Mathematics and Biosciences in Interaction

Managing Editor

Wolfgang Alt
Division of Theoretical Biology
Botanical Institute
University of Bonn
Kirschallee 1
D-53115 Bonn
e-mail: wolf.alt@uni-bonn.de

Editorial Board

Mathematics and Biosciences in Interaction is devoted to the publication of advanced textbooks, monographs, and multi-authored volumes on mathematical concepts in the biological sciences. It concentrates on truly interdisciplinary research presenting currently important biological fields and relevant methods being developed and refined in close relation to problems and results relevant for experimental bioscientists.

The series aims at publishing not only monographs by individual authors presenting their own results, but welcomes, in particular, volumes arising from collaborations, joint research programs or workshops. These can feature concepts and open problems as a result of such collaborative work, possibly illustrated with computer software providing statistical analyses, simulations or visualizations.

The envisaged readership includes researchers and advanced students in applied mathematics – numerical analysis as well as statistics, genetics, cell biology, neurobiology, bioinformatics, biophysics, bio(medical) engineering, biotechnology, evolution and behavioral sciences, theoretical biology, system theory.

POLYMER and CELL DYNAMICS

Multiscale Modeling and Numerical Simulations

Wolfgang Alt
Mark Chaplain
Michael Griebel
Jürgen Lenz
Editors

Springer Basel AG

Editors:

Prof. Dr. Wolfgang Alt
Abteilung Theoretische Biologie
Botanisches Institut
Universität Bonn
Kirschallee 1
D-53115 Bonn
Germany

Prof. Dr. Michael Griebel
Institut für Angewandte Mathematik (IAM)
Abteilung Wissenschaftliches Rechnen und
Numerische Simulation
Universität Bonn
Wegeler Str. 6
D-53115 Bonn
Germany

Prof. Dr. Mark Chaplain
Department of Mathematics
University of Dundee
23 Perth Road
Dundee DD1 4HN
UK

Dr. Jürgen Lenz
Bioreact GmbH
Botanisches Institut
Kirschallee 1
53115 Bonn
Germany

A CIP catalogue record for this book is available form the Library of Congress, Washington D.C., USA

Bibliografische Information der Deutschen Bibliothek
Die Deutsche Bibliothek verzeichnet diese Publikation in der Deutschen Nationalbiografie; detaillierte bibliografische Daten sind im Internet über http://dnb.ddb.de abrufbar.

ISBN 978-3-0348-9417-3 ISBN 978-3-0348-8043-5 (eBook)
DOI 10.1007/978-3-0348-8043-5

Originally published by Birkhäuser Verlag Basel, Switzerland in 2003
Softcover reprint of the hardcover 1st edition 2003
Member of the BertelsmannSpringer Publishing Group
Printed on acid-free paper produced from chlorine-free pulp. TFC ∞
Cover design: Armando Losa, graphic designer
Cover illustration: see p. 12, with the friendly permission of D.C. Rapaport

ISBN 978-3-0348-9417-3

9 8 7 6 5 4 3 2 1 www.birkhauser-science. com

Table of Contents

Preface vii

Introduction ix

Part I: Molecular Dynamics and Protein Configuration

Introduction to Part I 3

Dennis C. Rapaport
Molecular Dynamics Studies of Micelle and Capsid Self-assembly 7

Daniel Hoffmann
A Functional Study on Saposin B and C Using Experimentally Validated Models 19

Werner G. Krebs, Mark Gerstein
A Review of the Morph Server and the Macromolecular Motions Database: A Standardized System for Analyzing and Visualizing Macromolecular Motions in a Database Framework 29

Part II: Polymer Dynamics and Cell Motility

Introduction to Part II 45

Roland Rzehak, Andreas Arend, Diego Kienle, Walter Zimmermann
Brownian Dynamics of Flexible Polymers in Flow 49

Zénó Farkas, Imre Derényi, Tamás Vicsek
A Microscopic Model for the Dynamics of Actin Filaments in Motility Assays and Its Numerical Simulation 69

Jürgen Lenz, Dieter Felix
Multiparticle Modeling of Actin-Myosin Networks: From Molecular Interactions to Cell Motility 75

Angélique Stéphanou, Xavier Ronot, Philippe Tracqui
Analysis of Cell Motility Combining Cytomechanical Model Simulations and an Optical Flow Method 91

Jeanie L. Drury, Micah Dembo
Micropipet Aspiration of the Human Neutrophil 113

Emmanuel Promayon, Jean-Louis Martiel, Philippe Tracqui
Physical-object-oriented 3D Simulations of Cell Deformations and Migration 125

Part III: Multicellular Dynamics

Introduction to Part III 141

Aaron Fogelson, Haoyu Yu, Andrew Kuharsky
Computational Modeling of Blood Clotting: Coagulation and Three-dimensional Platelet Aggregation 145

Hans-Joachim Bungartz, Martin Kühn, Miriam Mehl, Stefan Würtz
Space- and Time-resolved Simulations of Processes in Biofilm Systems on a Microscale 155

Dirk Drasdo
On Selected Individual-based Approaches to the Dynamics in Multicellular Systems 169

András Czirók, Andreas Deutsch, Michael Wurzel
Individual-based Models of Cohort Migration in Cell Cultures 205

Wolfgang Alt, Till Bretschneider, Ralf Müller
Interactive Movement, Aggregation, and Swarm Dynamics 221

R. Brent Rice, Victor H. Barocas
A Discrete-cell Model of Tissue-equivalent Compaction 243

Alexander R.A. Anderson
A Hybrid Discrete-continuum Technique for Individual-based Migration Models 251

Alexander R.A. Anderson, Alan W. Pitcairn
Application of the Hybrid Discrete-continuum Technique 261

List of contributors and workshop participants 281

Index 285

Color Plates

Preface

It was at the inviting guest and seminar house 'Magdalena', run by the 'Franziskanerinnen von Nonnenwerth' (Fransiscan nuns) in Bad Honnef near Bonn that in June 2000 the International Workshop on *Numerical Simulations of Polymer and Cell Dynamics* took place. In the peaceful but inspiring atmosphere of a monastery-like cloister, the workshop collected ideas from more than 30 scientists from various fields and brought them into a fruitful week of interdisciplinary exchange and collaboration. Coming from several European countries, Israel, and the USA, the participants represented institutions or research groups in applied and numerical mathematics, scientific computing, theoretical physics, molecular biophysics, cell and molecular biology as well as chemical and biomedical engineering.

The Workshop was financially supported by a DFG research program (SFB 256) on 'Nonlinear Partial Differential Equations'. For the very effective local organization, we express our thanks to all those who enabled the successful meeting, in particular to our ever kind secretary Anke Thiedemann. At Bonn University, the two local organizing groups in some sense span the wide spectrum of our common research interests, namely the Division of *Scientific Computing* within the Institute of Applied Mathematics and the Division of *Theoretical Biology* within the Botanical Institute. In the meantime, these two groups are now even more focused on the central topic of this Workshop and of this book, namely, *Multiscale Modelling and Numerical Simulations* with applications to the physical chemistry of polymers and the biophysics of cells, and have implemented it into the new DFG research program at our University (SFB 611) on 'Singular Phenomena and Scaling in Mathematical Models'.

Moreover, during the (longer than anticipated) period between collecting the first contributions to this Volume and its final publication within the Birkhäuser book series *'Mathematics and Biosciences in Interaction'*, collaborations between workshop participants increased. Thus, the articles in the three parts of this book reflect the scientific discourse during the workshop and afterwards. The topics are of ongoing importance, since they address a series of principle problems in mathematical modeling and adequate methods for numerical simulation. All contributions have been carefully reviewed by expert referees in the field and by the editors.

Besides 16 color plates and a subject index at the end of the book, there is an updated address list not only of all corresponding authors but also of all workshop participants. The editors hope that their study of the presented scientific concepts and methods will stimulate the reader's own research ideas or activities.

Bonn,
September 2002

Wolfgang Alt, Mark Chaplain,
Michael Griebel, Jürgen Lenz

Introduction

The theme of these workshop proceedings, namely, *modeling and simulation of polymer and cell dynamics*, is, on the one hand, 'classical', as it concerns the description, explanation, and reproduction of the basic mechanisms of the application of force and the resulting motion. Hence, it essentially requires adequate methods from physical and mathematical theories on spatial and temporal scaling levels that are beyond quantum mechanics but are below systems theory of whole-body organisms and their evolution (see table on page xi).

On the other hand, the theme is quite 'modern', as it asks for quantitative solutions to basic questions in *Molecular and Cell Biology*, now the central field of life sciences, including relevant applications in biotechnology and medicine. In addition, the topics and concepts presented here are part of an ever-continuing attempt to understand the borderlines between the more 'passive' dynamics of pure matter, as molecules or polymers, and the more 'active' processes of structuring, organization, and regulation within living units, such as biological cells, tissues, or organs.

Therefore, the scientific contributions collected in this book, dealing with typical case studies of polymer and cell dynamics, are ordered into three parts of increasing 'biological complexity'. Part I on **Molecular Dynamics and Protein Configuration** shows how to represent the assembly and formation of elementary lipid or protein aggregates and how to study and classify geometric configuration changes in functional domains of enzymes or 'motor' proteins. Applications range from biomedical engineering to molecular pharmaceutics to a potential treatment of specific diseases.

Moving to the more complex realm of polymer organization, Part II on **Polymer Dynamics and Cell Motility** first deals with typical deformations of polymer chains, filaments, or networks and their interaction with surrounding media, ligands, cross-linkers, or molecular motors. These models of polymer dynamics, at least potentially, serve as ingredients for modeling and simulation on the next higher organizational level: shape changes and motility of single cells such as leukocytes, fibroblasts, or tumor cells. The aim of such theoretical analyses is not only to explain possible mechanisms of cell dynamics but also to offer quantitative tools for the characterization and classification of different cell types with potential applications to immunology and cancer research.

Finally, Part III on **Multicellular Systems** translates some of these concepts and methods, developed for multiparticle systems, to the level of cell-cell and cell-medium interactions and adds the necessary modeling elements for cell growth kinetics, cell adhesion, or cell-cell communication by chemical signalling and control loops or by mechanical force transduction via extracellular matrices. In particular, it presents simulation models and corresponding applications for the investigation of blood clotting, biofilm formation, which occurs, for example, in waste-water technology, and tissue formation and regeneration, relevant in wound healing, tumor invasion, or angiogenesis. Again, the quantitative analyses offered can help to improve data evaluation of 2D and 3D assays for cell and tissue behavior *in vitro* and *in situ*.

It is obvious that in order to explore the modeling of such phenomena in this wide range of biochemical, biotechnical, and biomedical applications – from molecular dynamics to tissue formation – very differing scales have to be considered: spatial scales reaching from nanometers to centimeters, temporal scales from microseconds to hours. Although in various recent biological research programs such tasks as 'exploring the whole genome and proteome of an organism' or 'remodeling and simulation of the whole cell' have been announced by referring to the increased computational power of new computer generations, it seems to be unrealizable to include the very microscale models directly into simulation models on very macroscale levels (both spatially and temporally). Rather, the modeling experience so far shows that it is already challenging enough to consider pairs or triples of particular submodels on certain selected micro-, meso- and macro-scale levels, relative to one another.

For example, the molecular dynamics approach used in Part I is based on 'mesoscopic' approximations of field potentials gained from more 'microscopic' quantum dynamics and finally leads to relatively 'macroscopic' geometric and dynamic descriptions of whole proteins or even of protein assemblies (such as micelles or capsoids).

Based on such 'molecular' submodels, the multiparticle models for polymer dynamics in Part II use lumped groups of protein-monomers and their approximate ('elastic') interaction properties as 'mesocopic' ingredients to simulate the more 'macroscopic' motion of single polymer chains. Furthermore, Part II presents a modeling approach for polymer network dynamics, where again lumped groups, now of polymer-filaments, are analogously used as units of a multiparticle model for simulating the spontaneous clustering and bundeling of an actin-myosin cortex near the model plasma membrane.

Finally, in Part III this whole story of multiscale modeling is repeated on the level of multicellular systems, namely, by developing individual-based multiparticle models using 'mesoscopic' interaction potentials or fields derived from more 'microscopic' submodels for chemical or mechanical cell-cell interactions, in order to simulate aggregation and tissue-formation dynamics.

On these different modeling levels, one common scaling problem appears: should a submodel be explicitly used, e.g., by complete calculation on the relatively 'microscopic' scale, or only in a 'coarse-grained' version, e.g., by implementing a rescaled and more easily computable approximation of it? Typical examples occur when polymers (Part II) or cells (Part III) are embedded in a viscous fluid or gel and thereby 'influence' each other indirectly by global flow or tension. Then high-power simulations can solve the full hydrodynamic or viscoelastic evolution equations (or realize the corresponding stochastic multiparticle process), and, alternatively, 'cheaper' simulations can use analytical or numerical approximations of pseudo-steady-state solutions (or of continuum versions of the discrete model) to compute the required expressions for flow or tension only at discrete positions of polymers or cells. The decision whether to choose the microscopic or the mesoscopic scaling variant usually depends on the desired or required output of the actual

Scaling of Simulation Models for Polymer and Cell Dynamics

Particles	**Space scale**	**Time scale**	**Interactions forces & motions**	**Models & methods**
Atoms	$\sim$ Å	$\sim \mu\,$sec	Gravity, electromagnetic, weak forces, (quasi-)periodic or chaotic, dynamic states	Schrödinger & Hartree-Fock equations, MOLECULAR DYNAMICS, interaction potentials
Molecules	$\sim$ nm	$\sim m\,$sec	Covalent binding, weak bonds, hydration, stochastic rotation, diffusion	(Stochastic) multi-particle systems, orientation & hinge dynamics, protonation dynamics, Poisson equation
Polymers	50 nm $\sim \mu$m	$\sim$ sec	Drag & elasticity, conformational changes, flickering, diffusion	(Semi-)flexible string model, Stokeslet approximations, THERMODYNAMICS, binding & rupture dynamics
Polymer systems	10 $\sim 100\,\mu$m	min	Steric restriction, cross-linking, shear forces, mixture & segregation, deformation, flow	(Stochastic) multi-particle models, cross-linking & contraction, two-phase flow equations, (an-)isotropic (non-)Newtonian fluids
Cells	50 μm $\sim$ mm	min $\sim$ h	Receptor kinetics, cell adhesion, tension & pressure, cytoplasm flow, deformation, translocation	'Tensegrity' model, adhesion kinetics & dynamics, free boundary problems, surface tension & viscosity
Cell collections	10 mm $\sim$ cm	*h* $\sim$ days	Exterior fluid, matrix, cell contact, stress & shear forces, individual motion, deformation, pushing & tugging	Immersed particle model, two-phase gel dynamics, multi-particle and lattice models, diffusion equations & hybrid models

macroscopic model, i.e., which averaged quantities or indices are going to be presented as simulation results.

Thus, on any level of dynamic phenomena in biophysics or biology, namely, for proteins, polymers, cells, tissues, organisms, populations, and ecosystems, mathematical modeling and numerical simulation have to find and use the appropriate strata of spatial and corresponding temporal scales. One of the most important future tasks will be to study more completely the relationship between submodels on the corresponding scaling levels.

This requires analytical and numerical work, i.e., formal derivations and proven approximation theorems as well as appropriate, compatible, and effective numerical algorithms. The algorithms appearing in these proceedings, either those used explicitly or merely mentioned, are mainly standard methods that have been developed for dynamic modeling systems in physics but, sometimes, are altered versions with suitable adaptations or generalizations. Typical examples are numerical approximation algorithms for stochastic ODE systems (Langevin equations) or lattice models, minimization of energy functionals, hyperbolic transport equations, elliptic or parabolic systems of partial differential equations (reaction, diffusion and taxis equations), and various combinations, in particular, the so-called hybrid methods combining discrete stochastic algorithms with discretization procedures for continnum models.

Part I

Molecular Dynamics and Protein Configuration

"At the atomic level, we have new kinds of forces and new kinds of possibilities, new kinds of effects. The problems of manufacture and reproduction of materials will be quite different. I am, as I said, inspired by the biological phenomena in which chemical forces are used in repetitious fashion to produce all kinds of weird effects (one of which is the author)."

[Richard Feynman's talk (1959) "There's Plenty of Room at the Bottom". Annual Meeting of the Amer. Phys. Soc., Caltech, Los Angeles]

Introduction to Part I

In principle, we can consider any matter, and thus the dynamics of polymers and cells, the interplay of particles. On a quantum mechanical level, Schrödinger's equation is the basis of description. Here, the particles are electrons and the nuclei of the overall system that interact by electrostatic forces. However, except for very simple systems, there is no analytical solution. Also, since the Schrödinger equation consists in $\mathcal{R}^{3(N+K)}$, where N is the number of electrons and K is the number of nuclei, we encounter the curse of dimensionality. This prohibits any numerical approach for reasons of complexity. Therefore, we have to resort to approximations. A typical approach follows the ideas of Born and Oppenheimer: Since the masses of electrons and nuclei differ by several orders of magnitude, the nuclei evolve classically by Newton's equations of motion in a potential field which is formed on the quantum mechanical level by the electrons. This results in the molecular dynamics methods of Born and Oppenheimer, Ehrenfest, or Car and Parinello [2], see also [8]. Here, in every time step, an electronic Schrödinger equation has to be solved approximatively, e.g., by the density functional method or the Hartree-Fock approach. Since this is still quite costly and can only be done for systems of moderate size, a further approximation is used: The effect of the electrons is expressed by analytical potential functions whose forms depend on a set of parameters. These parameters are fitted either to measurements or to the results of *ab initio* calculations for model situations. Thus, empirical force fields are obtained in which the nuclei now evolve over time according to Newton's equations. This is the basis of the classical molecular dynamics method.

For quite some time, this approach has been the 'work horse' next to the Monte Carlo technique in material science, biophysics, and biochemistry. Over the years, molecular dynamics packages like Amber [9], Charmm [1], Gromos [10], EGO [3] and others have been developed with specific force field parameters. Together with data from the Brookhaven protein database, they allow one to simulate, at least in principle, the dynamical behavior of peptides and most existing biopolymers. Due to the increasing power of modern computers and the availability of supercomputers with huge memory and large processor numbers, it is currently possible to handle systems with 100 million atoms and more for a simulation time up to several microseconds. However, many biological processes take place on larger time scales. Here, time step limitations prohibit the simulation of the underlying dynamics. Grubmüller and Schulten proposed to prescribe the macroscopic dynamics of a system by applying additional constraints such as outer forces. Thus, the effect of a (virtual) atomic force microscope is achieved [6] or the so-called steered or targeted molecular dynamics [7] results. This way, time step limitations can be overcome to a certain extent. Also, the Langevin dynamics approach [5, 11], where additional stochastic terms are added, allows for larger time steps and thus longer simulation times.

Macroscopic observables can be gained from the computed evolution of a system by means of averaging procedures from statistical mechanics. Furthermore, a particle itself can be considered as an average of a set of atoms or molecules. Here, again, proper potential functions have to be found empirically for these agglomerated particle models. Nevertheless, relatively simple potentials result in a rich variety of dynamical behavior ranging from simple pattern formation to complete self-assembly. This aspect is reflected in the article 'Molecular dynamics studies of micelle and capsid self-assembly' by **Rapaport**. Here, the molecular dynamics method is used to study supramolecular self-assembly processes as micelle formation and the growth of virus protein shells.

The *function* of a biomolecule is closely related to its structure in three-dimensional space. Therefore, knowledge of the structure is essential to further understand the function of a protein. The exploration of the structure of proteins is done in the laboratory with methods such as X-ray diffraction or nuclear magnetic resonance spectroscopy. This is a complicated, expensive, and time-consuming process. Here, numerical simulation techniques related to the molecular dynamics method can help to speed up this process substantially. In his article 'A functional study on saposin B and C using experimentally validated models (EVAMs)', **Hoffmann** describes the results of his EVAM approach for the fast generation of structural models of proteins. It starts from an NMR structure as a template and generates experimentally validated structural models with the help of homology techniques from bioinformatics and additional energy-minimization methods involving molecular dynamics. In this study two small proteins, saposin B and saposin C, were used as specific examples to demonstrate the feasibility of the method.

However, not only knowledge of the three-dimensional structure but also possible changes of conformation and their variability over time are important to further understand the function of proteins. Many macromolecules can be considered as molecular machines, and their basic movements contribute substantially to their function. An increasing amount of such conformal changes is being explored nowadays. The classification of typical movements of biomolecules, their database storage, and their visualization are discussed in the article 'The morph server and the macromolecular motions database: a standardized system for analyzing and visualizing macromolecular motions in a database framework' by **Krebs & Gerstein**. Here, the motion of a protein is described as a rigid body rotation of a small core part relative to a larger one using a set of hinges. A standardized coordinate system then allows comparisons of two motions and thus gives the possibility of a classification. Furthermore, an adiabatic mapping between two conformations by means of a restrained energy-minimization calculation similar to molecular dynamics allows first for a visualization of the motion and second for an indication of the energy barrier that must be surmounted in the motion.

In summary, molecular dynamics simulations permit a better understanding of the processes on the molecular and macromolecular level and are a helpful tool to explore the dynamics, the structure, the conformations, and thus the function of proteins and other biomolecules. For an introduction to the numerical techniques needed for molecular dynamics simulation, see, for instance, [4].

References

[1] B. R. Brooks, R. E. Bruccoleri, B. D. Olafson, D. J. States, S. Swaminathan and M. Karplus (1983) CHARMM: A program for macromolecular energy, minimization, and dynamics calculations. *J. Comput. Chem.* 4, 187–217

[2] R. Car and M. Parrinello (1985) Unified approach for molecular dynamics and density functional theory. *Phys. Rev. Lett.* 55, 2471–2474

[3] M. Eichinger, H. Heller und H. Grubmüller (2000) EGO – an efficient molecular dynamics program and its application to protein dynamics simulations. In: (R. Esser, P. Grassberger, J. Grotendorst and M. Lewerenz, eds.) *Workshop on Molecular Dynamics on Parallel Computers*, World Scientific, Singapore, 154–174.

[4] M. Griebel, A. Caglar, S. Knapek, G. Zumbush eds. (2003) *Numerische Simulation in der Molekulardynamik. Numerik, Algorithmen, Parallelisierung, Anwendungen.* Springer, Berlin.

[5] B. Hess, H. Bekker, H. J. C. Berendsen and J. G. E. M. Fraaije (1997) LINCS: A linear constraint solver for molecular simulations. *J. Comput. Chem.* 18, 1463–1472

[6] B. Heymann and H. Grubmüller (1999) AN02/DNP unbinding forces studied by molecular dynamics AFM simulations. *Chem. Phys. Lett.* 303, 1–9

[7] H. Lu, B. Isralewitz, A. Krammer, V. Vogel and K. Schulten (1998) Unfolding of titin immunoglobulin domains by steered molecular dynamics. *Biophys. J.* 75, 662–671

[8] M. Parrinello (2000) Simulating complex systems without adjustable parameters. *Comp. Sci. Eng.* 2, 22–27

[9] D. A. Pearlman, D. A. Case, J. W. Caldwell, W. R. Ross, T. E. Cheatham III, S. DeBolt, D. Ferguson, G. Seibel and P. Kollman (1995) AMBER, a computer program for applying molecular mechanics, normal mode analysis, molecular dynamics and free energy calculations to elucidate the structures and energies of molecules. *Comp. Phys. Comm.* 91, 1–41

[10] W. F. van Gunsteren and H. J. C. Berendsen (1990) Computer simulation of molecular dynamics: Methodology, applications and perspectives in chemistry. *Angew. Chem. Int. Ed. Engl.* 29, 992–1023

[11] G. Zhang und T. Schlick (1995) Implicit discretization schemes for Langevin dynamics. *Mol. Phys.* 84, 1077–1098

Polymer and Cell Dynamics – Multiscale Modeling
and Numerical Simulations, 7–17
ed. by W. Alt, M. Chaplain, M. Griebel und J. Lenz

Molecular Dynamics Studies of Micelle and Capsid Self-assembly

Dennis C. Rapaport

Physics Department, Bar-Ilan University, Ramat-Gan 52900, Israel

Summary. Molecular dynamics methods are used to study supramolecular self-assembly. The examples described here involve micelle formation and the growth of spherical virus protein shells (or capsids). The simulation techniques used are outlined and some of the results obtained using these methods are discussed.

1. Introduction

There are many processes – both natural and industrial – wherein self-assembly of multimolecular structures occurs. The size and morphology of the resulting entities depend on the systems involved but can range from atomic-scale polyhedra and spheres to layered and lattice-like structures orders of magnitude larger. A variety of experimental approaches have been employed to elucidate various features of these structures, but, despite a considerable investment of effort, theory is very limited in its ability to relate the characteristics of the structures and the manner in which they spontaneously assemble to the properties of the individual molecular units [7, 13].

The problem is a fundamental one, stemming from the fact that interacting many-body systems are theoretically intractable when the description is at the level of individual particles; indeed, theoretical progress is conditioned upon a major reduction in the number of degrees of freedom. Thus, concentrations of molecular species in a multicomponent system might be represented as continuous variables in a mean-field framework [1], or polymers that form a surfactant layer might be described as a continuous fluctuating membrane [22]. Such approximations often allow considerable progress, but the highly simplified description limits the questions that can be posed, such as, how does the surfactant envelope of a micelle actually form? Alternatively, elements of the discrete particle representation can be retained, but with the particles confined to a lattice [8, 28]; this is a generalization of the Ising problem – helpful but restricted in the kinds of behavior it can describe.

In this paper we examine two categories of supramolecular self-assembly, both addressed at a discrete particle level by means of molecular dynamics (MD) simulation [18, 19], namely, the formation of micelles and related structures in surfactant solutions and the growth of the polyhedral protein capsids that form the external shells of spherical viruses. Spontaneous self-assembly occurs in each case under suitable physicochemical conditions, and a complex interplay of molecular structure, forces, and temperature-driven dynamics underlies the intriguing behavior that is observed. The fact that nonequilibrium processes are involved limits theoretical analysis, leaving direct simulation as the only way to deal with these many-body systems. There is in fact no guarantee that such systems ever achieve 'true' thermodynamic equilibrium; any apparent long-term stability could be due to high free-energy barriers that are readily crossed during cooperative assembly, but not easily traversed in the reverse direction without a significant external perturbation (such as heating).

The MD approach is computationally intensive in the extreme. The question that must therefore be asked at the outset is, which of the molecular characteristics are essential for constructing a simplified model capable of reproducing the principal aspects of the behavior? The advantage of adopting a relatively simple model is that substantially larger systems can be simulated over longer periods of time than when using a highly detailed molecular representation. Because aggregate sizes are much larger than individual molecules, such simplifications are essential; only after extensive investigation of the properties of simplified models will it be possible to determine how much of the behavior is 'generic' and how much is likely to depend on specific molecular details.

2. Background

2.1. Micelle Assembly

Binary mixtures of simple liquids, such as oil and water, can form either a macroscopically homogeneous phase or a heterogeneous state in which the fluids are separated by an interface; the particular thermodynamic state depends on external variables such as relative concentration and temperature. The situation becomes much more complicated in ternary systems, one of the better-known examples being amphiphilic surfactants in an oil-water mixture; a multitude of distinct phases appear in such systems, including micelles (microemulsions), bilayer vesicles, stacks of bilayer membranes, and liquid crystals of various symmetries [5, 7, 14]. The origin of this rich polymorphism is rooted in the molecular structure and interactions of the constituents, and one of the goals of surfactant research is to quantitatively relate physicochemical behavior to molecular structure.

Self-assembly phenomena occur in a variety of molecular contexts, leading to the hypothesis that there are certain common features of the molecules that are responsible for the overall behavior. Hence it is reasonable to expect that simplified models ought to be capable of elucidating the equilibrium and dynamical behavior of such systems. In particular, such models should help reveal how the collective behavior of aggregates – such as the shapes and their fluctuations and the mechanisms for growth such as fusion and solubilization – is related to the molecular properties.

Simplified models have been used in previous MD studies of the nature of a preformed surfactant interface and the way the length of the hydrophobic tails affects the surface tension [24, 25]. Increasing system size from the original 1000 particles to almost 40,000 permits the observation of a spontaneously formed micelle [26]. Further studies [6, 11] aimed at exploring the mechanisms for transferring oil from a droplet to within a micelle and, to a limited extent, the manner in which micelle shape depends on the size of the surfactant head group. The effect of altering the relative interaction strengths on the nature of the micelles has been studied using small systems with about 5000 particles [15]. More complex surfactant 'topologies', as in the case of gemini surfactants, have been similarly explored [12], as has the role of surfactant spacer molecules [16].

2.2. Capsid Assembly

Viruses come in many forms, but there are two key facts about a virus that influence much of the analysis of its structure and function [2, 4]: the infective agent consists of genetic material (DNA or RNA), and this genome is contained in a protective protein shell – the capsid – whose task is to ensure that it arrives at its destination in a functionally intact state. The genetic information carried by the virus must be able, at a minimum, to ensure the formation of any new enzymes required in the viral reproduction process, as well as the proteins actually used for virus construction. Despite the apparently large range of possible geometrical forms a capsid might adopt, there are in fact only two basic designs in use, namely helical tubes and icosahedral ('spherical') shells.

An important limitation on the physical size of the genome, and thus on the amount of genetic information that it can contain, is the size of the capsid [10, 21]. The principles dictating the architecture of the capsid of a spherical virus recognize the need for a container of maximum volume that can be constructed with only a minimal amount of information provided by the genetic material: a near-spherical shape provides maximal volume for given surface area, with symmetry minimizing the engineering design information needed for construction. While there are capsids that consist of 60 elementary units, the volume is insufficient for significant amounts of nucleic acid, and most capsids actually consist of multiples of 60 units. These larger shells also have icosahedral symmetry [29], but because 60 subunits is the maximum possible under conditions of complete equivalence, the concept of quasi-equivalence [2, 3] is used to explain how such structures might be constructed with only minimal departure from the identical local environments experienced by each subunit of the 60-mer.

A great deal is known about assembled capsid structures and the nature of the stabilizing interactions, but because partially assembled intermediates are highly transient little is known about the mechanism of assembly [27]. The assembly pathway could involve intermediate combinations of the basic subunits that are added in a specific order (recent evidence suggests that growth can proceed by adding dimers to an initially nucleated triplet of dimers [30] or by starting with the formation of a pentamer of dimers [31]), or it could be based on more gradual growth by adding individual monomers; scaffolding proteins are also known to play a role in some assembly processes. Both *in vivo* assembly and in certain cases *in vitro* assembly and disassembly [17] are studied to learn more about the principles involved.

The fact that self-assembly is so robust [3] suggests that MD simulation of low-resolution models emphasizing 'shape' – which is independent of the molecular details of particular viruses [10] – might capture the essence of the process. While simple mechanical models have been used to explore steric aspects of assembly [3], the MD approach should permit exploration of the assembly pathways themselves. At the present time, the only attempt at such an approach to the study of capsid assembly is [20]; an alternative method based on spherically symmetric particles and highly directional interactions is described in [23].

3. Methodology

3.1. Micelle Simulation

A very simple class of model is available for modeling surfactant behavior that is based on the dynamics of two varieties of spherical particles. The particles are labeled W and O, corresponding to water and oil (the solvent and solute), respectively. The interaction between pairs of W particles is typically a Lennard-Jones interaction (see eq.6 in the following article by Hoffmann); a similar interaction occurs between O particles. The interaction between W and O particles is a short-range repulsion that ensures that W and O are immiscible under suitable conditions. This simplification ignores properties such as long-range electrostatic forces (justifiable for nonionic surfactants or when screening is strong) and the formation of highly directional hydrogen bonds (while these often play a role in surfactant behavior since they help explain hydrophobicity, their presence is not essential) and the fact that oil molecules are actually polymer chains. The amphiphilic surfactant molecules are represented as flexible linear chains, with a hydrophilic head group consisting of one or more W particles, and a hydrophobic (hydrocarbon) tail made from several O particles; these are linked together by spring-like interactions.

Even such simplified models allow ample scope for exploring a wide range of phenomena and examining the effect of modifying various characteristics. The relative particle sizes can be varied (e.g., by using larger particles to represent head groups or head groups with a broader cross-sectional area constructed by attaching a number of head particles to the first particle of the tail), as can the chain lengths and the relative interaction strengths; more complex surfactant structures can be represented, and bond- and dihedral-angle torques can be introduced to reduce the flexibility of the surfactant chains. By means of such modifications of the simple models and the consequent changes in, among other properties, the effective value of the critical packing parameter [9], it is possible to explore alternative preferred packing arrangements of the surfactant molecules and the kinds of structural morphologies that result. Improved computer performance permits calculations using systems substantially larger than those considered in earlier work, an essential element when phenomena embracing a range of length scales are involved.

3.2. Capsid Simulation

Modeling capsid assembly calls for an MD approach involving rigid body dynamics [18]. Each monomer subunit is represented by a rigid planar assembly of spheres, as shown in Fig. 1. These are responsible for the overall shape of the protein complex that forms the subunit, with spheres in different subunits interacting by means of short-range repulsion to prevent overlap. In addition, there are sites (denoted by the small spheres in Fig. 1) at which the attractive forces act. These forces (which are Coulomb-like but which merge smoothly into narrow harmonic wells at short range) draw the capsids together and eventually bind them into nearly rigid assemblies (by replacing the full force with only its harmonic component); three sites per face provide a simple way to achieve torsional stiffness. Two subunits are regarded as fully bonded when all three pairs of attraction sites are mutually bonded; torsion forces act between partially bonded subunits to accelerate the alignment process. The planes formed by the attraction site triplets are oriented so that the dihedral angles between bonded subunits are those of the desired polyhedral structure.

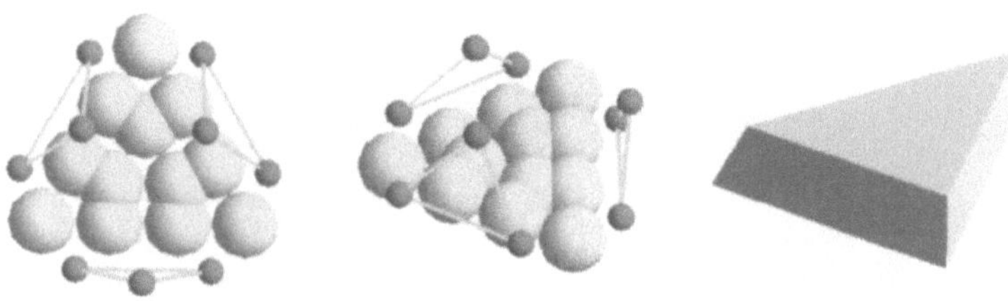

Figure 1. Views of the structure of the capsid subunit showing the large spheres that produce excluded volume and the small spheres denoting attraction sites; an alternative view shows the faces of the equivalent triangular block (see color plate 2).

This representation is a simplified paradigm that aims to capture the basic capability of self-assembly. In addition to the monomer shapes and interactions, ensuring an orderly assembly pathway requires additional assumptions. In particular, the initial version assumed that polyhedral structures form by the addition of a single monomer at a time and that once bonds form they are unbreakable (this represents kinetically limited rather than equilibrium assembly). Subsequently, this was extended, in view of experimental evidence, to allow the formation of intermediate subassemblies such as dimers or trimers, and the formation of larger assemblies was based on using these oligomers as building blocks.

3.3. Visualization

The MD approach provides unparalleled scope for measurements that correspond directly to what can be studied in the laboratory, as well as permitting a range of more detailed probes of features that are experimentally inaccessible. In the absence of a viable theoretical basis for analyzing the assembly processes, this detailed exploration is the key to understanding the underlying mechanisms. Visualization plays a particularly important role in developing the simulational approach, as some characteristics of the assembly process, while easy to follow visually, are not readily expressed in quantitative form. Thus, in addressing these and other aspects of the simulations, interactive 3D computer graphics have proved indispensable, and advanced techniques for visualizing different aspects of the structure and dynamics, including animation, are used to supplement more conventional quantitative methods of analyzing data generated by the simulations.

4. Measurements and Results

4.1. Micelle Formation

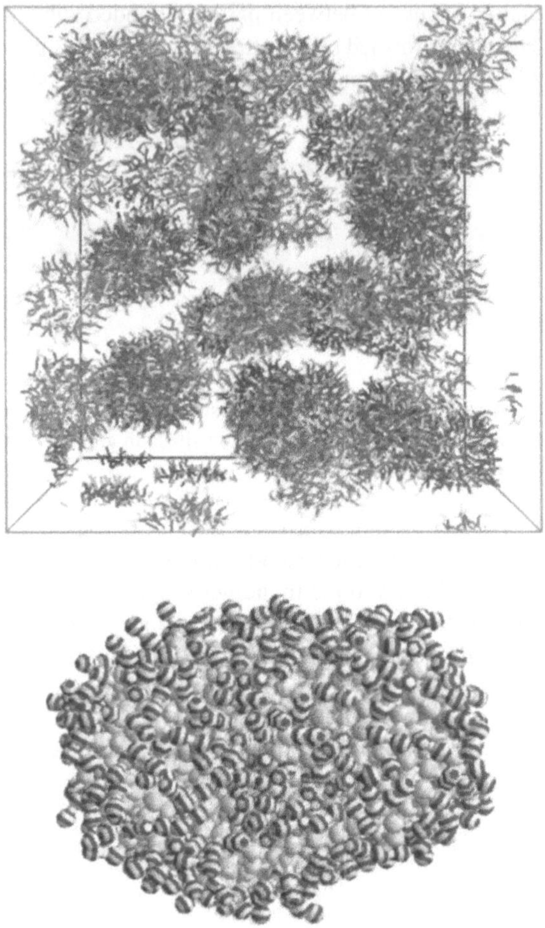

Figure 2. (a) Surfactant system (c = 0.05) after 5×10^5 time steps showing chains and solute (because of periodic boundaries, some micelles wrap around); (b) the largest micelle extracted from Fig. 2a (For colored picture and color scheme see color plate 1.)

Examples of results from preliminary studies of the problem are shown here. The systems contain up to 1.25×10^5 particles, of which a certain fraction c ($\approx$ 0.02-0.05) are solutes and a certain number are linked into surfactant chains of specified length and head-group size (for example, the chains might have five monomers, two of which form the hydrophilic head). This represents a relatively concentrated system that allows exploration of the mutual interactions between micelles.

Fig. 2a shows a state reached during a relatively long run, where solvent particles have been removed for clarity and only surfactant chains and solute are shown. The largest micelle in this system is shown separately in Fig. 2b; its hydrophobic core is mostly covered by the hydrophilic heads of the surfactant chains. Micelles are defined here on the basis of spatial proximity of solute (and surfactant tail) particles.

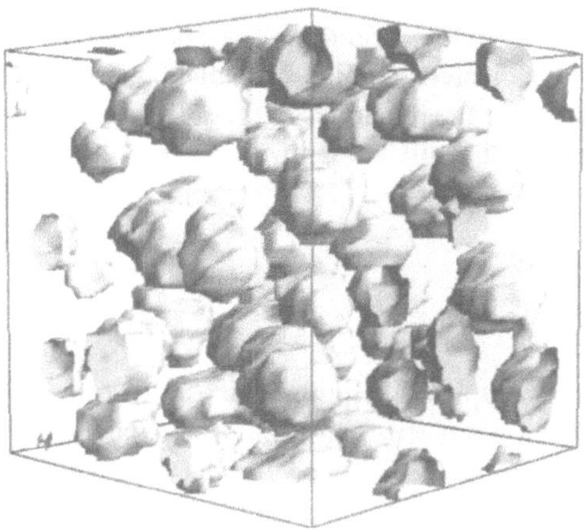

Figure 3. Surfactant system ($c = 0.025$) after 5×10^5 time steps showing the micelle boundaries (the fact that some surfaces appear open is an artifact due to periodic boundaries) (see color plate 2).

A state from a separate run with lower c is shown in Fig. 3; here, solute isodensity surfaces are rendered in a way that identifies the micelle boundaries (the particles themselves are not shown).

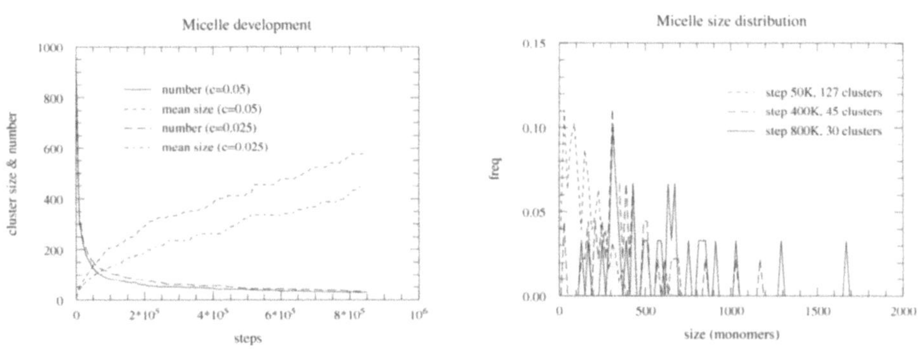

Figure 4. (a) Number of micelles and mean micelle size vs. time (in MD time steps); (b) distribution of micelle sizes ($c = 0.05$) at different stages of the run.

Fig. 4a plots the number of micelles and their mean size as functions of time; the decreasing rate of micelle mergers makes it difficult to determine the final state. Fig. 4b shows the micelle size distributions at various points in the run (due to reduced surfactant concentration, the trend here is the opposite of [16]); quantitative analysis of the mass distribution within each micelle suggests a preference for a near-spherical shape, a conclusion confirmed by visual examination.

These exploratory results serve to illustrate the potential of this approach. The underlying structure of the model can now be varied to establish how the various features impact the nature of the assembly products. The analysis can be extended to include a more detailed study of the clusters (and the surfactant layers that form at high solute concentrations) and the nature of the dynamics of the self-assembly process. These and other aspects of the problem are currently being investigated.

4.2. Capsid Formation

Fully assembled polyhedra typically require up to 5×10^5 time steps to form. Here we describe one example [20] of a run involving 1000 triangular subunits in which a maximum of 13 capsids are allowed to nucleate. Fig. 5 presents a series of images showing several growth stages; the monomers have been erased from all but the final picture for clarity. All polyhedra that are allowed to form assemble completely and correctly. Different colors are used to distinguish those subunits that are fully bonded from those that are only partially bonded. Examination of this and other simulation runs indicates that subassemblies tend to remain fairly compact, suggesting that incoming monomers seek locations where bonding to more than one attached subunit is possible. Similar results have been obtained with trapezoidal subunits, designed to produce polyedral shells that correspond to the so-called T=1 capsids [10], also with 60 faces.

Quantitative estimation of growth rate can be made by measuring subassembly size as a function of time. Runs such as the one above, where most monomers are eventually incorporated into the capsids, are not suitable for this, because gradual monomer depletion inhibits growth. Instead, only a single capsid is allowed to form in an environment containing several times (here 216) the required number of subunits; once complete, all bonds are broken and the process is repeated. Fig. 6 plots the mean number of steps required for each 5-subunit size increment; the growth rate is practically constant, except near completion. The reason for the slowdown is that some bonding sites in a nearly complete structure are less accessible; such behavior can be contrasted with the substantial retardation of the growth rate when monomer depletion dominates the final stages of the run. Note that this particular model has no favored intermediate states along the assembly pathway.

There is considerable scope for refining the model while still maintaining its highly simplified form, based on knowledge of the underlying protein structures and interactions; such information, which can eventually be incorporated into the force laws used in the simulations, is derived from both laboratory measurements and calculations employing detailed molecular representations.

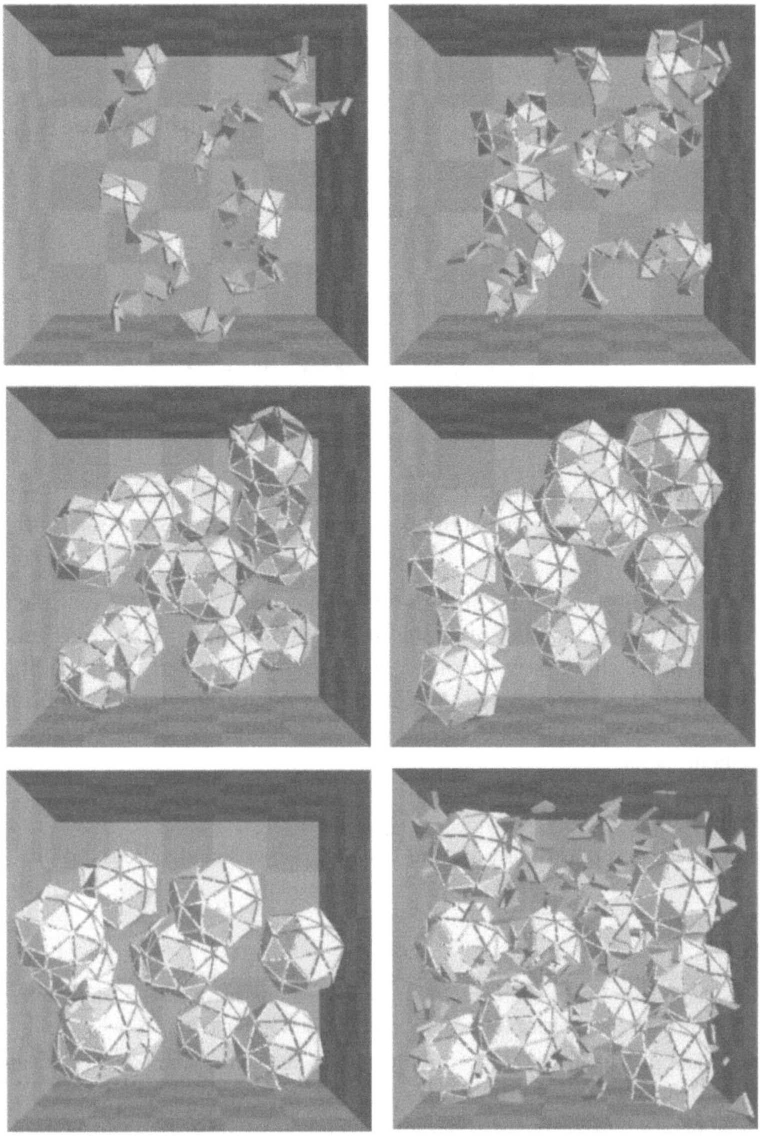

Figure 5. Stages in the capsid self-assembly process; unattached monomers are omitted for clarity, except in the final image (for colored picture see color plate 3).

5. Conclusions

By avoiding the complexity that accompanies a detailed molecular description in favor of high simplified representations (Occam's razor) it becomes possible to study larger systems over sufficiently long periods of time to allow truly interesting – and often surprising

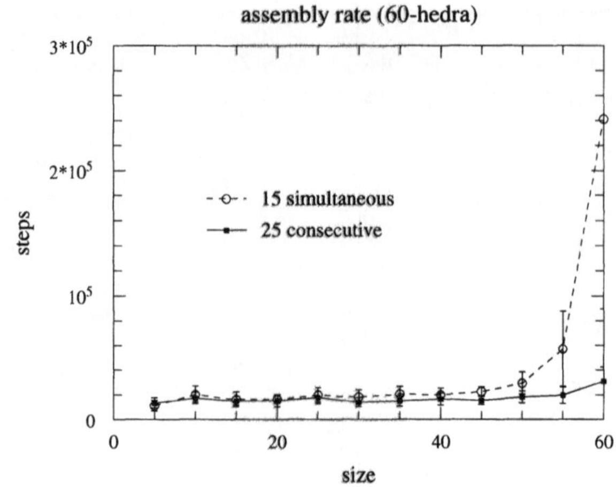

Figure 6. Assembly rates for multiple (15 assembled simultaneously) and single (25 assembled consecutively) capsids: mean numbers of time steps required to add an extra 5 subunits are shown.

– behavior to emerge. With the aid of high-performance computers it is feasible to extend the understanding of the factors governing molecular self-assembly. More generally, as the role of computer simulation becomes increasingly important in the biological and chemical sciences, it is important to understand the conditions under which major simplifications in the design of models can be tolerated due to the generic nature of certain classes of phenomena.

Acknowledgment

Part of this work was supported by a grant from the Israel Science Foundation.

References

[1] A. Ben-Shaul and W.M. Gelbart (1994) Statistical thermodynamics of amphiphile self-assembly: Structure and phase transitions in micellar solutions. In: [7], 1

[2] D. Caspar and A. Klug (1962) Physical principles in the construction of regular viruses. *Cold Spring Harbor Symp. Quant. Biol.* **27**, 1

[3] D. Caspar (1980) Movement and self-control in protein assemblies. *Biophys J* **32**, 103

[4] S. Casjens, ed. (1985) *Virus structure and assembly.* Jones and Bartlett, Boston

[5] K.W. Dillan, E.D. Goddard, and D.A. McKenzie (1979) Oily soil removal from a polyester substrate by aqueous nonionic surfactant systems. *J Am Oil Chem Soc* **56**, 59

[6] K. Esselink, P.A.J. Hilbers, N.M. van Os, B. Smit, and S. Karaborni (1994) Molecular dynamics simulations of model oil/water/surfactant systems. *Colloids and Surfaces A* **91**, 155

[7] W.M. Gelbart, A. Ben-Shaul and D. Roux, eds. (1994) *Micelles, membranes, microemulsions, and monolayers.* Springer, NY

[8] G. Gompper and M. Schick (1994) Lattice theories of microemulsions. In: [7], 395

[9] J.N. Israelachvili (1992) *Intermolecular and surface forces 2nd edn.* (Academic, London)

[10] J.E. Johnson and A.J. Fisher (1994) Principles of virus structure. In:*Encyclopedia of Virology* (Academic, NY, 1994), 1573

[11] S. Karaborni, N.M. van Os, K. Esselink, and P.A.J. Hilbers (1993) Molecular dynamics simulations of oil solubilization in surfactant solutions. *Langmuir* **9**, 1175

[12] S. Karaborni, K. Esselink, P.A.J. Hilbers, B. Smit, J. Karthäuser, N.M. van Os, and R. Zana (1994) Simulating the self-assembly of gemini (dimeric) surfactants. *Science* **266**, 254

[13] R.G. Laughlin (1994) *The aqueous phase behavior of surfactants.* (Academic, London)

[14] D. Myers (1988) *Surfactant science and technology.* (VCH Publishers, NY)

[15] B.J. Palmer and J. Liu (1996) Simulations of micelle self-assembly in surfactant solutions. *Langmuir* **12** 746, 6015

[16] B.J. Palmer, J. Liu, and J. Virden (1999) Simulations of the effect of diblock spacer molecules on micelle structure and function. *Langmuir* **15**, 7426

[17] P.E. Prevelidge, D. Thomas, and J. King (1993) Nucleation and growth phases in the polymerization of coat and scaffolding subunits into polyhedral procapsid shells. *Biophys J* **64**, 824

[18] D.C. Rapaport (1995) *The art of molecular dynamics simulation.* (Cambridge University Press, Cambridge)

[19] D.C. Rapaport (1997) An introduction to interactive molecular dynamics simulations. *Computers in Phys* **11**, 337

[20] D.C. Rapaport, J.E. Johnson, and J. Skolnick (1999) Supramolecular self-assembly: Molecular dynamics modeling of polyhedral shell formation. *Comp Phys Comm* **121**, 231

[21] M.G. Rossmann and J.W. Erickson (1985) Structure and assembly of icosahedral shells. In: [4], 30

[22] S.A. Safran (1994) Fluctuating interfaces and the structure of microemulsions. In: [7], 427

[23] R. Schwartz, P.W. Shor, P.E. Prevelige, and B. Berger (1998) Local rules simulation of the kinetics of virus capsid self-assembly. *Biophys J* **75**, 2626

[24] B. Smit (1988) Molecular-dynamics simulations of amphiphilic molecules at a liquid-liquid interface. *Phys Rev A* **37**, 3431

[25] B. Smit, A.G. Schlijper, L.A.M. Rupert, and N.M. van Os (1990) Effects of chain length of surfactants on the interfacial tension: Molecular dynamics simulations and experiments. *J Phys Chem* **94**, 6933

[26] B. Smit, P.A.J. Hilbers, K. Esselink, L.A.M. Rupert, and N.M. van Os (1991) Structure of a water/oil interface in the presence of micelles: A computer simulation. *J Phys Chem* **95**, 6361

[27] P.K. Sorger, P.G. Stockley, and S.C. Harrison (1986) Structure and assembly of turnip crinkle virus II. Mechanism of reassembly in vitro. *J Mol Biol* **191**, 639

[28] D. Sornette and N. Ostrowsky (1994) Lamellar phases: Effect of fluctuations (theory). In: [7], 251

[29] R. Williams (1979) *The Geometric Foundation of Natural Structure.* (Dover Publications Inc., NY)

[30] A. Zlotnick, J.M. Johnson, P.W. Wingfield, S.J. Sahl, and D. Endres (1999) A theoretical model successfully identifies features of Hepatitis B virus capsid assembly. *Biochemistry* **38**, 14644

[31] A. Zlotnick, R. Aldrich, J.M. Johnson, P. Ceres, and M.J. Young (2000) Mechanism of capsid assembly for an icosahedral plant virus. *Virology* **277**, 450

Polymer and Cell Dynamics – Multiscale Modeling
and Numerical Simulations, 19–27
ed. by W. Alt, M. Chaplain, M. Griebel und J. Lenz

A Functional Study on Saposin B and C Using Experimentally Validated Models

Daniel Hoffmann

Center of Advanced European Studies and Research (caesar), Ludwig-Erhard-Allee 2,
D-53175 Bonn, Germany

Summary. To speed up the structure-based analysis of protein function, we have developed the concept of Experimentally Validated 3D Models (EVAMs), i.e., structural models that can be generated relatively fast by a combination of bioinformatics and experiments. Here, we test the suitability of EVAMs for functional studies by applying the concept to two small proteins, saposin B and saposin C. We find that EVAMs are well able to explain some of the functional properties of these proteins, such as their pH-dependent association with lipid bilayers, that are known from independent biophysical experiments.

1. Introduction

Proteins are molecular machines, and in analogy to macroscopic machines, the 3D structure of a protein is closely related to its biological function. In fact, it is very difficult to thoroughly understand protein function without knowing the 3D structure. Unfortunately, to date the solution of protein 3D structures with methods like X-ray diffraction or nuclear magnetic resonance spectroscopy (NMR) is a laborious and time-consuming process[1]. We have devised a method [6] that promises to speed up this process significantly, although in general we probably will have to sacrifice some accuracy in comparison to high-quality structures measured with X-ray diffraction. A detailed account of our method is given elsewhere [5]. Application of the method to a protein results in a so-called Experimentally Validated Model (EVAM) of the protein 3D structure.

To investigate the suitability of our approach for the study of protein function, we generated EVAMs of two proteins, saposin B (SapB) and saposin C (SapC), and we used these structural models to study the biological functions of these proteins.

The saposins are a class of small homologous proteins of about 80 residues that have a conserved pattern of three disulfide bridges and that probably share a fold ("saposin fold", Fig. 1) consisting of five α-helices, as determined for the homologous protein NK-lysin (NKL) by NMR [9][2]. Four of the saposins – SapA, SapB, SapC, and SapD – are

[1] This may change with a wider use of robots for crystallization etc. However, robots provide no guarantee for a successful solution of structures in all cases.

[2] In the literature, other proteins having similar function but not the saposin fold, such as the GM2 activator protein, are also termed *s*phingolipid *a*ctivator *p*roteins or "saposins".

derived from a common precursor protein by proteolytic processing [7]. All proteins for which the saposin fold has been postulated seem to interact with membranes as part of their biological functions. Saposins in particular have been shown to stimulate the degradation of certain sphingolipid head groups in lysosomal membranes [13]. Consequently, the deficiency in saposins can lead to sphingolipid storage diseases [7]. Specifically, deficiencies in SapB and SapC have been shown to cause accumulation of the spingolipids sulfatide and glucosylceramide, respectively, that clinically manifest themselves as variants of metachromatic leukodystrophy and Gaucher disease [7], respectively.

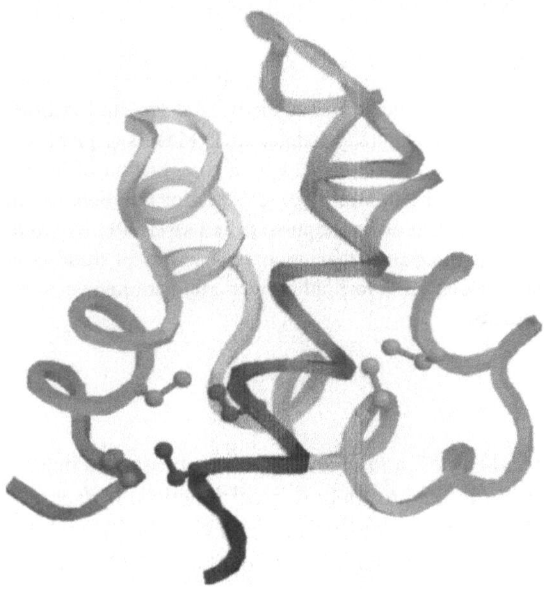

Figure 1. Saposin fold as given by backbone structure of the polypeptide chain of NKL (determined by NMR [9]). Although not a saposin itself, NKL has a significant homology to many members of the saposin family of proteins and probably shares the "fold" of these proteins, i.e., the global arrangement of the polypeptide chain shown here, consisting of five α-helices. Side chains of three pairs of cystein residues that form disulfide bridges in NKL and in most known saposins are shown in a ball-and-stick representation.

Despite the fact that many experimental results on saposins are available, the mechanism by which the saposins achieve their function is unknown. A hypothesis that could describe the mechanism of saposin action is that of a "liftase": the saposin may insert into the lysosomal membrane, bind to a sphingolipid, and lift it out of the membrane, such that the enzyme responsible for the actual degradation can access the head group more easily [7] (this is a general theme with variations between the saposins, e.g., SapC probably also stimulates its target enzyme by directly binding to it). From *in vitro* experiments it is known that saposins are most active at about pH 4.5, whereas at higher pH values they are inactive and do not partition into nonpolar media [12] and, hence, probably cannot insert into membranes. It has been speculated [12] that the molecular basis of this behavior may be a pH-triggered conformational transition of saposins, in the course of which hydrophobic areas of the proteins are exposed that promote insertion into the nonpolar membrane phase. On the other hand, saposins are heat stable and resistant to enzymatic proteolysis [14], indicating that they have a rigid fold that is unlikely to undergo large conformational transitions. Changes in polarity like those observed can be described in the physical framework of electrostatics. Hence, we have investigated how the electrostatic properties of saposins change with pH, and how this may explain experimental observations. We have based our study on EVAMs of SapB and SapC. High-resolution 3D structures of saposins from X-ray diffraction or NMR experiments have not been available, and it will be interesting to see whether our analyses will be confirmed once such experimental structures become accessible.

2. Methods

We proceeded in two steps: first, we set up 3D models of SapB and SapC using homology modeling with the NMR structure of NKL as template [9]; second, based on these models we calculated some properties of the proteins, in particular the protonation state of various groups and the electrostatic potential at the surface of the proteins, both at various pH values.

2.1. Generation of 3D Models

We first prepared a multiple sequence alignment [11] of six saposins, namely, SapA, SapB, SapC, SapD, the saposin domain of human acid sphingomyelinase, and NKL (Fig. 2 shows the part of the alignment that refers to SapB, SapC, and NKL). Only the 3D structure of the latter is known [9] and can be used as a structural template. The homology throughout all six sequences is significant and the alignment is compact, with only a few gaps in the loops of NKL, i.e., outside α-helical regions. The positions of cysteins are conserved. Because the disulfide bridges of SapB and SapC have been solved experimentally [13] and are in accordance with the disulfide pattern of NKL, we transferred the fold of NKL to SapB and SapC and considered this fold to be experimentally validated. Residues in gaps have been omitted, which means that we actually have generated models of deletion mutants. This is a source of error in the calculation of protonation and electrostatic properties, and if one compares computational results for the mutants and experimental results for the wild-type proteins, there may be differences originating from these omissions. However, none of the residues omitted is charged or can be expected to

```
SapB  GDVCQDCIQMVTDIQTAVRTNSTFVQALVEHVKEECDRLGPGMADICKNYISQYSEIAIQMMM HMQPKEICALVGFCDEV
SapC  DVYCEVCEFLVKEVTKLIDNNKT EKEILDAFDKMCSKLPKSLSEECQEVVDTYGSSILSILLEEVSPELVCSMLHLCSGT
NKL   GLICESCRKIIQKLEDMVGPQPN EDTVTQAASRVCDKMKIL RGVCKKIMRTFLRRISKDILTGKKPQAICVDIKICKE
         *  *  :       :  :      :       *  :       *   :  :              *   *  :  *
        hhhhhhhhhhhhhhhh      hhhhhhhhhhhhhh     hhhhhhhhhhhhhhhhhhhh     hhhhhhh
```

Figure 2. Sequence alignment of SapB, SapC, and NKL. Asterisks mark conserved residues, colons mark highly homologous residues, and 'h' indicates α-helical regions in the 3D structure of NKL.

change its protonation state under neutral to lysosomal pH. Hence, we judged the error resulting from these omissions to be of minor importance with respect to the properties of interest.

Using the ToPLign program [11], we modeled the residues of SapB and SapC upon the 3D structure of NKL [9], transferring as many corresponding atomic coordinates as available from the original structure of NKL into the structures of SapB and SapC. Missing parts of side chains were supplemented with CHARMM [3] in their default all-trans conformations. Structures then were subjected to 1500 steps of steepest-descent energy minimization [3], with strongly screened electrostatic interactions and polypeptide backbone position fixed, using a classical force field function of the following form as a model for the potential engergy [3]:

$$V(\mathbf{r_1}, \ldots, \mathbf{r_{atoms}}) = \tag{1}$$

$$\tfrac{1}{2}\textstyle\sum_i k_{r,i}(r_i - r_{0,i})^2 \quad \text{bonds} \tag{2}$$

$$+\tfrac{1}{2}\textstyle\sum_j k_{\alpha,j}(\alpha_j - \alpha_{0,j})^2 \quad \text{bond angles} \tag{3}$$

$$+\textstyle\sum_k k_{\phi,k}\left(1 - \cos\left(n_k(\phi - \phi_{0,k})\right)\right) \quad \text{dihedral angles} \tag{4}$$

$$+\textstyle\sum_l k_{\theta,l}(\theta_l - \theta_{0,l})^2 \quad \text{improper dihedrals} \tag{5}$$

$$+\textstyle\sum_{m>n} 4E_{0,mn}\left(\left(\frac{\sigma_{mn}}{r_{mn}}\right)^{12} - \left(\frac{\sigma_{mn}}{r_{mn}}\right)^{6}\right) \quad \text{Lennard-Jones} \tag{6}$$

$$+\textstyle\sum_{p>q} \frac{q_p q_q}{4\pi\epsilon_0\epsilon r_{pq}} \quad \text{Coulomb} \tag{7}$$

with equilibrium constants k, bond lengths r, bond angles α, dihedrals Φ, improper dihedrals θ (e.g., to fix chirality or planarity of certain groups), Lennard-Jones well-depths E, van der Waals radii σ, atomic (partial) charges q, and dielectric constant ϵ. In the minimization the electrostatic interaction was screened strongly by setting ϵ equal to a high value of $80r_{pq}$, which in practice almost corresponds to turning off the electrostatic interactions. In this way we avoided an artifical "mating" of groups of opposite charges that is often observed after *in vacuo* minimizations, and that can introduce significant errors in the electrostatics. Because of the homology of the saposins to NKL and the good quality of the alignment, we assumed that the backbone of NKL is a good model for the backbones of SapB and SapC, in contrast to the detailed side chain conformations, which are expected to be more diverse. Hence, to prevent distortions of the backbone during minimization, we fixed all positions of backbone atoms and only relaxed the side chains.

2.2. Calculation of Protonation and Electrostatics

3D models generated with the above procedure were used as input for the MEAD suite of programs [1, 2] that calculate pK_a values and pH-dependent protonation states of specified protonatable groups.

Briefly, MEAD considers protonation to be solely modulated by electrostatics. The protein and the aqueous solution surrounding it are treated as dielectric continua with dielectric constants ϵ of about 4 and 80, respectively. The two dielectric media are separated by the molecular surface of the protein. Additionally, inside the protein one assumes a charge density ρ given by the partial charges q in Eq. (7) and, in the solvent outside the protein, an ionic density n due to dissolved salt. The electrostatic potential φ is then computed by solving the so-called Poisson-Boltzmann equation:

$$-\epsilon_0 \nabla \epsilon \nabla \varphi = \rho - 2en \sinh(e\varphi/k_B T). \tag{8}$$

Under physiological conditions, the sinh term can be linearized to a good approximation, and the linearized version of Eq. (8) is then solved using a finite difference method. From the resulting electrostatic potential φ, the protonation probability can be calculated for each protonatable group, in the presence of all other protonatable groups. For details, see [1, 2].

Since we were most interested in groups that are expected to change their protonation state under acidic conditions like those found in lysosomes, we restricted our choice of protonatable groups to carboxylates (Asp, Glu, C-terminus) and imidazole rings (His). For these groups, pH-dependent protonation states were calculated using MEAD. Other groups were considered to have their usual protonation states at neutral pH, e.g., amido or guanidino groups were protonated and thus positively charged.

Experiments show that at neutral pH saposins are inactive and that they perform their function most effectively at a pH between 4 and 5 [12]. To gain insight into the structural basis of this behavior, we therefore prepared for more detailed inspection of 3D structures of SapB and SapC at pH 4, 5, and 7, applying the respective computed protonation states at these values of pH. To this end, protons were removed or added accordingly and structures were submitted to a second round of energy minimizations with CHARMM. Finally, using the minimized structures, the electrostatic potential on a cubic grid with a grid spacing of 0.1 nm around the proteins was calculated with the Poisson solver in the program CAMLab [4] and the potential was visualized with gOpenMol [8].

3. Results and Discussion

The total charge of SapB and SapC as a function of pH calculated as described above is depicted in Fig. 3. Both curves cross the zero-charge axis at a pH of about 4.3. This value is in good agreement with the pH value of about 4.5 at which saposins become most active according to experiments [12]. The calculated loss of net charge suggests that at this pH the proteins become most nonpolar in character and thus are most likely to insert into the nonpolar membrane phase.

The slope for SapC in Fig. 3 is much steeper around pH 4.3 than for SapB, meaning that the charge of SapC is more sensitive to changes of pH than is that of SapB. For low pH the total charge of SapC rises significantly above that of SapB. The increase in charge of both proteins is mainly caused by the protonation, i.e., neutralization, of carboxylates. Protonation of His residues plays no significant role at pH ≥ 4. The remaining positive net charge at pH ≤ 4.5 is located at amino and guanidino groups.

There are no experimental measurements available for the behavior of SapB and SapC at values of pH below 4.3. However, experimental results exist for the activity of SapB and SapC under the influence of various concentrations of negatively charged lipids such as bis(monoacylglycero)phosphate (BMP). Qualitatively, an increasing concentration of negative charges should have an effect similar to a decrease in pH, namely, it

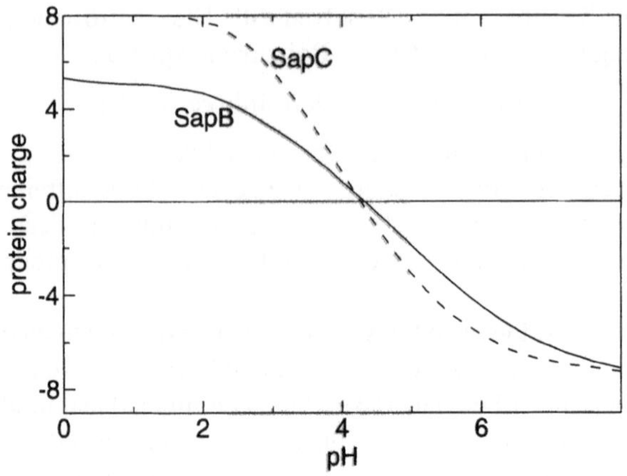

Figure 3. Dependence of total charge (in units of proton charge) of SapB and SapC upon pH as computed for the respective EVAMs.

makes protonation of the protein more likely. According to experiments [10], the stimulatory effect of SapC in the degradation of sphingomyelin is strongly affected by the presence of BMP, whereas BMP seems to have little effect on the activity of SapB. This agrees well with our findings that the slope in Fig. 3 is particularly steep for SapC and that SapC with dropping pH experiences a large increase in charge from about -5 e (proton charge) at pH 5 to +2 e at pH 4 (as compared to a change from about -2 e to +1 e for SapB). The larger positive charge of SapC will cause a stronger attraction by negatively charged lipids in the membrane, as observed experimentally.

Fig. 4 gives a more detailed view of the electrostatic properties at the surfaces of SapB and SapC at pH 4, 5, and 7. It can clearly be seen how the low potential (blue) patches – due to negative charges – dominate the surface at pH 7, but are replaced more and more by nonpolar (green) areas as the pH drops to 5 or 4, with some of the acidic groups becoming protonated. Especially for SapC, Fig. 4 also shows the emergence of several high-potential spots at low pH. It is plausible that the (green) nonpolar areas insert into the nonpolar phase of the lipid bilayer, whereas the high-potential (red) spots may favorably interact with negatively charged lipid head groups like that of BMP and with the aqueous solvent.

Despite its strongly increasing charge when going from pH 5 to pH 4, the number of negative charges remains high for SapC, dropping from about 12 to 8, whereas for SapB this number changes from about 8 to 5. This means that SapC is more able to attract positive groups such as the positively charged cholin in the head group of the substrate sphingomyelin. These negative charges cause distinct patches of low electrostatic potential (Fig. 4), as may be expected for a pattern responsible for a specific protein-ligand interaction.

At very low values of pH, or, following the above argument, at high concentrations of negatively charged lipids like BMP, more acidic groups will become protonated (data

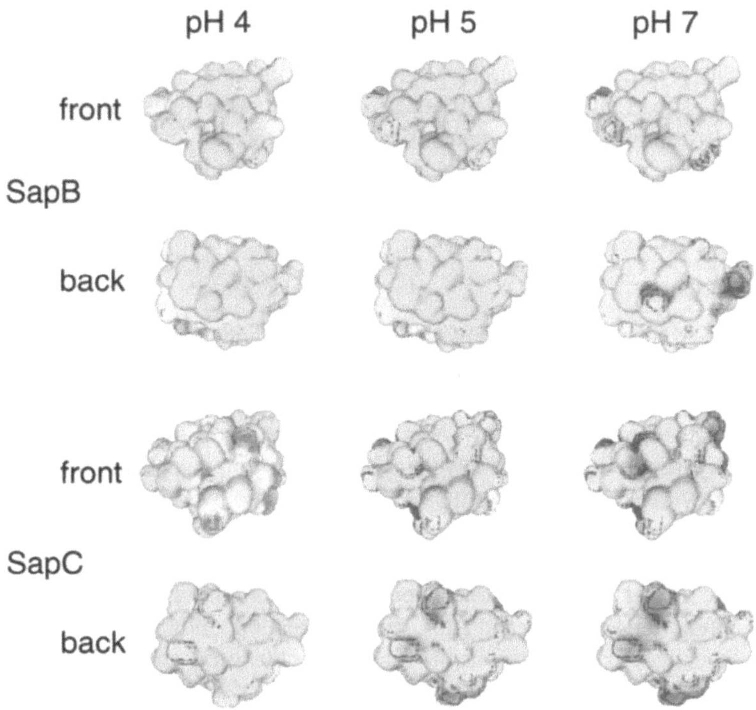

Figure 4. Electrostatic potential at the surfaces of EVAMs of SapB and SapC at pH 4, 5, and 7. (For colored picture see color plate 4).

not shown) and the negative patches will vanish. Under these conditions SapC will be "blind" for positively charged substrate groups and will no longer specifically interact with sphingomyelin and hence will fail to stimulate the degradation of this lipid. Together with the attractive interaction of BMP and SapC described above, we have a stimulation of SapC function at low concentrations of BMP because of this attraction between SapC and the negatively charged membrane, whereas at high concentrations of BMP, we expect a breakdown of specific interaction of SapC and sphingomyelin and hence a failure of SapC to stimulate the degradation of sphingomyelin. This dependence of the stimulating effect of SapC upon the concentration of BMP or other negatively charged lipids is indeed found experimentally [10].

In contrast to SapC, SapB at pH 4 shows no significant low-potential patches that could attract positively charged lipid head groups (Fig. 4), and at pH 5 it has much fewer low-potential patches than does SapC. Therefore, SapB seems unlikely to interact specifically with lipids like sphingomyelin in the presence of BMP or other negatively charged lipids. This is in agreement with experimental results, indicating that SapB does not stimulate degradation of sphingomyelin independent of the presence of BMP. However, SapB displays a high-potential patch (red area at lower left backside of SapB in Fig. 4) that is relatively stable in its extension over the considered values of pH 4, 5, and 7. Interestingly, SapB has been shown to support the decomposition of sulfatide, a sphingolipid with a negatively charged sulfate head group. The high potential patch in Fig. 4 could well mark the site of interaction between SapB and its sulfatide ligand. Since there is essentially only a single spot of high potential on the surface of SapB, sulfatide will have to compete with other negatively charged lipids for interaction with this binding site, and no stimulation of SapB action by BMP, etc., can be expected.

4. Conclusions

This study has shown that experimentally validated 3D models can provide insight into the mechanisms of protein function. In particular, the models suggest that a number of acidic residues of SapB and SapC acquire protons at values of pH between 4 and 5 (typical for their native lysosomal environment). As a consequence, at this pH large surface areas become more hydrophobic and are likely to insert into the nonpolar phase of the lipid bilayer, in agreement with experimental results. We do not need to hypothesize pH-triggered conformational changes that expose hydrophobic areas [12] to explain the observed behavior, but we also cannot exclude such changes on the basis of our study.

Despite the very plausible conclusions derived from this study, most questions concerning the functions of SapB and SapC remain open, e.g., how do the saposins interact with the respective enzymes responsible for the degradation of sphingolipids, and what is the basis of saposin specificity with respect to the various substrates and enzymes?

A thorough understanding of saposin function would require an adequate model comprising all parts involved, namely, saposin, membrane, substrate, aqueous solution, and enzyme. An analysis of experimental results may show that some of the interactions between components of this complex system could be considered independently, e.g., interactions among saposin, substrate, and enzyme. However, even the modeling of such a smaller system is difficult with the capabilities available today [4]. Hence, there is still need for the development of more powerful modeling tools as well as for more experiments.

Acknowledgments
The author gratefully acknowledges funding from the Center of Advanced European Studies and Research, the provision of computing facilities by the German National Research Center for Information Technology, and helpful comments from the group of Konrad Sandhoff, especially by Thomas Linke, Natascha Remmel, and Michaela Wendeler.

References

[1] Bashford D, Case DA, Dalvitt C, Tennant L, Wright PE (1993) Electrostatic calculations of side-chain pK_a values in myoglobin and comparison with NMR data for histidines. *Biochemistry* **32**, 8045–8056

[2] Bashford D, Karplus M (1990) pK_a's of ionizable groups in proteins: Atomic detail from a continuum electrostatic model. *Biochemistry* **29**, 10219–10225

[3] Brooks BR, Bruccoleri RE, Olafson BD, States DJ, Swaminathan S, Karplus M (1983) CHARMM: A program for macromolecular energy, minimization, and dynamics calculation. *J Comp Chem* **4**, 187–217

[4] Hoffmann D, Kramer B, Washio T, Steinmetzer T, Rarey M, Lengauer T (1999) Two-stage method for protein-ligand docking. *J Med Chem* **42**, 4422–4433

[5] Hoffmann D, Schnaible V, Wefing S, Albrecht M, Hanisch D, Zimmer R (2002) A new method for the fast solution of protein-3D-structures, combining experiments and bioinformatics. In: Coupling of Biological and Electronic Systems, *Proceedings of the 2nd caesarium*, K-H Hoffmann (ed.) Springer, Berlin, 59–78

[6] Hoffmann D, Zimmer R (1999) Chemical activation mediated by the transfer of fluorescent energy for elucidating the 3D-structure of biological macromolecules. *German Patent Office*, PCT WO 99/41607

[7] Kolter T, Sandhoff K (1999) Sphingolipids – Their metabolic pathways and the pathobiochemistry of neurodegenerative diseases. *Angew Chem Int Ed* **38**, 1532–1568

[8] Laaksonen L (2001) gOpenMol. `http://www.csc.fi/~laaksone/gopenmol/`

[9] Liepinsh E, Andersson M, Ruysschaert JM, Otting G (1997) Saposin fold revealed by the NMR structure of NK-lysin. *Nat Struct Biol* **4**, 793–795

[10] Linke T, Wilkening G, Lansmann S, Moczall H, Bartelsen O, Weisgerber J, Sandhoff K (2001) Stimulation of acid sphingomyelinase activity by lysosomal lipids and sphingolipid activator proteins. *Biol Chem* **382**, 283–290

[11] Mevissen H, Thiele R, Zimmer R, Lengauer T (1994) The ToPLign software environment – Toolbox for protein alignment. In *Bioinformatik '94*. IMB, Jena, Germany

[12] Vaccaro AM, Ciaffoni F, Tatti M, Savioli R, Barca A, Tognozzi D, Scerch C (1995) pH-dependent conformational properties of saposins and their interactions with phospholipid membranes. *J Biol Chem* **270**, 30576–30580

[13] Vaccaro AM, Salvioli R, Barca A, Tatti M, Ciaffoni F, Maras B, Siciliano R, Zappacosta F, Amoresano A, Pucci P (1995) structural analysis of saposin B and C. *J Biol Chem* **270**, 9953–9960

[14] Vaccaro AM, Salvioli R, Tatti M, Ciaffoni F (1999) Saposins and their interaction with lipids. *Neurochem Res* **24**, 307–314

Polymer and Cell Dynamics – Multiscale Modeling
and Numerical Simulations, 29–42
ed. by W. Alt, M. Chaplain, M. Griebel und J. Lenz

A Review of the Morph Server and the Macromolecular Motions Database: A Standardized System for Analyzing and Visualizing Macromolecular Motions in a Database Framework

Werner G. Krebs[1] and Mark Gerstein[2]

1) San Diego Supercomputer Center, Dept. 505, University of California, San Diego, 9500 Gilman Drive, La Jolla, CA 92093-0505, USA
2) Department of Molecular Biophysics & Biochemistry, Yale University, P.O. Box 208114, New Haven, CT 06520-8114, USA

Summary. The number of solved structures of macromolecules that have the same fold and thus exhibit some degree of conformational variability is rapidly increasing. It is consequently advantageous to develop a standardized terminology for describing this variability and automated systems for processing protein structures in different conformations. We have developed such a system as a 'front-end' server to our database of macromolecular motions, a database that classifies proteins into a limited number of categories, first on the basis of size (distinguishing among fragment, domain, and subunit motions) and then on the basis of packing. Our system attempts to describe a protein motion as a rigid-body rotation of a small 'core' relative to a larger one, using a set of hinges. The motion is placed in a standardized coordinate system so that all statistics between any two motions are directly comparable. We find that while this model can accommodate most protein motions, it cannot accommodate all; the degree to which a motion can be accommodated provides an aid in classifying it. Furthermore, we perform an adiabatic mapping (a restrained interpolation) between every two conformations. This gives some indication of the extent of the energetic barriers that need to be surmounted in the motion, and, as a byproduct, results in a 'morph movie.' We make these movies available over the Web to aid in visualization. Many instances of conformational variability occur between proteins with somewhat different sequences. We can accommodate these differences in a rough fashion, generating an 'evolutionary morph.' Users have already submitted hundreds of examples of protein motions to our server, producing a comprehensive set of statistics. So far the statistics show that the median submitted motion has a rotation of 10 degrees and a maximum C-alpha displacement of 17 Å. Almost all involve at least one large torsion angle change of >140 degrees. The server is accessible at `http://bioinfo.mbb.yale.edu/MolMovDB`.

1. Introduction

Function can often be thought of as being linked to structure by means of macromolecular motions (i.e., those of proteins and nucleic acids) that are often the essential link between structure and function. Because of their relationship to the principles of protein structure and stability, macromolecular motions, moreover, are of great intrinsic interest. By systematizing and analyzing many of the instances of protein structures solved in multiple conformations, it is now possible to study these motions within a database framework.

In this sense, our paper is less mathematical, since much of the mathematics (molecular dynamics and sieve-fitting) has already been worked out elsewhere, and we focus mainly on the application of existing mathematical methods into a database framework. In this sense, our work is complementary to that of other sections, as it illustrates how mathematical models may be integrated into a database framework. In this way, the database may be used as the computational bridge between empirical data and research (in the form of database entries) and mathematical models (software and tools integrated into the database).

Here we present a comprehensive database of macromolecular motions, intended to be useful to those studying structure-function relationships (in particular, rational drug design [27]) and also to those involved in large-scale protein or genome surveys. Such a database became feasible and favorable in the late 1990s for a number of reasons:

(i) The amount of raw data (known protein structures and sequences homologous to them) was exponentially increasing [21, 32], and an appreciable fraction of new structures had non-trivial motions.

(ii) The graphical and interactive nature of a database was particularly well suited for presenting macromolecular motions, which are often difficult to represent on a static printed page.

(iii) A loose federation of databases had emerged in the structural community, allowing the motions database to connect to a variety of information sources. There had been only one previous attempt made at the systematic classification of protein motions [4].

One of the best and most obvious ways to communicate protein motions is through "movies," especially when they are made available over the Web. Vonrhein et al. [48, 49], Sawaya et al. and other groups have made custom movies of protein motions available over the Web [33, 34, 40, 41, 43, 53, 54].

Here we present a perspective on how protein motions can be put into standardized, consistent terms. We develop a simple model for protein motions involving rigid-body motion of parts, apply our model to actual cases, and measure how well it fits. An integrated Web server provides tools to compare solved conformations of proteins involved in motion, generates statistics to characterize and classify them into a database, and automatically makes a morph movie to represent them.

The Web server is accessible at:

`http://bioinfo.mbb.yale.edu/MolMovDB/morph.`

It is integrated with the Database of Macromolecular Motions [13, 19]

(http://bioinfo.mbb.yale.edu/MolMovDB)

and is also connected with a variety of tools for aligning protein folds and studying their occurrence in genomes [12, 17, 18, 29].

2. The Database of Macromolecular Motions

2.1. Unique Motion Identifier

A single protein or nucleic acid can have a number of motions and the same essential motion can be shared among different macromolecules. For this reason, each entry is indexed by a unique motion identifier rather than around individual macromolecules.

2.2. Attributes of a Motion

Each entry has the following information in addition to the motion identifier:

- Classification. A classification number gives the place of a motion in the size and packing classification scheme for motions described below. In addition to its basic classification, a motion can also be annotated as being particularly "similar to" one in another or "part of" or "containing" another motion in the same protein.
- Structures. The identifiers have been made into hypertext links that link indirectly to the structure entries at the RCSB and to sequence and journal cross-references *via* the Entrez database [10, 42]. Links are also made to related structures *via* the Structural Classification of Proteins (SCOP) [31].

For most entries we describe the overall motion using standardized numeric terminology, such as the maximum displacement (overall and of only backbone atoms), the degree of rotation around the hinge, and residues with large torsion angle changes when these numbers are available from the scientific literature. (The morph server, described below, attempts to automatically compute these values from the structures.). Each entry has links to graphics and movies describing the motion, often depicting a plausible interpolated pathway (see below).

2.3. Size Classification

The classification scheme for proteins has a hierarchical layout shown in Figure 1. Protein motions are first ranked in order of their size (subunit, domain, and fragments). Domain motions, such as those in hexokinase or citrate synthase [3, 35], provide the most common examples of protein flexibility [2, 16, 23]. Usually, the motion of fragments smaller than domains refers to the motion of surface loops, such as the ones in triose phosphate isomerase or lactate dehydrogenase. It can also refer to the motion of secondary structures, such as of the helices in insulin [1, 7, 52]. Domain and fragment motions are important for a variety of protein functions and usually involve portions of the protein closing around a binding site, with a bound substrate stabilizing a closed conformation. Subunit motions are distinctly different and often involve allosteric effects.

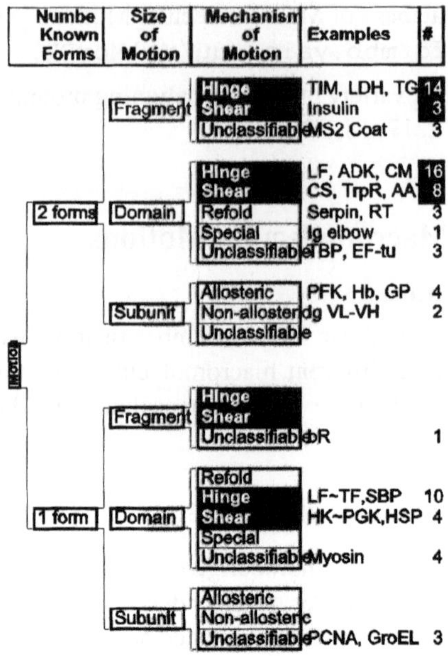

Figure 1. Schematic showing the overall classification scheme for motions. The database is organized around a hierarchical classification scheme, based on size (fragment, domain, subunit) and then packing (hinge or shear). Figure adapted from [13, 15, 16].

2.4. Packing Classification

For fragment and domain protein motions, we have systematized the motions on the basis of the packing of atoms inside of proteins, which is a fundamental constraint on protein structure [20, 30, 36, 37, 38]. Interfaces between different parts of a protein are usually packed very tightly. Consequently, two basic mechanisms for protein motions, hinge and shear, are proposed depending on whether or not there is a continuously maintained interface preserved through the motion. A complete protein motion can be built up from a number of these basic motions. For the database, a motion is classified as "shear" if it is predominately a shear motion and "hinge" if it is predominately composed of hinge motions.

The shear mechanism basically describes the special kind of sliding motion a protein must undergo if it wants to maintain a well-packed interface; these constraints mean that individual shear motions are constrained to be very small.

When no continuously maintained interface constrains the motion, a hinge motion occurs. Typically, these motions occur in proteins with two domains (or fragments) connected by linkers (i.e., hinges) that are relatively unconstrained by packing. The whole motion may be produced by a few large torsion angle changes.

Over 60% of the motions in the database are classified as domain motions, while the hinge mechanism is the most common mechanistic classification in the database, accounting for 45% of the entries. Reflecting the greater ease with which smaller motions can be studied experimentally, a greater percentage of fragment motions have structures for multiple conformations in the motion. Most of the fragment and domain motions in the database fall into the hinge or shear classification. There are some exceptions:

(i) A special mechanism that is clearly neither hinge nor shear accounts for the motion. An example of this sort of motion is what occurs in the immunoglobulin ball-and-socket joint [28], where the motion involves sliding over a continuously maintained interface (like a shear motion), but because the interface is smooth and not interdigitating, the motion can be large (like a hinge).

(ii) Motion involves a partial refolding of the protein. This usually results in dramatic changes in the overall structure.

(iii) Motion cannot yet be classified is a catch-all category. Subunit motions are classified differently as either allosteric, non-allosteric, or unclassifiable.

(iv) Complex motions. Finally, large protein motions that cannot easily be classified as subunit motions are classified as complex movements, e.g., the order-to-disorder transition that the headpiece domain undergoes when it binds DNA. Another example involves a molecule binding between two other domains in the protein, such as observed in the bacterial periplasmic binding proteins [50].

3. Annotation of Evidence Related to the Motion

We have tried to indicate the evidence behind its description and classification for every entry in the database. For example, the classification might be based on careful manual analysis of two conformations, based on automatic output of a conformation comparison program, inferred based on structure comparison, or inferred based on sequence comparison. We make a clear distinction between the carefully analyzed "gold-standard" motions, such as lactoferrin, and the much more tentatively understood motion in a protein that is a sequence homologue of another protein that is structurally similar to lactoferrin. We indicate the evidence behind a motion through listing information about the experimental techniques used, telling whether or not the motion is inferred, and giving a standardized "annotation level." At the time of this writing, the database describes 120 "gold standard" macromolecular motions that reference 241 PDB (Protein Data Base) structures, as well as larger set of 4000 motions automatically culled from the RCSB PDB.

4. Computer Implementation as a Relational Database

The database originally implemented as a free relational database server engine, called mini-SQL [22], has been used with a schema that contains $\sim$10 tables. Subsequently, the database was migrated to Informix Online Dynamic Server, a commercial product that traces its roots to the postgreSQL database project at Berkeley [25, 39, 44]. Although we plan to maintain pre-built pages on our Web server, we are investigating the use an Apache HTTP modperl system to allow instantaneous updates to the database's Web presence yet maintain a level of performance comparable to static pages.

5. The Morph Server: Representing Motion Pathways

We have begun a statistical characterization of macromolecular motions using the significant 'standardized values' that describe each motion, such as maximum atomic displacement or degrees of rotation. It is advantageous to develop a standardized terminology and automated systems for processing protein structures in different conformations and for determining these statistics. We have integrated such a system into our database of macromolecular motions. Our system attempts to describe protein motions as a rigid-body rotation of a small "core" relative to a larger one, using a set of hinges. To ensure that all statistics between any two motions are directly comparable, the motion is placed in a standardized coordinate. Although our model can accommodate most protein motions, it cannot accommodate all, and the degree to which a motion can be accommodated provides an aid in classifying it. We performed an adiabatic mapping (a restrained interpolation) between every two conformations. Users have already submitted hundreds of examples of protein motions to our server, producing a comprehensive set of statistics.

A major aim of the morph server is to illustrate how protein motions can be put into standardized, consistent terms. The morph server embodies a simple model for protein motions involving rigid-body motion of parts. Our approach is embodied in a Web server (the morph server) integrated into our main Database of Macromolecular Motions and provides tools to compare solved conformations of proteins involved in motion, generates statistics to characterize and classify them into a database, and automatically makes a morph movie to represent them. In addition, the server presents a database linking protein motions with custom movies of motions available at other sites, along with our own morphs generated automatically by the server upon request by members of the Internet community. Our server and database have been used by Internet users to analyze a number of recent structures including human interleukin 5 [47], bc1 complex [8, 9], glycerol kinase [6, 11], and lactoferrin [45, 46].

The database contains graphics showing the structures and some representation of the pathway for the motion, in addition to its textual elements. Without special techniques, such as high temperature simulation or Brownian dynamics [24, 51], normal dynamics simulations cannot approach the time scales of the large-scale motions in the database. Rather, a pathway movie is produced as an interpolation between known endpoints (usually two crystal structures). There are two principal methods:

(i) Straight Cartesian interpolation. The difference in each atomic coordinate (between the known endpoint structures) is calculated and then divided into a number of evenly spaced steps.

(ii) Adiabatic mapping. The database is home to a server that generates interpolated motion representations using this technique. This is a modification of straight Cartesian interpolation, adding the addition of energy minimization after each Cartesian interpolation step. This procedure produces interpolated frames with much more realistic geometry.

6. Information Flow

One starts by submitting to the server two or more conformations of a given protein. Then, the server classifies the motion in the database and produces an appealing movie through a wide variety of transformations.

6.1. Data Sources

Solved conformations analysis as performed by the server requires two kinds of information:

(i) three-dimensional atomic coordinates of protein conformations as solved structure files (such as those at the PDB) and, more importantly,

(ii) information relating two or more of these solved structures, thus selecting them for analysis. (This information has come from both automated searching of databases for proteins related by structure or sequence and from a simple user input form on the Web.) A selection scheme is important because the number of ordered pairs of PDB structures is rather large (more than $10,000^2$). Figure 2 diagrams the server in the larger context of data sources.

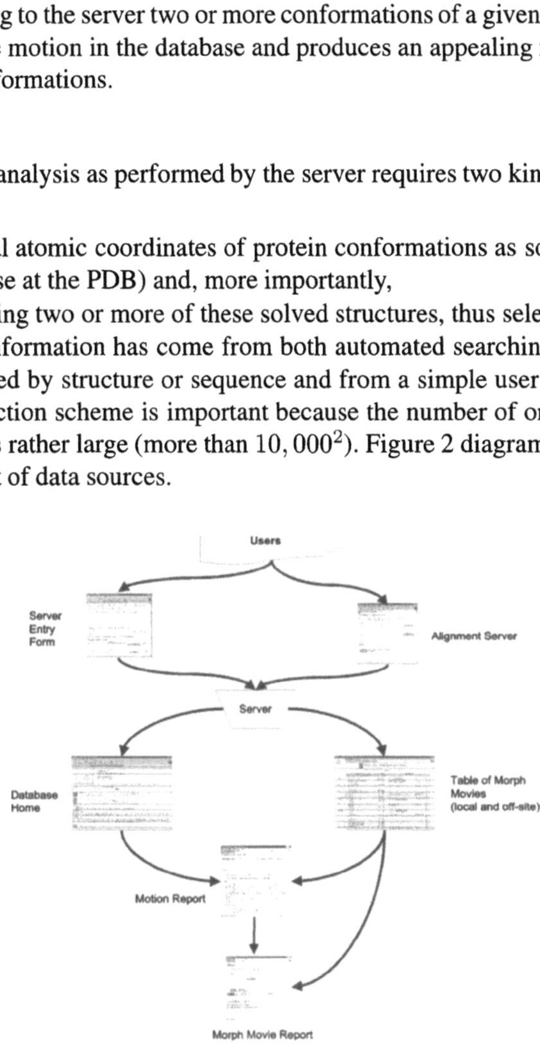

Figure 2. The morph server and motions database on the Web. This figure shows the interaction of the various Web pages to make up the morph server and Database of Macromolecular Motions. Figure adapted from [26].

6.2. Alignment

The latter method relies primarily on the use of 3D coordinates (i.e., solved PDB structures of proteins) to produce a sequence alignment otherwise analogous to an alignment produced purely from sequence information. Sequence alignment is used unless sequence similarity is below a user-defined cutoff, at which point structure alignment is used. The choice of approach (sequence or structural alignment) may also be forced by the user upon morph submission.

6.3. Superposition

One of the major aims of the server is to collect standardized statistics on the proteins involved in motions. Traditional "all-atom" RMS superposition minimizes the RMS difference between Cα atoms in the open and closed conformations. In a simple hinge motion, e.g., Calmodulin, such an alignment fits the closed conformation symmetrically inside the open conformation. Among other things, the maximum C-alpha displacement computed from such a superposition is considerably underestimated from the common sense alignment, and a morph movie made with such an alignment would give the impression of motion far more complicated than a simple opening of a hinge.

6.4. Orientation and Hinge Location

A comparison of the new position of the ending conformation following this latest fit with its position following the "sieve-fit" procedure yields a geometric transformation whose screw axis is (approximately) the axis of the hinge motion, i.e., the location of the hinge, as has been published elsewhere [14]. Straightforward calculations allow characterization of the angle of rotation around the hinge axis.

If a significant hinge motion is present, the software uses these transformations to align the Z-axis of the coordinate frame parallel to the hinge axis so that, when the motion is rendered, viewers will look down the screw-axis of the hinge motion.

6.5. Interpolation

Once the structures have been homogenized in terms of solved atomic coordinates, interpolation may proceed. When more than two solved conformations are present, the distance between the current structure and a solved intermediate conformation is evenly reduced instead. This remains the default interpolation engine for the server. We use a modified X-PLOR system to perform the adiabatic mapping. Brünger [5] provides an excellent reference on the mathematics of molecular dynamics in X-PLOR.

6.6. Visual Rendering

With the intermediate conformations morphed, the molecule is now visually rendered. The first two formats are also rendered into a small moving MultiGif icon to afford the database user with a quickly downloaded preview of the larger movies available.

6.7. Statistics

A server collects a number of statistics, include hinge angle rotation during motion. Of the over approximately 175 motions submitted for analysis, the median motion has a maximum rotation of 9.5° over a range of 0 through 150° as computed by our algorithm, whereas the 12 motions culled from the scientific literature had an average rotation of 24° over a range of 5–148°. Similarly, our algorithms found a median maximum C-alpha displacement of 17 Å ranging from 0 to 81 Å for the submitted motions, whereas 11 motions reported in the scientific literature average 12 Å over a range of 1.5–60 Å. Although most of the structures are very similar in sequence, the server has been able to accommodate sequence identity down to 8% for some motions. Most motions have at least one large torsion angle change. Table 1 presents a statistical description of motions in the database, a main scientific benefit of the server.

6.8. Integration with Database

The server produces a Web page for each successfully completed morph, allowing easy download of the coordinates or as movies (in a number of video formats), in addition to displaying the molecule in the moving VRML format. The page also includes the standardized statistics (previously mentioned) generated for the conformations used in the morph. A URL containing a special code that is e-mailed back to the submitting user when the morph is complete allows access to the morphs; for users seeking to keep their morphs private (for publication reasons), this URL serves as the user's password, allowing access to the morph page in the server. The public morphs are also accessible through an index `http://bioinfo.mbb.yale.edu/MolMovDB/movies`.

Each motion is linked either *via* a motion ID to an existing entry in the Macromolecular Motions Database or to a newly generated entry in the database. Each comparison is assigned a unique ID entered into a "comparison table" in the database that references the IDs of the PDB structures involved. These comparisons are, in turn, referenced by entries in the motions database. More than one comparison is often possible, and some database entries do reference multiple comparisons because many motions in the database are associated with more than two structures.

To illustrate the technique of adiabatic mapping as implemented by the server, Figure 3 depicts the frames in five automatically generated morphs produced by the server's adiabatic mapping interpolation engine. Comprehensive statistics for all five morphs can be found in Table 1.

6.9. Statistics

In most cases the structures of a given macromolecule involved in motions have been solved in two or more conformations, so that endpoint structures for the motion are available. This, in turn, means that automatic conformation comparison tools are possible, which, when applied en masse, allow the generation of a consistent, standardized set of statistics characterizing the motions in the database. In the process of analyzing the structures, pathway interpolation is performed as well.

6.10. What Constitutes an Optimal Morph?

One might define the "optimal morph" as the most likely (or most frequently taken) pathway between two conformations. Unlike the adiabatic interpolation engine used in the server, many interpolation engines on proteins take approaches that do not meet these criteria. With the exception of the simplest motions, a simple linear interpolation of atomic coordinates yields intermediates with clearly distorted geometry. The adiabatic mapping interpolation engine used by the server takes physical reality more into account, producing a much better morph.

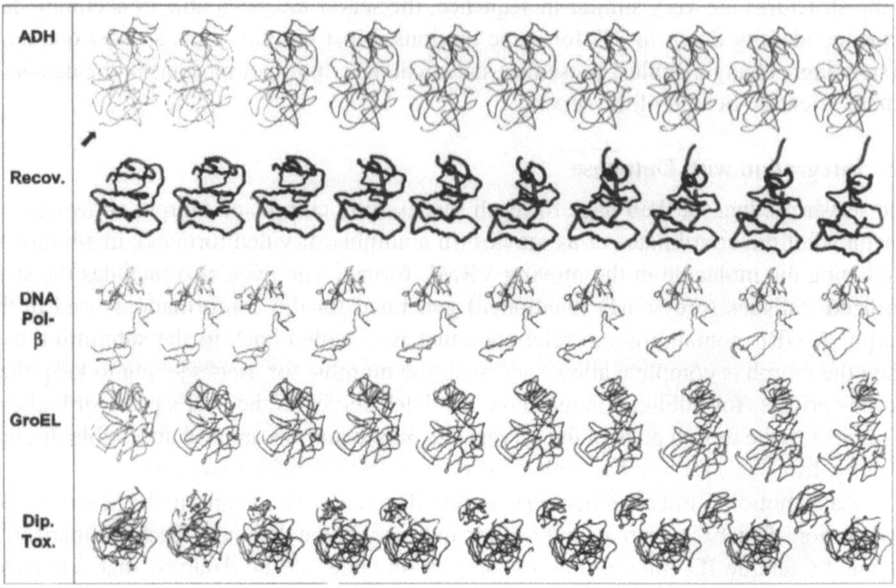

Figure 3. Interpolated motion pathways. This figure illustrates the morph "movies" produced by the server. The first morph at top is alcohol dehydrogenase, which has a stunning motion when viewed as a movie on the Web, but is very difficult to see on the static printed page. Figure adapted from [26].

7. Conclusions

We have developed a database of macromolecular motions along with an integrated set of protein conformation comparison tools on the Web for use in conjunction with the database or as a stand-alone, publicly accessible server. The server can produce a useful comparison of the structures involved in protein motions when solved endpoint structures are available. The server then uses an adiabatic mapping technique to generate a visually rendered interpolated pathway, or 'morph', of the motion or evolution of the protein.

The server also collects a number of statistics on the motion, including maximum C-alpha displacement and maximum rotation around the putative hinge. These are useful in both analyzing and classifying individual proteins and in generating a statistical picture of motions in the motions database as a whole. The software then presents the visual representation, statistics, orientation, alignment, and interpolated coordinates to the user. We have found the server useful in the analysis of protein motions and anticipate that use of the server will help standardize statistics and nomenclature for protein motions subsequently presented in the scientific literature.

	Statistic	[Code]	Easy	Typical		Large	Impossible
			ADH	Recoverin	DNA Pol-Beta	GroEL	Diphtheria Toxin
Input Structures	Motion ID	[ID]	adh	recvin	polbeta	groel	dt
	1st input frame	[inputframe0]	8ADH	1IKU	1BPD	1GRL	1DDT
	2nd input frame	[inputframe1]	6ADH	1JSA	2BPF	1AON	1MDT
	Size (Å) (in terms of window for rendering)	[max_x_or_y]	36	41	52	55	39
	Number of atoms	[natoms]	2887	1639	2697	3993	4110
	Number of residues	[nresidues]	374	201	335	548	535
Overall Motion	Overall RMS between first and last frames	[RMSoverall]	2.0	13	8.6	16	20
	Rotation (degrees)	[kappa]	4.9°	73°	9.9°	62°	62°
	Overall translation of centroid (Å)	[translation]	2.1	13	6.1	47	66
	X translation (Å)	[TransX]	1.1	-0.24	0.94	45	-45
	Y ""	[TransY]	-0.95	-9.14	4.1	-2.1	-0.54
	Z ""	[TransZ]	1.5	-9.78	-4.4	-10	48
1st Core	Number Cα's in 1st core	[AlignedCoreCAs]	187	95	160	259	262
	RMS of 1st core (Å)	[AlignedCoreRMS]	0.40	3.0	0.92	1.4	0.37
	Max Cα displacement in 1st Core (Å)	[MaxCore Deviation]	0.66	7.6	1.7	4.2	0.60
2nd Core	Num. Cα's in 2nd core	[2ndCoreCAs]	190	94	160	260	260
	RMS of 2nd core (Å)	[2ndCoreRMS]	2.9	18	12	23	29
	Max Cα displacement in 2nd core (Å)	[Max2ndCore Deviation]	7.1	38	28	49	60
	RMS of 2nd core (Å) after fitting on 1st core	[2ndCoreRMS postrefitting]	1.6	11	11	10	18
Hinge	Number of putative hinges detected	[NHinges]	0	0	0	1	1
	X position of 1st hinge (Å) rel. to centroid	[Hinge000X]	-	-	-	-4.7	-7.2
	Y position ""	[Hinge000Y]	-	-	-	11	-0.91
	Z position ""	[Hinge000Z]	-	-	-	3.3	-3.0
	1st Hinge Residue Selection	[Hinge000res]	-	-	-	380:403	352:375
	Sequence of 1st putative hinge	[Hinge000sec]	-	-	-	EVEMKE KKARVE DALHAT RAAVEE	NLFQVV HNSYNR PAYSPG HKTQP
Screw Axis	Distance betw. screw-axis (x0) & centroid (Å)	[x0ToCentroid Distance]	21	8.4	23	30	39
	X displacement centroid from screw axis (Å)	[x0X]	-0.16	-0.5	-2.5	17	-20
	Y ""	[x0Y]	-5.0	-8.2	-5.2	-16	-24
	Z ""	[x0Z]	-20	5.7	-22	19	-24
	Distance between screw axis and 1st hinge (Å)	[Hinge000x0dist]	-	-	-	26	45
Torsion Angles	Max phi change (Max of Abs. degrees, 0°-180°)	[MaxPhi]	180°	180°	180°	180°	180°
	Max psi change	[MaxPsi]	180°	180°	180°	180°	170°
	Max alpha change	[MaxAlpha]	150°	180°	180°	180°	170°

Table 1. Standardized statistics for the magnitude of the motions produced by the morph server. This table gives standardized statistics for five protein motions. The motions range from "easy" for a small motion to "impossible." The latter illustrates an example of domain swapping, rather than a true motion. Figure adapted from [26].

Acknowledgments

The authors gratefully acknowledge the financial support of the Keck Foundation and the U.S. National Science Foundation (Grant DBI-9723182). Numerous people have also either contributed entries or information to the database and morph server or have given us feedback on what the user community wants. The authors also wish to thank Informix Software, Inc. for providing a grant of its database software.

References

[1] Abad-Zapatero, C., J. P. Griffith, et al. (1987) Refined Crystal Structure of Dogfish M4 Apo-lactate Dehydrogenase. *J Mol Biol* **198**, 445–67

[2] Bennett, W. S. and R. Huber (1984) Structural and Functional Aspects of Domain Motion in Proteins. *Crit Rev Biochem* **15**, 291–384

[3] Bennett, W. S., Jr and T. A. Steitz (1978) Glucose induced conformational change in yeast hexokinase *Proc Natl Acad Sci USA* **75**, 4848–4852

[4] Boutonnet, N. S., M. J. Rooman, et al. (1995) Automatic analysis of protein conformational changes by multiple linkage clustering. *J Mol Biol* **253**, 633–647

[5] Brünger, A. T. (1993) X-PLOR 3.1, *A System for X-ray Crystallography and NMR.* New Haven, Yale University Press.

[6] Bystrom, C. E., D. W. Pettigrew, et al. (1999) Crystal structures of Escherichia coli glycerol kinase variant S58–>W in complex with nonhydrolyzable ATP analogues reveal a putative active conformation of the enzyme as a result of domain motion. *Biochemistry* **38**, 3508–3518

[7] Chothia, C., A. M. Lesk, et al. (1983) Transmission of conformational change in insulin *Nature* **302**, 500–505

[8] Crofts, A. R. and E. A. Berry (1998) Structure and function of the cytochrome bc1 complex of mitochondria and photosynthetic bacteria. *Curr Opin Struct Biol* **8(4)**, 501–509

[9] Crofts, A. R., S. Hong, et al. (1999) Pathways for proton release during ubihydroquinone oxidation by the bc(1) complex. *Proc Natl Acad Sci USA* **96**, 10021–10026

[10] Epstein, J. A., J. A. Kans, et al. (1994) WWW Entrez: A Hypertext Retrieval Tool for Molecular Biology. *2nd Ann Int WWW Conf*

[11] Feese, M. D., H. R. Faber, et al. (1998) Glycerol kinase from Escherichia coli and an Ala65–>Thr mutant: the crystal structures reveal conformational changes with implications for allosteric regulation. *Structure* **6**, 1407–1418

[12] Gerstein, M. (1997) A Structural Census of Genomes: Comparing Bacterial, Eukaryotic, and Archaeal Genomes in terms of Protein Structure. *J Mol Biol* **274**, 562–576

[13] Gerstein, M. and W. Krebs (1998) A Database of Macromolecular Movements. *Nucl. Acids Res* **26**, 4280

[14] Gerstein, M., A. Lesk, et al. (1995) *Structural Mechanisms for Protein Motions. Protein Motions.* S. Subbiah. Austin, TX, R G Landes.

[15] Gerstein, M., A. M. Lesk, et al. (1993) Domain Closure in Lactoferrin: Two Hinges produce a See-saw Motion between Alternative Close-Packed Interfaces. *J Mol Biol* **234**, 357–372

[16] Gerstein, M., A. M. Lesk, et al. (1994) Structural Mechanisms for Domain Movements. *Biochemistry* **33**, 6739–6749

[17] Gerstein, M. and M. Levitt (1998) Comprehensive Assessment of Automatic Structural Alignment against a Manual Standard, the Scop Classification of Proteins. *Protein Science* **7**, 445–456

[18] Gerstein, M. and C. Wilson (2000) Assessing Annotation Transfer for Genomics: Quantifying the relations between protein sequence, structure, and function through traditional and probabilistic scores. *J Mol Biol* **297**, 233–249

[19] Gerstein, M. B., R. Jansen, et al. (1999) Studying Macromolecular Motions in a Database Framework: From Structure to Sequence. *Rigidity theory and applications.* (M. F. Thorpe and P. M. Duxbury) Kluwer Academic/Plenum press, 401–442

[20] Harpaz, Y., M. Gerstein, et al. (1994) Volume Changes on Protein Folding. *Structure* **2**, 641–649

[21] Holm, L. and C. Sander (1996) Mapping the Protein Universe. *Science* **273**, 595–602

[22] Hughes, D. (1996) mini-SQL program. `http://Hughes.com.au.`

[23] Janin, J. and S. Wodak (1983) Structural domains in proteins and their role in the dynamics of protein function. *Prog Biophys Mol Biol* **42**, 21–78

[24] Joseph, D., G. A. Petsko, et al. (1990) Anatomy of Conformational Change: Hinged 'Lid' Motion of the Triosephosphosphate Isomerase Loop. *Science* **249**, 1425–1428

[25] Korth, H. and A. Silberschatz (1991) *Database system concepts*, 2nd edition. New York, McGraw-Hill

[26] Krebs, W. G. and M. Gerstein (2000) The morph server: a standardized system for analyzing and visualizing macromolecular motions in a database framework. *Nucleic Acids Res* **28**, 1665–1675

[27] Kuntz, I. D. (1992) Structure-Based Strategies for Drug Design and Discovery. *Science* **257**, 1078–1082

[28] Lesk, A. M. and C. Chothia (1988) Elbow Motion in the immunoglobulins involves a molecular ball and socket joint. *Nature* **335**, 188–190

[29] Levitt, M. and M. Gerstein (1998) A Unified Statistical Framework for Sequence Comparison and Structure Comparison. *Proceedings of the National Academy of Sciences USA* **95**, 5913–5920.

[30] Levitt, M., M. Gerstein, et al. (1997) Protein Folding: the Endgame. *Ann Rev Biochem* **66**, 549–579

[31] Murzin, A., S. E. Brenner, et al. (1995) SCOP: A Structural Classification of Proteins for the Investigation of Sequences and Structures. *J Mol Biol* **247**, 536–540

[32] Orengo, C. A., D. T. Jones, et al. (1994) Protein superfamilies and domain superfolds. *Nature* **372**, 631–634

[33] Pande, V. S. and D. S. Rokhsar (1999) Folding pathway of a lattice model for proteins. *Proc Natl Acad Sci USA* **96**, 1273–1278

[34] Pande, V. S. and D. S. Rokhsar (1999) Molecular dynamics simulations of unfolding and refolding of a beta- hairpin fragment of protein G [In Process Citation]. *Proc Natl Acad Sci USA* **96**, 9062–9067

[35] Remington, S., G. Wiegand, et al. (1982) Crystallographic refinement and atomic models of two different forms of citrate synthase at 2.7 and 1.7 Å resolution. *J Mol Biol* **158**, 111–152

[36] Richards, F. M. (1977) Areas, Volumes, Packing, and Protein Structure. *Ann Rev Biophys Bioeng* **6**, 151–176

[37] Richards, F. M. (1985) Calculation of Molecular Volumes and Areas for Structures ofKnown Geometry. *Methods in Enzymology* **115**, 440–464

[38] Richards, F. M. and W. A. Lim (1994) An analysis of packing in the protein folding problem. *Quart Rev Biophys* **26**, 423–498

[39] Rowe, L. A. and M. R. Stonebraker (1987). The POSTGRES data model. *Proceedings of the Thirteenth International Conference on Very Large Data Bases: 1987 13th VLDB*. (P. M. Stocker, W. Kent and P. Hammersley) Los Altos, CA, USA, Morgan Kaufmann, 83–96

[40] Rye, H., S. Burston, et al. (1997) Distinct actions of cis and trans ATP within the double ring of the chaperonin GroEL. *Nature* **388**, 792–798

[41] Sawaya, M. R., R. Prasad, et al. (1997). Crystal structures of human DNA polymerase beta complexed with gapped and nicked DNA: evidence for an induced fit mechanism. *Biochemistry* **36**, 11205–11215

[42] Schuler, G. D., J. A. Epstein, et al. (1996) Entrez: Molecular Biology Database and Retrievel System. *Meth Enz* **266**, 141–162.

[43] Sigler, P. B., Z. Xu, et al. (1998) Structure and function in GroEL-mediated protein folding. *Annu Rev Biochem* **67**, 581–608

[44] Stonebraker, M. R. and L. A. Rowe (1986). *The Design of POSTGRES.* Proc. 1986 ACM-ACM-SIGMOD Conf. on Management of Data Int. Conf. on Mgt. of Data.

[45] Sykes, J. A., M. J. Thomas, et al. (1982) Plasma lactoferrin levels in pregnancy and cystic fibrosis. *Clin Chim Acta* **122**, 385–393

[46] Thompson, A. B., M. B. Mueller, et al. (1992) Aerosolized beclomethasone in chronic bronchitis. Improved pulmonary function and diminished airway inflammation. *Am Rev Respir Dis* **146**, 389–95

[47] Verschelde, J. L., C. Ampe, et al. (1998) Analysis of three human interleukin 5 structures suggests a possible receptor binding mechanism. *FEBS Lett* **424**, 121–126

[48] Vonrhein, C., H. Bonisch, et al. (1998) The structure of a trimeric archaeal adenylate kinase. *J Mol Biol* **282**, 167–179

[49] Vonrhein, C., G. J. Schlauderer, et al. (1995) Movie of the structural changes during a catalytic cycle of nucleoside monophosphate kinases. *Structure* **3**, 483–490

[50] Vyas, N. K., M. N. Vyas, et al. (1991) Comparison of the periplasmic receptors for L-arabinose, D-glucose, and D-ribose – structural and functional similarity. *J Biol Chem* **266**, 5226–5237

[51] Wade, R. C., M. E. Davis, et al. (1993) Gating of the active site of triose phosphate isomerase: Brownian dynamics simulations of flexible peptide loops in the enzyme. *Biophys J* **64**, 9–15

[52] Wierenga, R. K., M. E. M. Noble, et al. (1991) The crystal structure of the "open" and the "closed" conformation of the flexible loop of trypanosomal triosephosphate isomerase. *Proteins* **10**, 93

[53] Xu, Z., A. L. Horwich, et al. (1997) The crystal structure of the asymmetric GroEL-GroES-(ADP)7 chaperonin complex. *Nature* **388**, 741–750

[54] Xu, Z. and P. B. Sigler (1998) GroEL/GroES: structure and function of a two-stroke folding machine. *J Struct Biol* **124**, 129–41

Part II

Polymer Dynamics and Cell Motility

"Wir haben ausreichenden Grund, anzunehmen, dass wir lediglich in der Gruppirung der unter sich verschiedenen Micellen des Protoplastins, ihrem engern Zusammen- und weiteren Auseinandertreten einstweilen den letzten, rein atomistischen Grund der chemischen und plastischen Thätigkeit des Protoplasten – also des ganzen Organismus – suchen dürfen."

[Johannes von Hanstein (1880) "Das Protoplasma als Träger der pflanzlichen und thierischen Lebesverrichtungen". C. Winter, Heidelberg, p. 291]

English translation: *"We have enough reason to assume that only in the grouping of various micelles of the protoplastin, their closer aggregation and wider dispersal we might, for the time being, search for the ultimate, purely atomistic basis of the chemical and plastic activity of the protoplast – i.e., of the whole organism."*

Introduction to Part II

The dynamic behavior of biopolymers strongly depends on their *in situ* configuration and mutual interactions. These mostly 'molecular' properties of biopolymers obviously have evolved in strong relation to their eventual *in vivo* functions, which appear within a broad range of biochemical, biophysical, and cell biological applications. The series of situations addressed and exemplified here in Part II extends from single polymer motion (thereby extending the molecular dynamics approach of Part I), to polymer cross-linking and network formation, and, finally, to single cell deformation and motility, both essentially depending on passive or active mechanical properties of actin filaments, the main constituent of the cytoplasm. Therefore, mathematical modeling and numerical simulations of these phenomena have to cover a wide spectrum of mechanical aspects on different spatial and temporal scales. One desired effect of the contributions in Part II (partially assembled in a 'cooperative' manner) is that some of the presented models that operate at a microscopic level may serve as submodel 'ingredients' for other models on a more macroscopic level (also in Part III).

The first three chapters present stochastic multiparticle models for the dynamic behavior of single polymers or polymer networks, partially applying newly developed numerical simulation algorithms for corresponding stochastic ODE systems of generalized Langevin type. On the other hand, the subsequent three chapters rely on continuum models of cell motion, either described by viscoelastic (or contractile) fluid dynamics in a deformable domain with suitable numerical discretizations or, in the last chapter, by a directly discretized dynamic system for elastic body deformation and motion.

The first contribution by **Rzehak, Arend, Kienle & Zimmermann** uses a simplified 'bead-spring' model in order to simulate the dynamics of *flexible polymer chains* that are tethered at one end and are exposed to an outer hydrodynamic flow. Model beads representing a coarse-grained discretization of the elastic polymer chain are connected by (nonlinear) springs. After adding hydrodynamic interactions (for an approximating steady state) and by regarding a non-overlap condition for the polymer beads, the resulting stochastic multiparticle system of Langevin type is appropriate to answer the two basic questions of interest: how does the outer flow deform the polymer chains, and, how does their motion alter the flow pattern. One biological application mentioned concerns possible hydrodynamic effects during ligand motion towards or away from a polymer binding site (cf. also [13]). Another application might occur in cytoplasmic streaming, where the more fluid-like cytosol can flow through a network of cytoskeletal polymers, possibly inducing their deformation or alignment. For a corresponding two-phase flow continuum model (cf. [4]), see also the alternative models below.

However, cytoskeletal polymer chains like actin filaments, for example, should be modeled as *semi-flexible polymer chains* since they are not compressible or stretchable but undergo elastic bending only. This well-known type of dynamic model description with an overdamped stochastic Langevin system is used in the contribution by **Farkas, Derényi & Vicsek**. They model an experimental motility assay, where actin filaments, at specific binding sites along their chain, dynamically interact with myosin monomers fixed on a planar substratum. As in muscle motion, the myosin heads can stochastically exert shifting forces to the polarized filament. The resulting gliding motion or rotation of the actin filament is simulated, and the corresponding speeds are compared to experimental data.

The same mechanism of myosin-induced force interaction, which contracts muscle sarcomeres – with their highly ordered and aligned filamenteous structure – is effective in the contractile motion of less-ordered actin-myosin polymer networks occuring in almost all non-muscle cells. Simulations of a corresponding 3D stochastic multiparticle model are presented in the first part of the contribution by **Lenz & Felix**. Here, dynamic interactions between two neighboring actin filaments, described as stiff rods, occur if dimers of myosin molecules by chance connect two specific binding sites and, as before, excert a 'pulling force' on both filaments. Depending on their polarization and relative position, the simulated actin filaments tend to align either in a parallel or anti-parallel manner. Furthermore, simulations of 3D actin-myosin polymer networks, with an embedded magnetic bead experiencing an oscillating outer force, can be used to reproduce experimental measurements of magnetic bead micro-rheometers and deduce macroscopic properties such as the elastic and viscous moduli of the 'polymer fluid' (cf. [8, 9]).

The second part of this contribution mimics the contractile and bundling properties of such actin-myosin networks by a mesoscopic 'coarse-grain model' introducing interacting 'polymer beads' as mechanochemical elements of actin filament bundles, which are able to cross-link, align, contract, and interact with proteins embedded in a flexible membrane. A corresponding 2D stochastic multiparticle model reveals simulations of (planar) whole cell dynamics, which qualitatively reproduce observed phenomena such as local and transient cortex formation, protrusion, and retraction of cell membrane and characteristic cell shape deformations. Many questions and controversies remain concerning, for example, the molecular mechanisms and energetics of actin-polymerization and plasma membrane motion (see [12], for example), and as a famous analogue problem, the motion of A-protein coated beads or infective *Listeria* bacteria within cells. A new and elegant stochastic simulation model using the idea of a molecular 'clamp motor' has recently been published by Dickinson and Purich [6].

In summary, the stochastic multiparticle models presented in the first three chapters are supplementary to each other and, therefore, call for a more detailed and complete modeling framework and a corresponding simulation tool that can quantitatively reproduce the chemical kinetics and mechanical dynamics of actin filament systems as reactive and interacting semi-flexible polymers, as they are experimentally observed in living cells [10] or artificial vesicles [3]. Moreover, such a general modeling framework could offer different possibilites to include the dynamics of an embedding fluid and surrounding membranes on various temporal and spatial subscales. Finally, there is an obvious need to link such 'microscopic' (or 'mesoscopic') stochastic interaction models for polymer and membrane dynamics to corresponding 'macroscopic' continuum models for two-

or multi-phase flow dynamics with free boundaries, see [2, 4, 5]. Suitable approximation procedures and limit theorems have to be developed that could generalize techniques used in hydrodynamics [11] or in simpler situations of interactive polymer or cell motion (cf. [1, 14] and also the relevant chapters in Part III).

A simple one-dimensional version of such two-phase flow models has been used and adapted in the article by **Stéphanou, Ronot & Tracqui**. Depending on the choice of critical parameters, most importantly the actin-myosin network contractility coefficient, the peripheral cortex of an adherent (fibroblast) cell, modeled by differential equations for actin density and lamellipod extension, can destabilize and develop standing oscillatory modes of protrusions and retractions. Circular mode numbers and temporal periods are explicitely related to a series of model paramters by linear stability analysis and numerical simulations. These results qualitatively reproduce experimentally observed data of cell shape deformations for a certain fibroblast cell line. Two different methods of video image data acquisition and processing lead to consistent results in time series and spectral analysis. One is the newly developed 'optical flow method' by which spatiotemporal gradients in optical gray values of the cell for subsequent pictures are captured by a suitable matrix model for 2D 'linear elastic' deformations.

The case of 3-dimensional mechanical models for cell deformation and motility is treated in the last two chapters. In the contribution by **Drury & Dembo** three fluid dynamical and cell surface models are presented and tested for fitting the data from classical micropipet aspiration experiments with leukocytes. It turns out that the best model to reproduce the observed deformation of a leukocyte is a non-Newtonian viscous fluid (with shear thinning bulk viscosity) surrounded by a viscous surface with area-dependent surface tension (and shear thinning dilatational viscosity). The authors point out that an even better fit could be reached by using the above-mentioned class of viscoelastic, two-phase fluid models (see also [9]).

The same problem of simulating cell deformations under exterior load forces is also treated in the final contribution by **Promayon, Martiel & Tracqui**, but by the direct use of a discrete mechanical ('elastic') modeling device describing 3D 'cell objects', in a similar way as known 'tensegrity' models, called the physical-object-oriented (POO) approach. In contrast to usual 'spring-mass' models, elastic properties of the object are implemented by simpler geometrical shape memory forces trying to restore the former shape. Anisotropic stress distribution due to cytoskeletal fiber organization can also be represented. Under further constraints (e.g., incompressibility, obstacle conditions, or exclusion in cell-cell interactions), the authors are able to simulate single cell deformations as they occur, for example, in experiments with optical tweezers (cf. [7]), and also the migration and accumulation of two types of cells on a substratum in a chemotactic gradient, mimicking aggregation processes in *Dictyostelium* slime molds.

This latter modeling result leads to the topic of the last part of the book (Part III): modeling the interactive dynamics in multicellular systems.

References

[1] W. Alt (2002) Nonlinear hyperbolic systems of generalized Navier-Stokes type for interactive motion in biology. In: (S. Hildebrandt, H. Karcher eds.) *Geometric Analysis and Nonlinear Partial Differential Equations*. Springer-Verlag, Berlin, 431–459

[2] W. Alt and M. Dembo (1999) Cytoplasm dynamics and cell motion: Two-phase flow models. *Math. Biocs.* **156**, 207–228

[3] A. Boulbitch, R. Simon, D.A. Simon, R. Merkel, W. Häckl, M. Bärmann, E. Sackmann (2000) Shape instability of a biomembrane driven by local softening of the underlying actin cortex. *Phys. Rev. E* **62**, 3974–3985

[4] Micah Dembo (1989) Field theories of the cytoplasma. *Comments on Theoretical Biology* **1**, 159–177

[5] M. Dembo (1994) On free boundary problems and amoeboid motion. In: (N. Akkas, ed.) *Biomechanics of Active Movement and Division of Cells, NATO ASI Ser. H 84*. Springer, Berlin et al., 231–283

[6] R.B. Dickinson and D.L. Purich (2002) Clamped-filament elongation model for actin-based motors. *Biophys. J.*, **82**, 605–617

[7] J. Guck, R. Ananthakrishnan, H. Mahmood, R.J. Moon, C.C. Cunningham and J. Käs (2001) The optical stretcher: A novel laser tool to micromanipulate cells. *Biophys. J.*, **81**, 767–784

[8] D. Humprey, C. Duggan, D. Saha, D. Smith and J. Käs (2002) Acitve fluidization of polymer networks through molecular motors. *Lett. to Nature*, **416**, 413–416

[9] R.E. Mahaffy, C.K. Shih, F.C. MacKintosh and J. Käs (2002) Scanning probe-based frequency dependent microrheology of polymer gels and biological cells. *Phys. Rev. Lett.* **85**, 880–883

[10] R. Merkel, R. Simson, D.A. Simon, M. Hohenadl, A. Boulbitch, E. Wallraff and E. Sackmann (2000) A micromechanical study of cell polarity and plasma membrane cell body coupling in *Dictyostelium. Biophys. J* **79**, 707–719

[11] K. Öhlschläger (1991) On the connection between Hamiltonian many-particle systems and the hydrodynamical equations. *Arch. Rational Mechanics and Analysis* **115**, 297–310

[12] R. Sambeth, A. Baumgaertner (2001) Anutocatalytic polymerization generates persistent random walk of crawling cells. *Phys. Rev. Lett.* **86**, 5196–5199

[13] U. Seifert (2000) Rupture of multiple parallel molecular bonds under dynamic loading. *Phys. Rev. Lett.* **84**, 2750–2753

[14] A. Stevens (2000) The derivation of chemotaxis equations as limit dynamics of moderately interacting stochastic many-particle systems. *SIAM J. Appl. Math.* **61**, 183–212

Polymer and Cell Dynamics – Multiscale Modeling
and Numerical Simulations, 49–68
ed. by W. Alt, M. Chaplain, M. Griebel und J. Lenz

Brownian Dynamics of Flexible Polymers in Flow

Roland Rzehak[1], Andreas Arend[2], Diego Kienle[1] and
Walter Zimmermann[2]

1) Institut für Festkörperphysik, Forschungszentrum Jülich, D-52425 Jülich, Germany
2) Theoretische Physik, Universität des Saarlandes, D-66041 Saarbrücken, Germany

Summary. The shape and the dynamics of flexible polymers in flow are described by bead-spring models taking into account hydrodynamic interactions and excluded volume effects. The Brownian dynamics of these models with many degrees of freedom and nonlinear interactions is simulated by algorithms that achieve a high efficiency. We consider a single tethered polymer in uniform flow as a simple model problem that has been studied recently both theoretically and experimentally. Here, we focus especially on hydrodynamic screening effects without employing commonly used averaging approximations. Potential applications to situations of biological interest are discussed.

1. Introduction

Flexible polymers are large chain-like molecules that consist of many parts with statistically independent orientation, the so-called Kuhn segments [13, 17]. They can assume a great variety of different arrangements in space, which are termed conformations. The time scales for conformational changes of flexible polymers cover a very wide range. There are modes of the polymer dynamics with relaxation times that are as slow as the macroscopic hydrodynamic modes. Therefore, the microscopic and macroscopic degrees of freedom do not decouple in a simple manner as for simple fluids like water. The rich and at times surprising visco-elastic behavior of polymer solutions is a consequence of this coupling between individual polymer molecules and the flow field [4, 12, 13, 26].

The interplay between polymer molecules and a flow seems to be governed by two elementary processes: i) How does a flow field deform a polymer and ii) how does the motion of the polymer perturb the flow field? In the past, quantitative analysis of the polymer-flow interaction causing the deformation of the polymer has been hampered by two difficulties. On the experimental side, the classic measurement techniques such as rheometry, birefringence, light scattering, and small angle neutron scattering provide only volume averaged quantities from which only little can be deduced about the behavior of individual polymer molecules, let alone the flow field. On the theoretical side, the large number of polymer degrees of freedom, which are coupled by various nonlinear interactions, makes treatment of the polymer-flow interaction on a fundamental level a formidable task.

Only recently a huge step forward in understanding flow-induced polymer deformation has been achieved by studying single DNA molecules. DNA is much larger than synthetic polymers and can be manipulated with optical tweezers. When decorated by fluorescent dyes, the action of flows on the polymer can be followed under an optical microscope [29, 35–37]. Thus, DNA provides an ideal model system to study conformations of deformed polymers.

On the theoretical side, modern high-speed supercomputers allow us to attack nonlinear problems with many degrees of freedom by means of numerical simulation. An advantage of computer simulation is that all observables are accessible in full detail, which facilitates the investigation of features, such as the flow around the polymer, that have not yet been resolved experimentally. Furthermore, this approach allows us to compare different models in order to assess the importance of various interactions in different parameter regimes. This provides guidelines for developing more simplified models such as effective dumbbell [4] or blob models [7, 8, 23, 43, 44], which neglect a large number of internal degrees of freedom of the polymer. Such models may then be used as building blocks for the description of more complicated systems (see, e.g., [31]).

A variety of bead-spring polymer models that give a faithful representation of chain flexibility and are well suited for computer simulation is presented in Section 2. These models include several important interactions: a finite extensibility of the springs, the excluded volume of the beads (**EVI**), and hydrodynamic interactions between the beads (**HI**). We put special emphasis on the latter, which is responsible for the perturbation of the flow field. So far, bead-spring models have been investigated mostly with various averaging approximations for the HI [27, 28, 33]. In contrast to these works, we keep the full conformation dependence of the HI by employing the so-called Oseen-approximation without any averaging.

A central issue in computer simulation is to devise efficient algorithms such that problems of interesting size and complexity can be tackled. At first sight, Brownian dynamics simulations with fluctuating HI require $\mathcal{O}(N^3)$ operations, but with an approximation due to Fixman [15], the operation count is reduced to $\sim \mathcal{O}(N^{2.25})$ operations. For the EVI, bookkeeping techniques familiar in molecular dynamics can be adapted to Brownian dynamics. These methods are described in Section 3.

As an illustrative application, we consider in Section 4 a simple model problem, namely, a single tethered polymer that is subjected to a uniform flow. This problem has been studied recently both experimentally [36, 37] and theoretically [7–9, 30]. Here, we discuss some advances that have been made using simulations [43, 44].

2. Bead-Spring Models for Polymer Dynamics

To study the long-time dynamics of polymers in dilute solution, coarse-grained models are used because atomistic models of long polymer chains are intractable even numerically with present-day computing equipment. A common class of such models is the so-called bead-spring models [5] where coarse-graining (cf. Fig. 1) is achieved by replacing a subchain of a real polymer with a bead and a spring with a suitable force-elongation

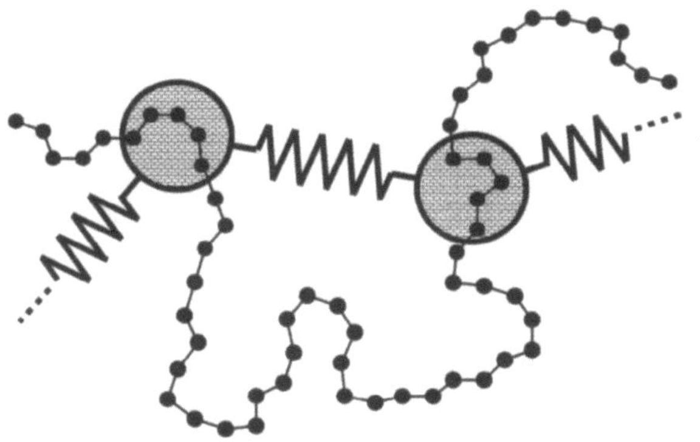

Figure 1. Sketch of the coarse-grain procedure leading to the bead-spring model. The small solid circles represent chemical groups in the backbone of the polymer, with the solid lines indicating chemical bonds between them. The large shaded circles are the beads that are connected by springs reflecting the entropic elasticity of the subchain.

law [13, 49]. Friction and mass of the subchains are lumped into the beads as depicted in Figure 1. For a flexible chain, each subchain corresponds to one or more Kuhn segments [13]. The solvent acts as a heat bath, which causes a stochastic motion of the polymer.

Usually, one considers the motion on the diffusive timescale only, i.e., bead inertia are neglected [46]. The equation of motion for the position of the ith bead ($i = 1 \ldots N$) is then obtained from a balance between all forces acting on the beads. These forces comprise viscous drag forces $\mathbf{F}^{\mathrm{H}}$ on one side and spring- and stochastic forces $\mathbf{F}^{\Phi}$, $\mathbf{F}^{\mathrm{S}}$ on the other:

$$-\mathbf{F}_i^{\mathrm{H}} = \mathbf{F}_i^{\Phi} + \mathbf{F}_i^{\mathrm{S}} . \tag{1}$$

The drag forces are proportional to the difference between the velocity of a bead $\mathbf{V}_i$ and the flow velocity at its position $\mathbf{u}^0(\mathbf{R}_i)$:

$$-\mathbf{F}_i^{\mathrm{H}} = \zeta \left(\mathbf{V}_i - \mathbf{u}^0(\mathbf{R}_i) \right) . \tag{2}$$

The single bead friction coefficient ζ is given by Stokes law, i.e., $\zeta = 6\pi\eta a$, where η is the solvent viscosity and a is the effective hydrodynamic radius of a bead. In thermal equilibrium, of course, $\mathbf{u}^0 \equiv \mathbf{0}$. The stochastic forces are related to the dissipative drag by the fluctuation-dissipation theorem in order to ensure the correct equilibrium distribution. We have

$$\mathbf{F}_i^{\mathrm{S}} = \sqrt{2k_{\mathrm{B}}T\,\zeta}\;\boldsymbol{\xi}_i , \tag{3}$$

where T is the solvent temperature, k_{B} is the Boltzmann constant, and $\boldsymbol{\xi}_i$ is an uncorrelated Gaussian white noise with zero mean, i.e.,

$$\begin{aligned} \langle \boldsymbol{\xi}(t) \rangle &= 0 , \\ \langle \boldsymbol{\xi}(t)\, \boldsymbol{\xi}^T(t') \rangle &= \delta(t-t')\,\mathbf{1} . \end{aligned} \tag{4}$$

Taking velocities and positions of *all* beads together as single supervectors for the velocity $\mathbf{V}$ and position $\mathbf{R}$, the equations of motion may be written in the following general form:

$$\mathbf{V} \equiv \frac{\partial}{\partial t}\mathbf{R} = \mathbf{u}^0(\mathbf{R}) + \mathbf{H} \cdot (-\nabla_{\mathbf{R}}\, \Phi) + \sqrt{2k_{\mathrm{B}}T\, \mathbf{H}}\; \xi\,. \tag{5}$$

Here we introduced a potential Φ for the spring forces and the mobility $\mathbf{H}$, which is the inverse of the friction coefficient, ζ^{-1}. In the most general case to be discussed later on, $\mathbf{H}$ will no longer be a simple scalar but a tensor that couples all beads. Boundary conditions are implemented by introducing additional beads with indices $i = 0, N+1$ that do not participate in the dynamics. For free chain ends we have $\mathbf{R}_0 = \mathbf{R}_1$ viz. $\mathbf{R}_{N+1} = \mathbf{R}_N$, while for fixed ends $\mathbf{R}_0$ viz. $\mathbf{R}_{N+1}$ are constant. Choosing $k_{\mathrm{B}}T = 1.0$ and $\zeta = 1.0$ basically fixes units for energy and time. The remaining unit for length is the bondlength b to be discussed below.

The Rouse model [42] is the simplest conceivable polymer model that embodies only the chain connectivity by assuming harmonic springs, i.e., by using in Equation (5)

$$\Phi \equiv \Phi_{\mathrm{H}} = \sum_i \frac{1}{2}\, k_{\mathrm{H}}\, |\mathbf{R}_{i+1} - \mathbf{R}_i|^2\,. \tag{6}$$

For the force constant k_{H} we use a value of 3.0 such that the root mean square bond length in equilibrium b becomes unity. The mobility in the Rouse model is simply a scalar, namely, the inverse of the single bead friction coefficient, i.e., $\mathbf{H} = \zeta^{-1}\,\mathbf{1}$. All of the above assumptions are tacitly introduced in order to make the equation of motion Equation (5) linear so that it may be solved analytically.

There are three obvious refinements, each of which renders the equation of motion Equation (5) nonlinear: a finite extensibility of the springs, the excluded volume of the beads (EVI), which may be chosen such that the chain cannot cross itself, and hydrodynamic interactions between the beads (HI).

Since chemical bonds have a fixed length, real polymers are inextensible. This can be modeled by a nonlinear spring law which keeps the stretching of the springs small even for large forces. Rather common in simulations is the phenomenological FENE (Finitely Extensible Nonlinear Elastic) spring law [48] with the potential

$$\Phi \equiv \Phi_{\mathrm{F}} = -\sum_i \frac{1}{2}\, k_{\mathrm{F}}\, R_{\mathrm{F}}^2 \ln\!\left(1 - \frac{|\mathbf{R}_{i+1} - \mathbf{R}_i|^2}{R_{\mathrm{F}}^2}\right). \tag{7}$$

Another approach is to make the chain completely unstretchable by replacing the springs with rigid rods. In numerical simulations this may be achieved to a good approximation by augmenting the FENE spring law with a nearest-neighbor repulsion of the form described in Equation (8) below [25]. The values $k_{\mathrm{F}} = 30.0$ for the force constant and $R_{\mathrm{F}} = 1.5$ for the maximum extension of the spring then result in a bond length of $b = 0.961$. Obviously, the effects of finite extensibility are most important when the polymer is driven far from equilibrium where the chain may be strongly stretched.

The excluded volume effect arises because different beads cannot occupy the same region in space. This is modeled by a repulsive interaction between any pair of beads as described, e.g., by a truncated Lennard-Jones potential [25]:

$$\Phi_{\mathrm{LJ}} = \begin{cases} \sum_{i,j} 4\epsilon \left(\left(\frac{\sigma}{|\mathbf{R}_j - \mathbf{R}_i|} \right)^{12} - \left(\frac{\sigma}{|\mathbf{R}_j - \mathbf{R}_i|} \right)^{6} + \frac{1}{4} \right) & \text{for } |\mathbf{R}_j - \mathbf{R}_i| < R_{\mathrm{LJ}} \\ 0 & \text{for } |\mathbf{R}_j - \mathbf{R}_i| \geq R_{\mathrm{LJ}} \,. \end{cases} \tag{8}$$

Here ϵ and σ define energy and length scales of the excluded volume interaction, and $R_{\mathrm{LJ}} = 2^{1/6}\sigma$ is the minimum of the conventional 6-12 Lennard-Jones potential. This term is included in the potential Φ in Equation (5) along with the spring potential. When combined with harmonic springs, the choice $\epsilon = \sigma = 1.0$ gives an effective equilibrium bond length of $b = 1.33$ as discussed in the Appendix. Together with the FENE potential for the springs, the bead-spring chain becomes self-avoiding like a real polymer for the parameters chosen [25]. The effects of excluded volume are most prominent in or close to thermal equilibrium where the polymer chain assumes a coil shape. Far from equilibrium, where the polymer chain is strongly stretched and the beads are far apart from each other, the excluded volume force tends to zero.

The hydrodynamic interaction (HI) is an effective interaction between any pair of beads which is mediated by the solvent. It arises in the following way: If the solvent exerts a drag force on a bead, then by virtue of Newton's third law there must be a drag reaction force of the same strength but with reverse direction acting as a driving force for the solvent. According to Equation (1), the average of this force is given by the potential force $-\nabla_{\mathbf{R}}\,\Phi$. Neglecting the finite bead diameter – i.e., idealizing the drag reaction force as a point force – and using the linear equations of Stokes flow for the solvent dynamics, one can derive a general expression for the perturbation $\mathbf{u}'$ of the imposed flow field resulting from the drag reaction forces of all beads [13, 51]:

$$\mathbf{u}'(\mathbf{R}_i) = \sum_{j \neq i} \boldsymbol{\Omega}(\mathbf{R}_i - \mathbf{R}_j) \cdot (-\nabla_{\mathbf{R}_j} \Phi) \,. \tag{9}$$

The ith bead is excluded from the sum on the right-hand side of Equation (9) in order to suppress unphysical hydrodynamic self-interactions. The Stokes friction force on bead i as given by Equation (2) now arises from the bead velocity relative to the *perturbed* flow field $\mathbf{u}^0 + \mathbf{u}'$. Using this perturbed flow field instead of $\mathbf{u}^0$ alone in Equation (5) and collecting all terms containing $-\nabla_{\mathbf{R}}\,\Phi$, one identifies the components of the mobility, which has now become a conformation-dependent supermatrix, as

$$\mathbf{H}_{ij} = \begin{cases} \frac{1}{\zeta}\mathbf{1} & \text{for} \quad i = j \\ \boldsymbol{\Omega}(\mathbf{R}_i - \mathbf{R}_j) & \text{for} \quad i \neq j \,. \end{cases} \tag{10}$$

Here the Oseen tensor [13] is nothing but the Green's function for Stokes flow,

$$\boldsymbol{\Omega}(\mathbf{r}) = \frac{1}{8\pi\eta|\mathbf{r}|} \left(\mathbf{1} + \hat{\mathbf{r}}\hat{\mathbf{r}}^T \right) . \tag{11}$$

It suffers from the deficiency that it makes the mobility non-positive at small bead separations. Therefore it is necessary to introduce EVI along with HI [43] or to use a regularization of the Oseen tensor [52]. So far we have considered deterministic forces only. In order to complete the equation of motion for the case with HI, we need to specify the stochastic forces. Fortunately, it turns out that the fluctuation-dissipation theorem for this case has precisely the same form as before [32].

The choice $\eta = 0.2$ for the solvent viscosity, with the bead radius a determined from $6\pi\eta a = \zeta = 1.0$, results in a value of $h^* = \sqrt{\frac{3}{\pi}}\frac{a}{b} \approx 0.25$ for the dimensionless parameter h^* measuring the strength of the HI. This value is compatible with other work [34].

The collective effect of the hydrodynamic interactions is often condensed in the so-called non-draining assumption, i.e., the assumption that the polymer coil may be replaced by a sphere with some effective radius into which the external flow does not penetrate. This assumption is inspired by Zimm's calculation of the diffusion coefficient of a polymer coil with HI [50], which is given by an expression similar to that of a hard sphere [13]. Apart from additional approximations in the calculation (see below), this evidence seems to be of rather scarce support for the conclusion drawn. Only recently has the perturbed flow field been calculated directly [44], and this calculation revealed that the flow is weakened inside the polymer coil but not completely suppressed as shown in Figure 2 below. Furthermore, there is a significant penetration depth where the flow remains strong and a large long-range effect that are both neglected in the simple traditional picture.

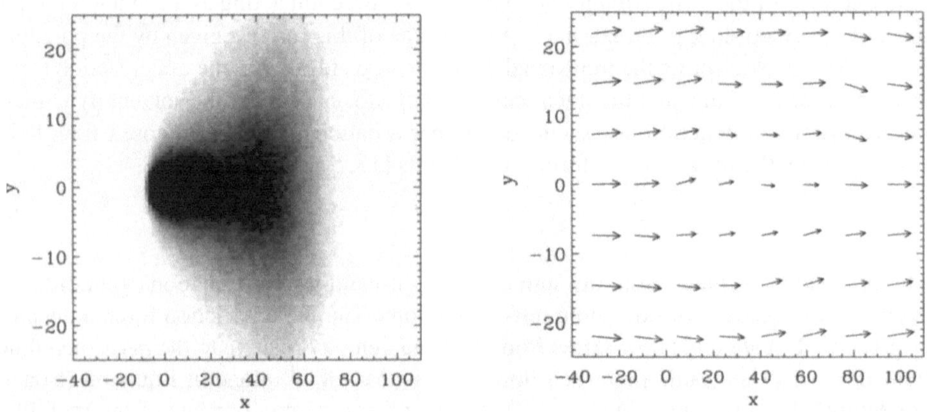

Figure 2. (a) Segment density $\rho(x, y, z = 0)$ and (b) time-averaged perturbed flow field $\mathbf{u}(x, y, z = 0)$ (right) for a chain with $N = 200$ segments which is fixed at the origin with one end and subjected to a uniform flow in the x-direction with $v = 0.02$. Both EVI and HI are included and harmonic springs are used. The streamlines go around the region where the density of polymer segments is high, but the flow does not vanish there.

If any of the three effects discussed above is to be included in the model without further approximations, one has to rely on numerical simulation in order to solve the equation of motion. Only by restricting ourselves to the Rouse model with HI does an approximate treatment become possible by replacing $\mathbf{H}(\mathbf{R})$ with its equilibrium average. This so-called preaveraging approximation has the effect of linearizing the equation of motion again. This procedure was proposed by Zimm [50] and therefore is referred to as the Zimm model. Far from equilibrium, the simple preaveraging approximation obviously breaks down. Various other approximation schemes have been proposed that avoid the use of the equilibrium distribution but that still lead to linear equations [27, 28, 33].

3. Efficient Brownian Dynamics Algorithms

The Brownian dynamics method pioneered by Ermack and McCammon [14] is nothing but the numerical integration of the Langevin equation Equation (5) with one realization of the stochastic process modeling the thermal noise. It yields one sample of polymer conformations $\mathbf{R}(t_n)$ from which all quantities of interest can be obtained as averages. Since several textbooks on the finite difference approximation of stochastic differential equations like Equation (5) are available [24, 34], we just briefly summarize our scheme and then move on to describe the efficient calculation of the various forces.

An important but not widely appreciated point in Brownian dynamics simulation concerns the accurate numerical reproduction of the equilibrium distribution [21]. Therefore, the usual procedure of trading accuracy for speed by choosing a time step so large that the algorithm is just stable is questionable here, because this might violate a fundamental physical principle. In order to cure this unfavorable situation, Hendriks et al. [21] suggested introducing an artificial inertial term, i.e., adding $m\ddot{\mathbf{R}}$ on the left-hand side of Equation (5). A stochastic velocity-Verlet scheme is then used to advance the solution in time. The final form of the numerical integration scheme with time step δt is

$$\begin{aligned} \mathbf{R}(t_n+\delta t) &= \mathbf{R}(t_n)+\mathbf{V}(t_n)\,\delta t+\frac{\delta t^2}{2m}\,\mathbf{F}(t_n)\,, \\ m\mathbf{V}(t_n+\delta t) &= m\mathbf{V}(t_n)+\frac{\delta t}{2}\left(\mathbf{F}(t_n+\delta t)+\mathbf{F}(t_n)\right) \\ \mathbf{F}(t_n) &= \zeta\left(\sqrt{2k_{\mathrm{B}}T\,\mathbf{H}(\mathbf{R}(t_n))}\;\Xi(t_n)\,\frac{1}{\sqrt{\delta t}}\right. \\ &\qquad \left. -\left(\mathbf{H}(\mathbf{R}(t_n))\,\frac{\partial\Phi}{\partial\mathbf{R}}(t_n)+\mathbf{V}(t_n)-\mathbf{u}^0(\mathbf{R}(t_n))\right)\right). \end{aligned} \tag{12}$$

The random numbers Ξ at each time t_n are drawn from a uniform distribution on the interval $\left[-\sqrt{3},+\sqrt{3}\right]$, which has zero mean and unit variance. This replacement of Gaussian random numbers approximating the noise in Equation (5) is justified if only moments of the stochastic process $\mathbf{R}(t_n)$ and not individual sample paths are sought, since in this case it suffices to approximate only the moments of the noise [24, 34].

As long as the mass is chosen small enough so the period of oscillations due to the added degrees of freedom is smaller than the diffusive time scale of the phenomena of interest, the results remain unaffected by the artificial inertia. As shown by the results in the Appendix, this condition is met for a value of $m = 1.0$. The time step δt then can be chosen up to a factor of 10 larger than in conventional BD (where the mass vanishes) if we require an error of less than 1% in the second moment of the equilibrium distribution for the positional degrees of freedom, which can be calculated from the potentials given in Section 2 (cf. Appendix 6). A value of $\delta t = 0.02$ has been found to be suitable for all models.

The numerical algorithm Equation (12) was coded in standard FORTRAN90. A machine-optimized version of the LAPACK library was used for the linear algebra calculations involved in the treatment of HI (see below). The simulations were performed on a single processor of a Cray T90 vector computer. For the 200-bead chain with harmonic springs and both EVI and HI taken into account (cf. Section 4), which is the most time-consuming case we considered, a maximal speed of 900–1000 MFLOPS was achieved. For shorter chains, of course, the vectorization is less efficient. Longer chains could not be treated for this model because the total requirement of CPU time – which was already $\mathcal{O}(100\,h)$ for the flow field calculations in Section 4 – becomes prohibitively large.

Several tests of the numerical scheme with respect to the equilibrium distribution of bond length for linear and nonlinear springs and the scaling of the root-mean-square end-to-end distance as a function of the number of beads are described in the Appendix. The agreement with analytical solutions for these equilibrium properties is nearly perfect.

3.1. Evaluation of Stochastic Forces in the Presence of Hydrodynamic Interactions

A key issue for the present study with emphasis on HI is the calculation of the matrix $\sqrt{\mathbf{H}}$ in Equation (5) and Equation (12), which is a function of all bead positions. To make the whole algorithm practical, an efficient way to evaluate the square root of the mobility matrix $\mathbf{H}$ must be developed. The straightforward calculation of such an expression [19] involves a diagonalization of $\mathbf{H}$ which numerically requires an effort of $\mathcal{O}(N^3)$ machine instructions. A second standard method for the calculation of matrix functions is *via* series expansion of the desired function [19]. A Taylor series will contain only powers of the matrix argument that are easily evaluated numerically. However, matrix multiplication also requires $\mathcal{O}(N^3)$ operations. A third method that again needs $\mathcal{O}(N^3)$ operations but that offers the most favorable prefactor becomes possible by noting that the square root is not precisely what is needed. Instead, one can also use the Cholesky decomposition of $\mathbf{H}$. This idea was exploited in the classic work by Ermak & McCammon [14]. In all cases the numerical effort of $\mathcal{O}(N^3)$ makes the computation prohibitive for long chains. Therefore, to our knowledge all previous Brownian dynamics studies of polymer dynamics that took HI into account were limited to chains with $N \leq 20$ beads with one exception, the work of Fixman [16], where a few results for a chain of 56 beads are given.

An approximate method that requires only $\sim \mathcal{O}(N^{2.25})$ operations was proposed later by Fixman [15]. The starting point for this method is an expression of the square root in terms of a complete set of polynomials as in method two above, i.e.,

$$\sqrt{\mathbf{H}} = \sum_{\mu=1}^{M} \mathcal{P}^{\mu}(\mathbf{H}) \,. \tag{13}$$

A reduction of the computational effort becomes possible by noting that the knowledge of the matrix $\sqrt{\mathbf{H}}$ is actually much more than what is really needed since once it is known it could be applied to *many different* random vectors $\boldsymbol{\xi}$. For the simulation, however, it needs to be applied to *one single* realization only. A scheme that takes advantage of this is obtained by multiplying both sides of Equation (13) with $\boldsymbol{\xi}$. Taking $\boldsymbol{\xi}$ into the sum on the right-hand side, one obtains a series expression for $\sqrt{\mathbf{H}}\,\boldsymbol{\xi}$

$$\sqrt{\mathbf{H}}\,\boldsymbol{\xi} = \sum_{\mu=1}^{M} \mathcal{P}^{\mu}(\mathbf{H})\,\boldsymbol{\xi}\,. \tag{14}$$

This expression contains only matrix-vector products, and thus its evaluation requires an effort of $\mathcal{O}(N^2)$ only. Furthermore, the individual terms in the sum may be calculated recursively, keeping the number of these operations low, too.

The polynomials $\mathcal{P}^{\mu}(x)$ may be taken from *any* complete set in function space. The most economic choice is not simple powers $\mathcal{P}^{\mu}(x) = x^{\mu}$ but Chebychev polynomials $C^{\mu}(x)$ [6, 39]. These can be evaluated by means of the recursion relation

$$\begin{aligned} C^{\mu+1}(x) &= 2\,x\,C^{\mu}(x) - C^{\mu-1}(x)\,, && (15)\\ \text{with}\quad C^{1}(x) &= x\,, \\ \text{and}\quad C^{0}(x) &= 1\,. && (16) \end{aligned}$$

Since the Chebychev polynomials are defined on the interval $[-1, 1]$, which is not suitable in the present context, one applies a transformation of the independent variable

$$x = \frac{2y}{b-a} - \frac{b+a}{b-a}\,, \tag{17}$$

which maps the domain of the problem $y \in [a, b]$ to the domain $x \in [-1, 1]$ of the Chebychev polynomials. $C^{\mu}(y)$ appears frequently in numerical analysis and is referred to as a shifted Chebychev polynomial [6, 39].

If, as in the problem under consideration, the argument x is a matrix, not a simple scalar, then $[a, b]$ is the range of eigenvalues of x. An estimate of the range of the eigenvalues of $\mathbf{H}$ is furnished by a simple physical argument: If two nearby beads experience a force in the same direction, the induced perturbations of the velocity field will have a large degree of coherence and thus add up to a larger perturbation. If, on the other hand, the forces are in opposite directions, the induced perturbations will cancel out to a large extent. Since beads that are neighbors along the chain are likely to be also close in space, an estimate for the largest eigenvalue is obtained by using a force vector with equal forces for all beads as a test vector $\mathbf{F}$ to form the Rayleigh quotient [19] $\mathbf{F}^T\,\mathbf{H}\,\mathbf{F}/\mathbf{F}^T\,\mathbf{F}$. Similarly, an estimate of the smallest eigenvalue is obtained by using a force vector with alternating forces for all beads as a test vector. In order to compensate for deviations of these estimates from the true values of largest and smallest eigenvalue of $\mathbf{H}$, one takes a somewhat larger interval for the shifted Chebychev polynomials.

The order of truncation of the series, M, has to be determined empirically and increases somewhat with N, from which the final effort grows with a somewhat higher power than 2. To monitor the accuracy of the approximation, we compute the exact square root of $\mathbf{H}$ *via* the spectral theorem using a QR algorithm [19] for the diagonalization whenever the conformation is saved. This happens only every 100–10,000 time steps of the integration and is thus acceptable in terms of computer time.

3.2. Calculation of Excluded Volume Interactions

The most obvious way to calculate the excluded volume force on bead i is to loop over all other beads j, checking whether they are within the interaction range R_{LJ} and if so calculating the force $-\nabla\Phi_{LJ}$. This leads to an algorithm where the total computational effort for calculating all EVI increases with the square of the number of beads. The short-range (in real space) nature of the forces leads to possible improvements since most pairs of beads i, j do not interact. Bookkeeping methods that exploit this fact by keeping track of particles within the range of interaction are well known from the molecular dynamics literature [1, 18, 41]. One approach is to maintain for each bead i a so-called neighbor list giving the indices of beads j within a distance that is somewhat larger than the interaction range, i.e., $|\mathbf{R}_i - \mathbf{R}_j| \leq R_{LJ} + R_s$. In the actual force calculation, only the beads in this list have to be considered. As soon as the first bead has traversed the extra shell of thickness R_s, the list must be reconstructed from scratch. This step, which takes a time of order $\mathcal{O}(N^2)$, ultimately dominates the force computation such that an improvement of only a constant factor independent of N can be achieved. A second possibility is the cell list method, where the space occupied by the beads is partitioned into cubic cells with edge length R_{LJ}. Having established which beads fall in each cell, the calculation of the force on bead i needs to consider only beads j in the same and adjacent cells. This method has the desirable property that the effort is only of order $\mathcal{O}(N)$. To achieve full vectorization, it is necessary to use a layered data structure [40]. A comparison [20] showed that in the context of molecular dynamics, a combination of both approaches, where a neighbor list is built from a cell list, gives the fastest method.

Some preliminary experimentation indicated that this may be different in the case of nonequilibrium Brownian dynamics simulation. For a single stretched polymer, the density of beads is extremely inhomogeneous. When the chain is placed in a quadrilateral box to set up a cell list, many cells will not contain any bead at all. Therefore, we use a simple neighbor list with a shell thickness of $R_s = 0.5R_{LJ}$, the value of which has been determined empirically. To implement a criterion when the list must be updated, we accumulate the displacement $\delta\mathbf{R}$ each bead has undergone since the last update. The first bead has traversed the shell when $max_i(|\delta\mathbf{R}_i|) \geq R_s$. Then an update of the neighbor list is necessary, which turns out to be the case approximately every 6th time step for the tests described in the Appendix. On a workstation larger values of $R_s \geq 1.0R_{LJ}$ turn out to be more advantageous, leading to a list update only every 10th time step.

For $N \gtrsim 1000$ the dominating part is the list update. The effort then continues to grow as $\sim \mathcal{O}(N^2)$ (actually the power is somewhat less than 2 on the Cray because of gains due to better vectorization for longer chains). Compared to the force calculation without neighbor list, this gives a speed-up of a constant factor of about 10 in an equilibrium situation. For a stretched polymer a somewhat larger gain is possible.

4. Results for a Simple Model Problem

In this section we describe results obtained in simulations of a bead-spring chain which is fixed at the origin with one end and subjected to a uniform flow in the x-direction.

The chain has $N = 200$ beads connected by harmonic springs and both EVI and HI are taken into account. For an unperturbed flow velocity $\mathbf{u}^0 = (v, 0, 0)$ with $v = 0.02$, we show in Figure 2a the temporally averaged segment density ρ as a measure of the chain localization. The full flow field $\mathbf{u}$ including the perturbation due to the HI between the beads is shown in Figure 2b. The flow velocity assumes its smallest values near the x-axis in the interval $0 < x < 40$, where the beads are located with the largest probability. The important result emerging from these simulations is that the flow at the average location of the polymer coil does not vanish. This partial draining is a superposition of two effects: The flow penetrates the polymer coil at any moment and, due to thermal fluctuations, the polymer also does not stay at a fixed location. The y-dependence of u_x shown in Figure 3 reveals a magnitude of the flow penetration of approximately one-third of the unperturbed flow velocity. Furthermore, there is a smaller decrease in the flow velocity which has a very long range, i.e., the unperturbed flow velocity is approached only very far away from the polymer coil.

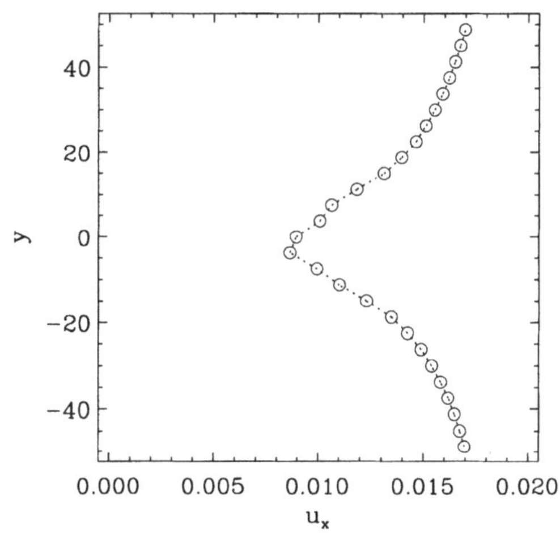

Figure 3. Spatially averaged x-component of the velocity field $\mathbf{u}$ shown in Figure 2 as a function of y. The average was taken over the range $0 \leq x \leq 40$, where the segment density is large. The smallest flow velocity is approximately one-third of the unperturbed value $v = 0.02$.

The partial penetration of the deformed polymer coil also depends on the imposed flow velocity. Quantitative information about this velocity dependence is furnished by the overall drag force acting on the polymer, which may be measured in our simulation by the tension T_0 in the first spring at the tether point. Even more revealing is the drag coefficient $\Gamma = T_0/v$ for the whole coil. Its velocity dependence is plotted in Figure 5a. Without HI (upper curve) where the flow can freely penetrate the polymer coil – the so-called free-draining case – each bead makes the same contribution ζ to the coil drag coefficient independent of the flow velocity v, i.e., $\Gamma = \zeta N$. With HI (lower curve) there is only partial draining, i.e., the flow velocity inside the coil is reduced. Thus, the inner beads experience smaller drag forces and the total drag on the coil becomes smaller than

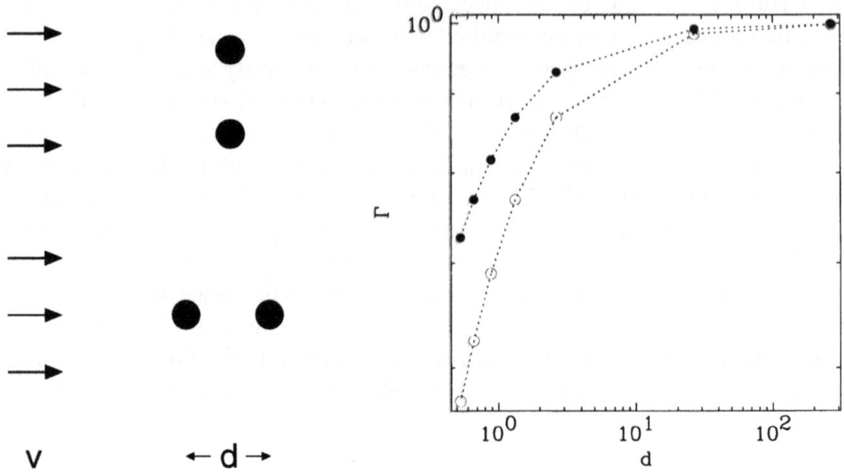

Figure 4. (a) Two beads in a uniform flow are held at fixed positions where the relative position vector of length d is oriented either perpendicular (top) or parallel (bottom) to the flow. (b) Drag coefficient Γ per bead as a function of the distance d for configurations with orientation perpendicular (solid circles) and parallel (open circles) to the flow. The flow velocity is $v = 1.0$, all other parameters have values as given in Section 2.

in the free-draining case. As the chain is stretched with increasing flow velocity, the flow penetration becomes stronger and finally the free-draining value of Γ is approached.

This effect of the screening of the flow field due to HI and the related reduction of the drag coefficient is also observed in the more didactic example of a two-bead system sketched in Figure 4a. The force required to keep the two beads fixed in a flow of velocity v is shown in Figure 4b as a function of their distance d. The drag coefficient per bead Γ is seen to decrease with d. This mimics in a simple manner the situation of the polymer, where the average distance between the beads is smaller in the coiled than in the stretched state. The dependence of the drag coefficient on the shape of the polymer coil and therefore also on the flow velocity is in sharp contrast to the constant drag coefficient of a solid sphere given by Stokes law. In more complex systems such as polymer-coated colloidal particles, similar deviations from Stokes' behavior can be caused by a variety of other effects [22].

The elongation of the polymer is the result of a balance between the viscous drag exerted on the polymer by the flow and the thermal forces acting on the polymer. The stronger the screening of the flow field, the smaller are the drag force and also the elongation of the polymer. Hence, complementary to the overall drag coefficient (plotted in Figure 5a) is the polymer elongation which may be measured by the root-mean-square end-to-end distance $R_E(v)$ (shown in Figure 5b). Without HI (upper curve), where each bead experiences the same velocity-independent viscous drag, we find in the velocity range $v = 0.001$–0.01 a behavior consistent with a power law $R_E \propto v^\mu$ with $\mu = 0.5$ and a transition to a linear dependence $R_E \propto v$ for $v > 0.1$. Beyond this crossover, EVI

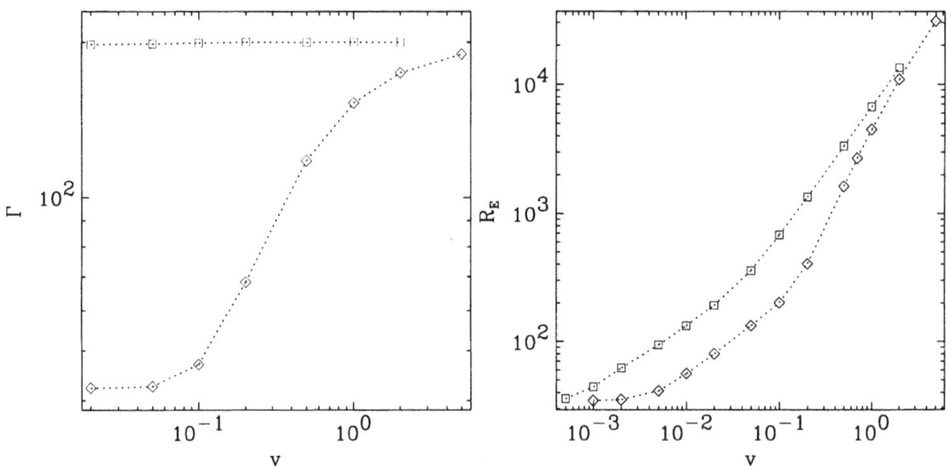

Figure 5. (a) Total drag Γ coefficient for the coil as a function of the imposed velocity v both with HI (diamonds) and without HI (squares). Otherwise the same model as in Figure 2 was used. (b) Root-mean-square end-to-end distance R_E for the same two models.

becomes negligible because the polymer is uncoiled at large flow velocities. Including HI (lower curve), the onset of stretching is shifted to larger flow velocities. Around $v = 0.5$ we find a maximal exponent of $\mu = 1.5$, which is larger than without HI such that free-draining behavior is approached for very large flow velocity. The shifted onset of chain stretching is another consequence of the screening of the flow inside the coil due to strong HI. With increasing flow velocity and polymer elongation, the average distance between the beads increases and therefore the strength of the HI is reduced. This weakening is more pronounced close to the tethered end than near the free end as long as the chain is not fully stretched. At this stage of deformation, the effects of HI vary significantly along the polymer chain. Finally, for chains at very large v which are already strongly stretched, the screening becomes negligible and the free-draining behavior is approached.

$R_E(v)$ may be compared with analytical scalings for blob models, where the chain is described by a sequence of impenetrable [7] or free-draining blobs [43]. For free-draining chains with harmonic springs, there is reasonable agreement with results for a blob model [43]. However, for impenetrable blobs, $R_E(v) \propto v^\mu$ with $\mu = 2$ is predicted [7], which is considerably larger than the maximal exponent $\mu \sim 1.5$ of the lower curve in Figure 5b. One reason for deviations between simulations and completely free- or non-draining blob models is that when the polymer uncoils with increasing flow velocity, EVI and HI effects are continuously "switched off" such that the degree of penetration varies along the chain. This inhomogeneous process, which sets in at the tether point and with increasing flow velocity extends to the free chain end, is not described by the blob models considered so far. As shown in [23, 44], it may be incorporated at least qualitatively in a generalized blob model where the blobs are composed of a non-draining inner core and a fully penetrable outer shell (f-shell).

The above considerations in principle carry over to the more realistic model with FENE springs. In that case, however, scaling regimes can be expected only for extremely long chains. An estimate shows that $N \approx 10,000$ is needed [43], which is not even reached in recent experiments [27, 37] where the number of Kuhn segments is less than 2000.

5. Conclusions

We have introduced a modified Brownian dynamics scheme, cf. Equation (12), by adding an artificial inertial term to the usual first-order Langevin equation. This allows the integration time step to be chosen by a factor of 10 larger as for the usual case corresponding to vanishing mass, while maintaining a prescribed accuracy of the second moment of the equilibrium distribution for the positional degrees of freedom even for the steep potentials occurring in the models with finite extensibility or excluded volume interactions. In addition, we have implemented efficient algorithms for the treatment of excluded volume and hydrodynamic interactions that allow us to overcome the chain length limitations of previous Brownian dynamics studies of polymers.

Formally the scheme Equation (12) is quite similar to that used in molecular dynamics simulations with a Langevin thermostat, whereas its interpretation is different [25, 47]. In molecular dynamics simulations, the solvent is modeled explicitly as a collection of particles. Thus, in that case noise and dissipation are a purely artificial means that is used to simulate a canonical ensemble while the masses of the particles represent true physical quantities. In our case in contrast, the masses of the particles are introduced as a computational device to speed up the simulation, while the stochastic and dissipative forces represent the solvent degrees of freedom that do not appear explicitly in the equations of motion.

In either case the length of the integration time that must be covered by the simulation is determined by the longest time scale of the problem, which increases with the system size as N^2. The computational effort for the force calculations at each time step in our scheme is $\sim \mathcal{O}(N^{2.25})$ (cf. Section 3.1), i.e., the required CPU time increases as $\sim \mathcal{O}(N^{4.25})$. In comparison, for state-of-the-art molecular dynamics simulations [20] that take full advantage of the short-range nature of the interaction potentials, the required CPU time increases only with the third power of the system size. However, in the case of molecular dynamics simulations, the size of the system is the number of both solvent particles and polymer beads, which is beyond 10^4. For Brownian dynamics simulations, in contrast, only the beads appear explicitly in the model and the system size is only on the order of 10^2. Comparing the chain lengths $N \leq 60$ attained in recent nonequilibrium molecular dynamics simulations [3, 38] to our work, where $N \leq 200$, (500 if only EVI is considered), we find that saving the large overhead of solvent particles still outweighs the unfavorable scaling with the system size in current applications. Furthermore, the Brownian dynamics approach we chose allows easy inclusion or removal of various effects and arbitrary external flows. Both features may be difficult to realize in molecular dynamics simulations.

Simulation of the Brownian dynamics of bead-spring models for tethered polymers allows a detailed monitoring of various interactions between chain segments at different stages of the polymer deformation. Thus, a test of the validity ranges of previous more coarse-grained model approaches becomes possible, as described in more detail in [43]. Moreover, a thorough study of the polymer conformations and of the flow field provides a starting point for generalization and modification of these models, such as the f-shell blob model for tethered polymers as introduced in [23, 44]. But not only static properties become amenable to investigation. The spectrum of relaxation times characterizes the dynamics of conformational fluctuations about a stationary average conformation. The complete spectrum together with the associated modes can be extracted from simulation data, as shown in [45]. This furnishes an important rheological function and gives further insight into the major motions that have to be accounted for in coarse-grained models. When the flow undergoes a sudden change, there is no stationary average polymer conformation anymore. Instead, average quantities such as the end-to-end distance become time dependent. This time-dependent behavior is also investigated in [45], where single-mode models like the dumbbell have been challenged. In summary, the study of the behavior of tethered polymers in uniform flow is a first step in bridging the gap between the microscopic scale at which the interaction of single polymers and the flow takes place and the macroscopic continuum dynamics of polymer solutions.

In the context of biology, bead-spring-type models similar to the one studied here are used to model systems on very diverse levels of description ranging from individual protein molecules to entire tissues (see other contributions in Parts II and III of this book). Consequently, the techniques developed here may be of use in a much broader context than just polymer science. Furthermore, hydrodynamic interactions are often important for the biological function of single macromolecules or more complicated aggregates. For example, a hydrodynamic steering effect has been found to be crucial for the reaction rates of protein-ligand binding [11]. Previously, the protein was considered as a rigid object while the methods described here allow us to include the effects of its internal motions [10]. As a second interesting example, we mention the computer study of a vesicle in Poiseuille flow using rather similar modeling approaches [2]. The vesicles were found to migrate towards the center of the flow, which may be a relevant factor to inhibit mechanical degradation of red blood cells flowing through a blood vessel. Finally, on a macroscopic level, objects like vesicles or cells also behave much like viscoelastic fluids. Thus, a basic question is again to determine the rheological functions from a more microscopic approach, a first step towards which was made here for polymer solutions.

6. Appendix: Program Tests

At thermal equilibrium, the simulations can be compared with several well-known analytical results. The equilibrium distribution of the bond lengths can be calculated *via* the Boltzmann factor $\exp(-\Phi/k_BT)$ using the spring potentials given in Section 2. These analytically calculated distributions are compared in Figure 6 with the distributions obtained from simulations using these potentials for the springs.

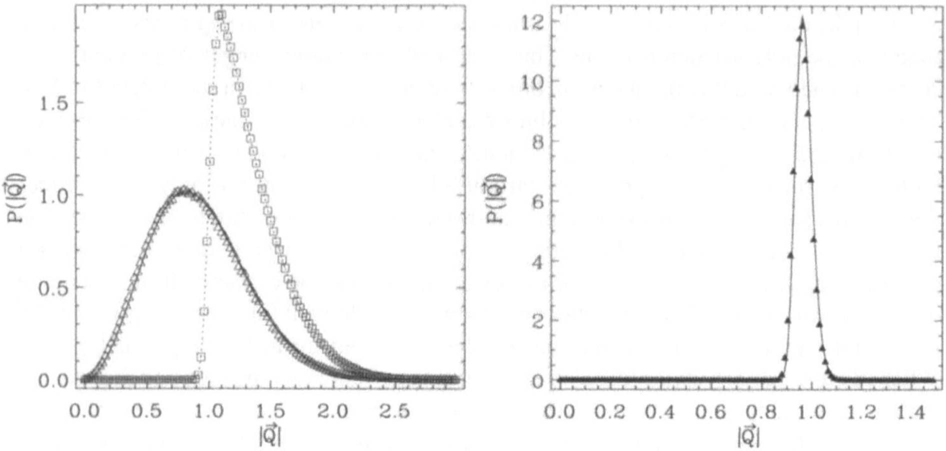

Figure 6. Probability distribution for the bond length $|\mathbf{Q}| = |\mathbf{R}_{j+1} - \mathbf{R}_j|$ at thermal equilibrium for models with (a) harmonic springs, both with (squares) and without (triangles) EVI, and (b) FENE springs. In the latter case the addition of EVI has no effect on the bond length distribution since nearest-neighbor repulsion is included in the spring potential anyways. The solid lines are obtained by evaluating the Boltzmann factor $\exp(-\Phi/k_B T)$ with the potentials given in Section 2. The symbols give the results of the corresponding simulations for a chain with $N = 100$ beads.

For the Rouse model, the bond length distribution is of course Maxwellian. For the FENE model, the distribution is sharply peaked around its maximum value so that it can be regarded as a good approximation to a freely jointed chain model. The bond length distribution for the Rouse model with excluded volume interactions reveals that the repulsive potential acts as a rather hard wall. Hence, the distribution is deformed and its mean value is shifted to larger values. The addition of the excluded volume interactions to the FENE model of course leaves the bond length distribution unchanged because nearest-neighbor repulsion is included as described in Section 2. As the analysis of the scaling of the end-to-end distance with the number of segments shows (see below), the effective bond length is rather close to the distance $b = 0.961$ where the bond length distribution has its maximum.

A second more global test is the comparison of the numerically obtained end-to-end distance R_E as a function of the number of segments N with the scaling result due to Flory [17]:

$$R_E = bN^\nu . \tag{18}$$

Here ν is the scaling exponent, while b gives an effective bond length. For any model with purely local interactions between the beads, it is well known that the exponent is $\nu = 1/2$ [13]. In Figure 7a, we verify this behavior for the pure Rouse and FENE models. In these cases, the bond lengths are known so that we can contrast the numerical values with an analytical result without any adjustable parameters. The small difference between both models comes from the fact that the bond lengths for the Rouse ($b = 1.0$) and FENE

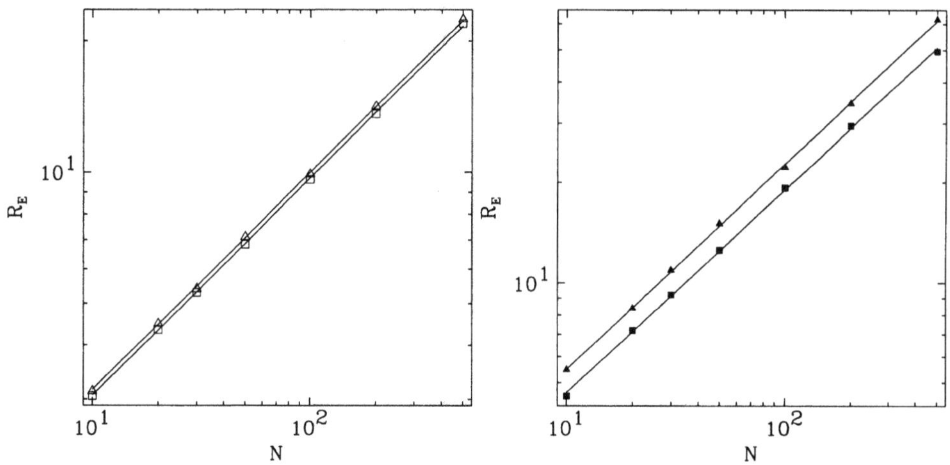

Figure 7. (a) Equilibrium scaling of the end-to-end distance R_E with the number of segments N for the Rouse (open triangles) and FENE models (open squares). The solid lines are due to the scaling relation $R_E = b\sqrt{N}$, with the mean bond length $b = 1.0$ for the Rouse chain and $b = 0.961$ for the FENE chain. (b) Equilibrium scaling of the end-to-end distance with the number of segments for the Rouse (solid triangles) and FENE models (solid squares) including the excluded volume effect. The solid lines are fits with a power law $R_E = bN^{\nu}$. The fit parameters are $\nu = 0.617$, $b = 1.33$ for the Rouse model and $\nu = 0.610$, $b = 1.15$ for the FENE model.

models ($b = 0.961$) are slightly different. For both models there is very good agreement with the numerical data even for rather short chains.

The excluded volume interactions change the scaling exponent for R_{E} from $\nu = 1/2$ to $\nu = 3/5$, neglecting a small correction to mean field theory [17]. In Figure 7b, we verify this behavior for the Rouse and FENE models with EVI. Since the prefactor is not known, we fit a power law of the form of Equation (18) to the numerical data. The values of the exponents are $\nu \sim 0.6$ for the models with either harmonic or FENE springs, which is in quite good agreement with the theoretical values. For the model with harmonic springs, the repulsive potential due to the excluded volume interaction leads to a considerable shift of the mean bond length to $b = 1.33$.

For the models with hydrodynamic interactions, a test is possible since all static properties at equilibrium are determined solely by the potential Φ, whereas the HI affects only the mobility. One can simply compare the numerical values of the above quantities for the models with and without the hydrodynamic effects. We find agreement within the statistical errors.

References

[1] M.P. Allen and D.J. Tildesley (1987) *Computer Simulation of Liquids*. Clarendon Press, Oxford

[2] A. Arend, J. Leonhard, D. Kienle, G. Ristow, R. Rzehak, and W. Zimmermann (2002) Cross-streamline migration of deformable objects in poiseuille flow. In preparation.

[3] C. Aust, M. Kröger, and S. Hess (1999) Structure and dynamics of dilute polymer solutions under shear flow via nonequilibrium molecular dynamics. *Macromolecules*, **32** 5660

[4] R.B. Bird, R.C. Armstrong, and O. Hassager (1987) *Dynamics of Polymeric Liquids, Vols. i, ii*. John Wiley, New York, second edition

[5] R.B. Bird and H.C. Öttinger (1992) Transport properties of polymeric liquids. *Annu Rev Phys Chem* **43**, 371

[6] A. Björck and G. Dahlquist (1979) *Numerische Methoden* (R. Oldenbourg) München, second edition

[7] F. Brochard-Wyart (1993) Deformation of one tethered chain in strong flows. *Europhys Lett* **23**, 105

[8] F. Brochard-Wyart (1995) Polymer chains under strong flows: Stems and flowers. *Europhys Lett* **30**, 387

[9] F. Brochard-Wyart, H. Hervet, and P. Pincus (1994) Unwinding of polymer chains under forces or flows. *Europhys. Lett.* **26**, 511

[10] C.L. Brooks, M. Karplus, and B.M. Pettitt, editors (1988) Proteins: A theoretical perspective of dynamics, structure and thermodynamics volume 71 of *Adv. Chem. Phys.* New York, 1988, John Wiley.

[11] D. Brune and S. Kim (1994) Hydrodynamic steering effects in protein association. *Proc Natl Acad Sci USA* **91** 2930

[12] P.G. deGennes (1981) *Scaling concepts in polymer physics* (Cornell University Press, Ithaca NY)

[13] M. Doi and S.F. Edwards (1986) *The theory of polymer dynamics* (Clarendon Press, Oxford)

[14] D.L. Ermak and J.A. McCammon (1978) Brownian dynamics with hydrodynamic interactions *J Chem Phys* **69**, 1352

[15] M. Fixman (1986) Construction of Langevin forces in the simulation of hydrodynamic interaction. *Macromolecules* **19**, 1204

[16] M. Fixman (1986) Implicit algorithm for Brownian dynamics of polymers. *Macromolecules* **19**, 1195

[17] P.J. Flory (1969) *Statistical Mechanics of Chain Molecules* (Interscience Publishers, New York)

[18] D. Frenkel and B. Smit (1998) *Understanding Molecular Simulation* (Academic Press, New York, third edition)

[19] G.H. Golub and C.F. van Loan (1989) *Matrix Computations* (John Hopkins University Press, Baltimore MD, second edition)

[20] G.S. Grest, B. Dünweg, and K. Kremer (1989) Vectorized link cell FORTRAN code for molecular dynamics simulation for a large number of particles. *Comp Phys Commun* **55**, 269

[21] J. Hendriks, T. Kawakatsu, K. Kawasaki, and W. Zimmermann On semiflexible polymers in an anisotropic medium: A phenomenological model. Unpublished.

[22] D. Kienle, R. Rzehak, and W. Zimmermann (2002) Nonlinear Stokes law and hydrodynamically enhanced fluctuations of polymers in flow. Submitted to *Phys Rev Lett*

[23] D. Kienle and W. Zimmermann (2001) F-Shell Blob-Modell for a Tethered Polymer in Strong Flows. *Macromolecules* **34**, 9173

[24] P.E. Kloeden and E. Platen (1992) *Numerical Solution of Stochastic Differential Equations* (Springer, Heidelberg)

[25] K. Kremer and G.S. Grest (1990) Dynamics of entangled linear polymer melts. *J Chem Phys* **92**, 5057

[26] R.G. Larson (1988) *Constitutive Equations for Polymer Melts and Solutions* (Butterworth, Stoneham MA)

[27] R.G. Larson, T.T. Perkins, D.E. Smith, and S. Chu. (1997) Hydrodynamics of a DNA molecule in a flow field. *Phys Rev E* **55**, 1794

[28] J.J. Magda, R.G. Larson, and M.E. Mackay (1988) Deformation-dependent hydrodynamic interaction in flows of dilute polymer solutions. *J Chem Phys* **89**, 2504

[29] S. Manneville, Ph. Cluzel, J.L. Viovy, D. Chatenay, and F. Caron (1996) Evidence for the universal scaling behavior of a freely relaxing DNA molecule. *Europhys Lett* **36**, 413

[30] Y. Marciano and F. Brochard-Wyart (1995) Normal modes of stretched polymer chains. *Macromolecules* **28**, 985

[31] M. Murat and K. Kremer (1998) From many monomers to many polymers: Soft ellipsoid model for polymer melts and solutions. *J Chem Phys* **108**, 4340

[32] H.C. Öttinger and Y. Rabin (1989) Diffusion equation versus coupled Langevin equations approach to hydrodynamics of dilute polymer solutions. *J Rheol* **33**, 725

[33] H.C. Öttinger (1987) Generalized Zimm model for dilute polymer solutions under Θ-conditions. *J Chem Phys* **86** 3731

[34] H.C. Öttinger (1996) *Stochastic Processes in Polymeric Fluids* (Springer, Heidelberg)

[35] T.T. Perkins, D.E. Smith, and S. Chu (1997) Single polymer dynamics in an elongational flow. *Science* **276**, 2016

[36] T.T. Perkins, S.R. Quake, D.E. Smith, and S. Chu (1994) Relaxation of a single DNA molecule observed by optical microscopy. *Science* **264**, 822

[37] T.T. Perkins, D.E. Smith, R.G. Larson, and S. Chu (1995) Stretching of a single tethered polymer in a uniform flow. *Science* **268**, 83

[38] C. Pierloni and J.P. Ryckaert (1995) Deformation and orientation of flexible polymers in solution under shear flow: A new picture for intermediate reduced shear rates. *Macromolecules* **28**, 5097

[39] W.H. Press et al (1990) *Numerical Recipes in C* (Cambridge University Press, Cambridge UK, third edition)

[40] D.C. Rapaport (1990) Vectorized molecular dynamics for systems with shortrange interactions. In M. Mareschal, editor, *Microscopic Simulations of Complex Flows* **page 57** (Plenum Press, New York)

[41] D.C. Rapaport (1995) *The Art of Molecular Dynamics Simulation* (Cambridge University Press, Cambridge UK)

[42] P.E. Rouse, Jr (1953) A theory of the linear viscoelastic properties of dilute solutions of coiling polymers. *J Chem Phys* **21**, 1272

[43] R. Rzehak, W. Kromen, T. Kawakatsu, and W. Zimmermann (2000) Deformation of a tethered polymer in uniform flow. *Eur Phys J E* **2**, 3

[44] R. Rzehak, D. Kienle, T. Kawakatsu, and W. Zimmermann (1999) Partial draining of a tethered polymer in flow. *Europhys Lett* **46**, 821

[45] R. Rzehak and W. Zimmermann (2002) Dynamics of strongly deformed polymers in solution. *Europhys Lett* **59**, 779

[46] J.D. Schieber and H.C. Öttinger (1988) The effects of bead inertia on the Rouse model. *J Chem Phys* **89**, 6972

[47] T. Schneider and E. Stoll (1978) Molecular dynamics study of a three-dimensional one-component model for distortive phase transitions. *Phys Rev B* **17**, 1302

[48] H.R. Warner, Jr (1972) Kinetic theory and rheology of dilute suspensions of finitely extendible dumbbells. *Ind Eng Chem Fundam* **11**, 379

[49] H. Yamakawa (1971) *Modern Theory of Polymer Solutions*. (Harper & Row, New York)

[50] B.H. Zimm (1956) Dynamics of polymer molecules in dilute solution. *J Chem Phys* **24**, 269

[51] R. Zwanzig (1969) Langevin theory of polymer dynamics in dilute solution. *Adv Chem Phys* **15**, 325

[52] W. Zylka and H.C. Öttinger (1989) A comparison between simulations and various approximations for Hookean dumbbells with hydrodynamic interaction. *J Chem Phys* **90**, 474

Polymer and Cell Dynamics – Multiscale Modeling
and Numerical Simulations, 69–74
ed. by W. Alt, M. Chaplain, M. Griebel und J. Lenz

A Microscopic Model for the Dynamics of Actin Filaments in Motility Assays and its Numerical Simulation

Zénó Farkas, Imre Derényi and Tamás Vicsek

Dept. of Biological Physics, Eötvös University, Pázmány P. stny 1A,
H-1117 Budapest, Hungary

Summary. We investigate a microscopic model of motility assays in which actin filaments move over a layer of myosin heads attached to a glass plate. The actin filaments are regarded as unstretchable semi-flexible polymer chains, and we use a standard method to numerically solve their equations of motion. The results of our computer simulations are in good agreement with those of the experiments, indicating that possibly this model can be used to predict behavior under modified conditions or determine some of the parameters that cannot be measured directly by finding the best fit to the experimentally observed quantities.

1. Introduction

Motor proteins such as actin or kinesin play an essential role in the motion of living organisms. Much effort has been made so far to understand how they function. An important experimental tool is the *in vitro* motility assay [2, 11], in which the associated filaments are placed on a surface coated with motor proteins (see Fig. 1). The filaments may produce various kinds of movements, e.g., they may slide on the surface, rotate around a fixed point, or make a flapping motion [1, 10]. In these experiments individual filaments can be analyzed; therefore, valuable information can be obtained regarding their elastic properties or the nature of the actin-myosin interaction. The motion of motor protein filaments has also been a subject of theoretical studies [3, 4]. We have developed a model for the motility assay experiments, and the results of the numerical simulation of this model are presented here.

2. The Model

The subject of our investigation is the following motility assay: an actin filament moves over a lawn of randomly distributed myosin heads attached to a horizontal glass plate. We model the actin filament as a semi-flexible polymer chain in two dimensions following the technique described in detail in [6–8]. The polymer chain is discretized as a sequence of beads, and the neighboring beads are connected to each other by weightless

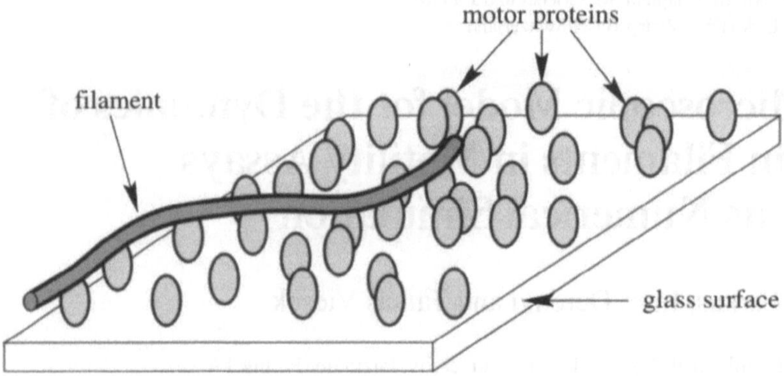

Figure 1. Example of the *in vitro* motility assays where a lawn of immobilized motor proteins moves a complementary filament.

stiff rods with constant length of a. The polymer is elastic in the transverse direction with *persistence length* L_p (see Fig. 2). In our model we assume that the myosin molecules can be attached to the actin filament only at certain places, called *binding sites*. These binding sites are practically the beads, although this is not necessary. Since the actin filaments move in liquid in the experiments, viscous friction forces with viscosity coefficient Γ act on the beads. The whole system is in a thermal bath of temperature T; therefore, a random force η_i is also present, satisfying the fluctuation-dissipation relation $\langle \eta_i(t)\eta_j(t')\rangle = 2kT\Gamma\delta_{ij}\delta(t-t')$. The equation of motion of the ith filament bead is the first-order Langevin equation

$$\Gamma \frac{d\mathbf{r}_i}{\delta t} = \mathbf{F}_i^{el} + \mathbf{F}_i^{constr} + \mathbf{F}_i^{metr} + \mathbf{F}_i^{myo} + \eta_i, \qquad (1)$$

where the forces in the r.h.s. are the following: $\mathbf{F}_i^{el}$ is the elastic force due to transversal elasticity; $\mathbf{F}_i^{constr}$ is a constraint force that keeps the length of the segments constant; $\mathbf{F}_i^{metr}$ is the *metric* or *pseudo* force, which makes a model polymer, defined by rigid segments, behave like a polymer being elastic in the normal direction without the need for taking into account the normal oscillations (for a detailed explanation, see [8, 7, 6, 5]); $\mathbf{F}_i^{myo}$ is the force exerted by an attached myosin molecule (0 if no myosin is attached); and the last term corresponds to the thermal fluctuations. The only difference from the algorithm described in [8, 7, 6] is that we used the fifth-order Runge-Kutta algorithm with fixed step size [9] because we found that this allows larger time steps in the computer simulation. Our model is microscopic in the sense that the actions of individual myosin heads are taken into account in addition to the microscopic perturbations resulting from thermal fluctuations.

The combination of the most effective algorithms available allows us to carry out simulations on a time scale compatible with the experimental observation times in spite of the microscopic nature of the model. A more simplistic approach (direct integration of the equations of motion for the beads that are connected to each other by springs instead of

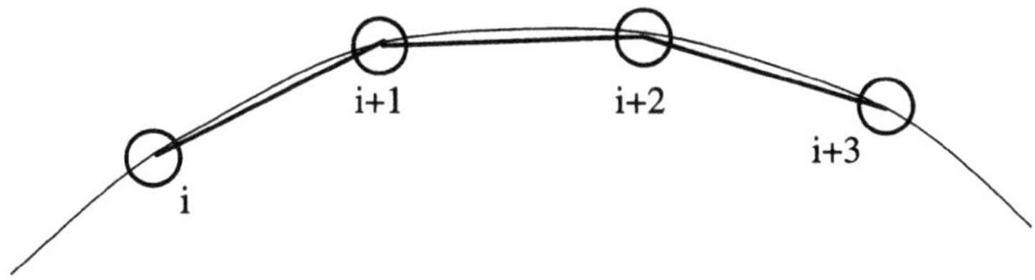

Figure 2. The actin filament is modeled as a sequence of beads that are connected to each other by weightless rods and threaded onto a flexible string.

rods) resulted in a simulation that was 10^3 to 10^4 times *slower* due to the high-frequency normal oscillation modes.

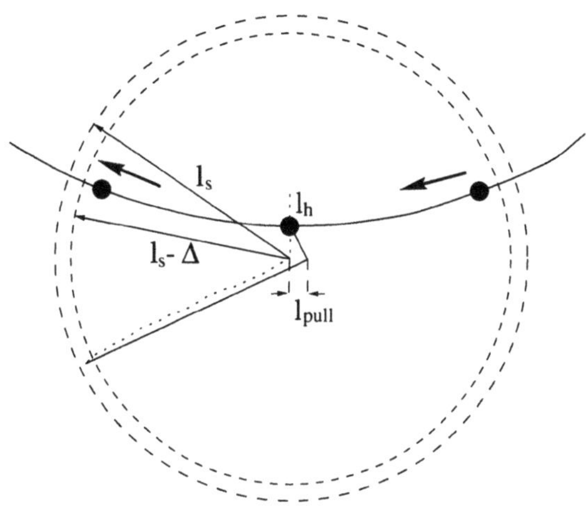

Figure 3. A myosin can be attached to an actin unit if it is located in a proper ring (see text). Once bound, its head changes its configuration by bending backwards a distance l_{pull} in a way that the point where the head and the string join each other moves parallel to the actin filament. If the sting stretches, a pulling force is exerted to the actin by the myosin string.

The myosin molecules are randomly distributed over the simulation area with surface density σ. In the model, a myosin molecule consists of a rigid head connected to a stretchable string (with length l_{h} and l_{s}, respectively). One end of the string (which is not connected to the head) is attached to the base. The myosin can freely rotate around this point, and no constraint is imposed concerning the angle of the head-string connection. A myosin can attach to a binding site if the point where it is attached to the base is within a ring with outer radius l_{s}, width Δ, and a center that is l_{h} far from the binding site in perpendicular direction from the actin filament (see Fig. 3). Once a myosin head is attached to an actin unit, it leans backwards in a way that the point where the head and the string join each other is moved by a distance of l_{pull} parallel to the actin filament (i.e., the length of the head increases); therefore, in most cases the string stretches, and a pulling force is exerted on the actin unit. The absolute value of this force is $D_{\text{m}}(l - l_{\text{s}})$, where l

is the length of the stretched string, and its direction is the same as that of the string. It is possible (geometrically) that the string becomes loose and, therefore, no force is exerted. The myosin heads can attach to and detach from the binding sites obeying the following rules: (i) no more than one myosin head can be attached to a binding site at any time; (ii) in case a myosin is located in the proper ring, its head can attach to the binding site with "attachment" time constant τ_a (throughout this paper, the probabilities are characterized by time constants: the probability of an event with time constant τ to occur during a δt step is $\delta t/\tau$); (iii) the myosin head detaches from the binding site with time constant τ_d when it is not stretched too much; and (iv) the myosin head detaches with time constant $\tau_d^{stretch}$ if the string (of length l) is stretched too much, i.e., when $l > l^{max}$.

3. Results of the Numerical Simulations

The following values were used in the simulations: $\Gamma = 6 \times 10^{-10}$ kg/s, $L_p = 16\ \mu$m, $a = 50$ nm, $D_m = 3 \times 10^{-4}$ kg/s^2, $l_h = 15$ nm, $l_s = 60$ nm, $l_{pull} = 11$ nm, $\Delta = 5$ nm, $\tau_a = 10 \times 10^{-3}$ s, $\tau_d = 4 \times 10^{-3}$ s, $\tau_d^{stretch} = 10^{-6}$ s, $\delta t = 10^{-8}$s.

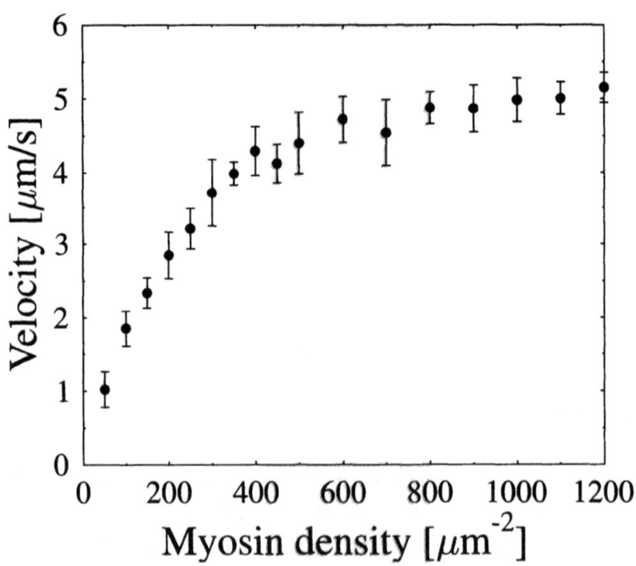

Figure 4. The velocity of sliding actin filaments rapidly saturates to about 5 $\mu m/s$ as the myosin density is increased.

First we discuss the qualitative results. In the simulation, the motion of the actin filaments is similar to that observed experimentally: (i) if a filament is free to move, it follows a winding path; (ii) if its leading endpoint gets stuck, the filament starts to rotate around that point and takes on a whirling spiral shape; and (iii) when the filament is attached to the base at two points, it executes an undulating motion. Furthermore, in our simulation we observe occasional reversals of the direction of rotation in case (ii). This was also observed in the experiments but cannot be described by continuum approximations [1].

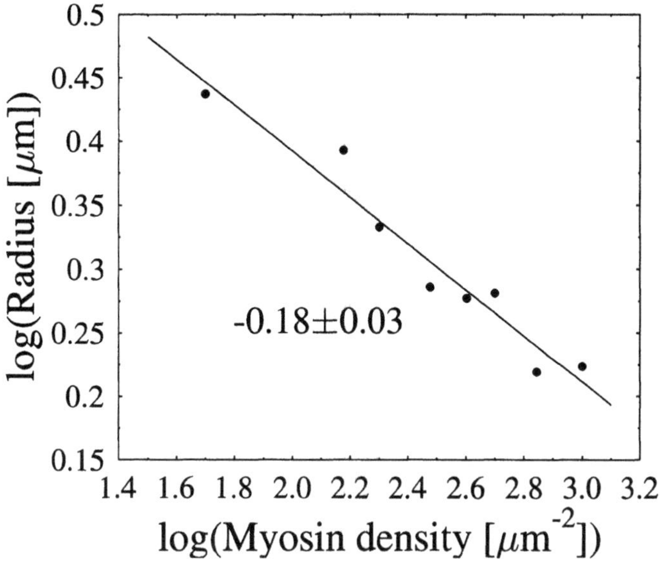

Figure 5. The radius of a rotating filament decreases linearly with the myosin density on log-log scale. This is in agreement with the experimental results; however, the slope is different.

As far as the quantitative results are concerned, first we present the velocity dependence on the myosin density of the sliding actin filaments (Fig. 4). Two regimes can be clearly distinguished: below myosin density 400 μm^{-2} the curve has a steep slope, while beyond this density value it is much less steep; we can say that the velocity is saturated to a value close to 5 $\mu m/s$. This result is in very good agreement with the experiments [2].

Another result presented here is the radius of the rotating filament as a function of the myosin density (Fig. 5). The values of the radii are close to the measured values and fall onto a line when plotted on log-log scale. The slope is -0.18 ± 0.03, which is, however, different from the experimental data (-0.36 ± 0.05) [1].

We measured other quantities as well, e.g., the undulation/rotation frequency or the occupation of actin-binding sites as functions of myosin density (not shown here). The conclusion is that with a simple microscopic model the experimentally observable quantities can be reproduced and other, experimentally not observable quantities can be determined, which could be important in understanding the mechanisms of molecular motors.

Supplemental material is available at
`http://biol-phys.elte.hu/~zeno/poldyn.`

References

[1] Bourdieu, L. *et al.* (1995) Spiral defects in motility assays: A measure of motor protein force. *Phys. Rev. Lett.* **75**, 176–179

[2] Bourdieu, L., Magnasco, M.O., Winkelmann, D.A., and Libchaber, A. (1995) Actin filaments on myosin beds: The velocity distribution. *Phys. Rev. E* **52**, 6573–6579

[3] Camalet, S., Jülicher, F., and Prost, J. (1999) Self-organized beating and swimming of internally driven filaments. *Phys. Rev. Lett.* **82**, 1590–1593

[4] Duke, T., Holy, T.E., and Leibler, S. (1995) "Gliding assays" for motor proteins: A theoretical analysis. *Phys. Rev. Lett.* **74**, 330–333

[5] Fixman, M. (1978) Simulation of polymer dynamics I. General theory. *J. Chem. Phys.* **69**, 1527–1537

[6] Grassia, P. and Hinch, E.J. (1996) Computer simulations of polymer chain relaxation via Brownian motion. *J. Fluid. Mech.* **308**, 255–288

[7] Grassia, P.S., Hinch, E.J., and Nitsche, L.C. (1995) Computer simulations of Brownian motion of complex systems. *J. Fluid. Mech.* **282**, 373–403

[8] Hinch, E.J. (1994) Brownian motion with stiff bonds and rigid constraints. *J. Fluid. Mech.* **271**, 219–234

[9] Press, W.H., Teukolsky, S.A., Vetterling, W.T., and Flannery, B.P. (1992) *Numerical Recepies in C.* Cambridge: Cambridge University Press, Section 16.2

[10] Sekimoto, K., Mori, N., Tawada, K., and Toyoshima, Y.Y. (1995) Symmetry breaking instabilities of an *in vitro* biological system *Phys. Rev. Lett.* **75**, 172–175

[11] Sekimoto, K. and Tawada, K. (1995) Extended time correlation of *in vitro* motility by motor protein *Phys. Rev. Lett.* **75**, 180–183

Polymer and Cell Dynamics – Multiscale Modeling
and Numerical Simulations, 75–89
ed. by W. Alt, M. Chaplain, M. Griebel und J. Lenz

Multiparticle Modeling of Actin-Myosin Networks: From Molecular Interactions to Cell Motility

Jürgen Lenz[1] and Dieter Felix[2]

1) Bioreact GmbH, Kirschallee 1, D-53115 Bonn, Germany
2) Abteilung Theoretische Biologie, Institut für Zelluläre und Molekulare Botanik, Universität Bonn, Kirschallee 1, D-53115 Bonn, Germany

Summary. Active motility and locomotion of animal tissue cells are based on the interaction of actin and myosin, which form a contractile network within the cell. We present two related stochastic multiparticle models of actin-myosin dynamics on different spatial scales: first, a model on the mesoscopic scale, focusing on some basic aspects of the behavior of small and medium-sized actin-myosin II systems *in vitro*, and second, a model on the macroscopic scale, which relies on some results of the first model and which describes whole cell shape changes *in vivo*. Computer simulations show how simple rules for molecular interaction can result in a quite complex and naturally appearing behavior in both model systems. We also stress the advantage of modeling certain phenomena within the same context on different scales.

1. Introduction

Movement of animal tissue cells is fundamental for many biological processes, e.g., during early embryonic development, wound healing, or cancerogenesis. Therefore, understanding of the molecular mechanisms and regulatory control of cellular movement is valuable for basic research, as well as for medical and biotechnological applications. Experimental results over the last 20 years have confirmed the outstanding role of the intracellular actin-myosin system in active cellular motility. Despite the huge complexity with respect to regulation, which might reflect the physiological and ontogenetic importance of these processes, there exists only a limited set of force-generating physical mechanisms. Hence, it seems that not force generation itself is complex but rather the spatial and temporal orchestration and network dynamics. It is therefore important to know which kinds of static and dynamic patterns emerge from basic self-organizing properties of actin-myosin networks *in vitro* and *in vivo*. This is the topic of the present paper. We will present two related stochastic multiparticle models for the dynamic behavior of actin-myosin networks on different length scales.

Actin filaments that are placed on a myosin-coated surface show a sliding motion [18, 34] due to directed forces exerted by myosin molecules. In the first model we replace the myosin-coated surface with a second actin filament to get a motion of two filaments parallel to each other. In addition, we investigate more complex types of filament motion, namely, changes in their relative positions.

Actin gels exhibit a wide range of complex behavior *in vivo* and *in vitro*: Bundles of actin filaments are present in cultured fibroblasts [9]; these stress fibers are capable of contraction, as are nonpolar actin bundles *in vitro* [31].

Understanding the mechanical and rheological properties of actin solutions is important to interpret the behavior of cells under shear stress and to elucidate the mechanisms of cell movement [20, 24]. The rheological properties are characterized by shear moduli, measured at different deformation frequencies; their values strongly depend on experimental conditions (see [36] and references therein). Because single actin filaments are easy to visualize by fluorescence labeling and video microscopy, they are widely used as model polymers for studying the dynamics of semiflexible chains [21] in polymer physics.

The first model [14] describes, on a mesoscopic scale, the spatial arrangement of two and up to a few hundred actin filaments interacting with myosin II molecules. Moreover, the rheological properties of this model system can be investigated. This model is mesoscopic in the sense that we follow a coarse-grained approach to simulate the behavior of actin filaments and myosin molecules.

The second model [22] was designed to explain active amoeboid shape changes of animal tissue cells migrating on flat, solid surfaces. Experiments performed by Alt et al. [2] and Libotte et al. [23] on human keratinocytes revealed a variety of dynamic shape changes with the appearance of stochastic, oscillatory or drifting lamellipodes (flat protrusions) or blebs/blisters (spherical protrusions that appear frequently under physiologically unstable conditions) [10].

From a methodological point of view, the two models are intended to complement the well-established continuum approaches in the field [3], especially by adding inhomogeneity and stochasticity. Here, the model systems dynamics are ascribed to interactions of small virtual entities, constituting the system as a whole.

In the following sections, we will present some case studies that show how both complex and natural appearing dynamic behavior can result from simple rules in our model systems.

2. Multiparticle Model of Actin-Myosin Systems

2.1. Model Description

The first model [14] describes the dynamic behavior of actin-myosin complexes in aqueous solutions *in vitro*. The model system is three-dimensional and consists of actin filaments and dumbbell shaped myosin molecules.

Actin filaments are modeled as stiff rods with fixed length. Filament lengths are distributed exponentially within the ensemble. A single filament consists of a given number of spherical monomers with radius R and well-defined position X_{ij} (position of monomer j in polymer i). Each monomer has a single myosin-binding site. The structural polarity of actin filaments is included in this model by allowing a bound myosin molecule to move only from monomer M_{ij} to monomer $M_{i(j+1)}$.

A myosin molecule is modeled as a hookean spring with two actin-binding sites, one at each end of the spring. Each actin monomer, with its binding site unoccupied, can react with a free myosin molecule. Free, i.e., unbound, myosin molecules are not modeled individually but are represented by a pool with constant concentration throughout the system. Therefore, the reaction probability per time depends only on an association rate proportional to the concentration of free myosin molecules. The dissociation probability is assumed to be constant. If a myosin molecule has already bound to an actin monomer, the probability of binding to a second one located on a different actin filament depends, in addition, on some geometrical factors, namely, the distance between the two monomers that will be bridged by the myosin molecule and the angle between the axes of the myosin molecule and each of the two actin filaments. The more this distance deviates from the equilibrium length of the myosin molecule (the spring), the smaller the association probability is. In a similar manner, a bound myosin molecule prefers a certain orientation with respect to the actin filament. Deviation from this orientation results in a smaller association probability. If a myosin molecule is bound to two actin monomers, the probability of breaking the first bond depends on a certain dissociation rate and on the same geometrical factors. An additional aspect of myosin dynamics is their capability to move actively towards one end of the actin filament. This motion is a step-by-step process, consisting of jumps from monomer M_{ij} to monomer $M_{i(j+1)}$ with a probability that depends on the increase or decrease of the length of the myosin molecule.

To compute the diffusional motion of actin filaments, we model them as long cylindrical particles. There are two types of diffusional motion: translational diffusion (which can be further subdivided into diffusion along the main axis and diffusion perpendicular to the main axis) and rotational diffusion. The constants of diffusion for each type were taken from [11].

In addition to diffusion, actin filaments undergo motion due to external forces. Monomers exert repulsive forces on each other. These forces depend inversely on the distance between an interacting pair of monomers, i.e., they approach infinity for zero distance and a value of zero for a finite distance r_{cut}. Myosin molecules bridging two actin monomers exert a repulsive or an attractive force depending on their length.

All forces $\vec{F}_{ij}$ that act on the monomers M_{ij} of a single filament i at a certain distance $\vec{r}_{ij}$ from the center of mass cause two effects: a translational motion due to the sum of forces $\vec{F}_i = \sum \vec{F}_{ij}$ acting on the center of mass of the filament and a rotational motion due to the rotation moment $\vec{M}_i = \sum \vec{r}_{ij} \times \vec{F}_{ij}$ perpendicular to the filament axis.

Neglecting inertial forces, translational motion of a filament due to $\vec{F}_i$ is computed using the following Langevin-type equation:

$$\vec{v}_i = \frac{1}{\gamma_{i\|}}\vec{F}_{i\|} + \frac{1}{\gamma_{i\perp}}\vec{F}_{i\perp} \quad , \tag{1}$$

where $\vec{v}_i$ is the velocity of filament i and $\vec{F}_{i\|}$ and $\vec{F}_{i\perp}$ are the components of $\vec{F}_i$ in direction of and perpendicular to the filament axis, respectively. $\gamma_{i\|}$ and $\gamma_{i\perp}$ are the corresponding friction coefficients [11].

Under the same conditions, the angular velocity $\vec{\omega}_i$ is given by

$$\vec{\omega}_i = \frac{\vec{M}_i}{\gamma_{\mathrm{rot}}}\,, \tag{2}$$

with γ_{rot} as the rotational friction coefficient [11].

The actual displacement (new position $\vec{x}_i$ at time $t+\Delta t$ and rotation angle δ_i during time interval Δt) of each actin filament is calculated from the velocity $\vec{v}_i$ and the angular velocity $\vec{\omega}_i$ using an explicit Euler scheme:

$$\vec{x}_i(t+\Delta t) = \vec{x}_i(t) + \vec{v}_i(t)\cdot\Delta t \tag{3}$$

$$\delta_i = \omega_i \cdot \Delta t \tag{4}$$

2.2. Results and Discussion of Computer Simulations

To understand the elementary behavior of actin-myosin systems, simulations with two actin filaments in the presence of myosin molecules were performed. At the beginning of each run, filaments were arranged antiparallel, parallel, or crosswise:

(i) Filaments antiparallel to each other: If the initial distance between the filaments allows myosin molecules to bridge them, they slide along each other (the vector of motion points to the minus end) until they have reached a position with only a slight overlap at their plus ends.

(ii) Filaments parallel to each other: In this case, the filaments do not show any net movement.

(iii) Filaments arranged at an angle: If the simulation starts with an angle of 45° or 67.5°, the filaments always move into antiparallel orientation; with a starting angle of 90° or 135°, they move into parallel orientation. As an example of the latter, see Figure 1 showing positions of two filaments during a simulation at different time steps.

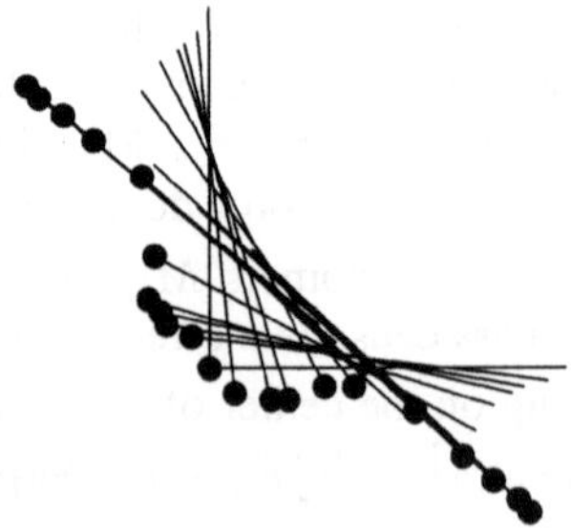

Figure 1. Projection of the positions of two filaments at 10 time steps. The simulation starts with an angle of 90° and overlap of the filaments at their minus ends (marked by a large dot). The filaments arrange into antiparallel orientation.

The phenomenon of (anti)parallelization in systems of two actin filaments reappeared in simulations with many actin filaments. The upper part of Figure 2 shows the initial configuration of 100 actin filaments and myosin molecules (myosin molecules are not displayed). The mass centers of the actin filaments were Poisson distributed. The lower part of this figure shows the same actin gel after 10^9 simulation steps. The mass centers of the filaments show a clustered distribution with parallelization within their groups.

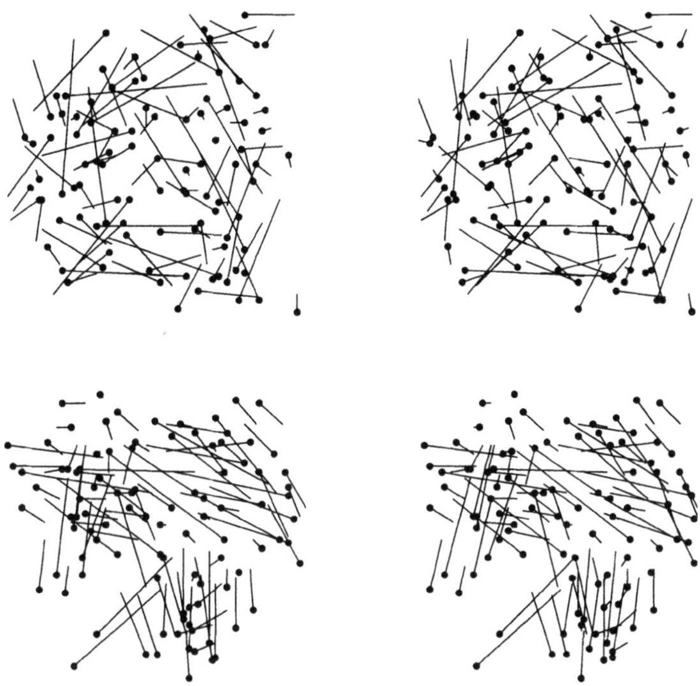

Figure 2. Stereo representation of an actin gel consisting of 100 filaments. Top: initial configuration. Bottom: configuration after 10^9 timesteps.

Viscoelastic properties of actin networks can be measured *in vitro* by means of an oscillating magnetic bead micro-rheometer (see, for example, [17, 37]). For an oscillating force $F_0 \cdot \cos(\omega t)$ acting on the magnetic bead with a certain amplitude, bead motion is analyzed by an image-processing system. The storage modulus $G'(\omega)$ and the loss modulus $G''(\omega)$ are obtained by analyzing amplitude $x_0(\omega)$ and phase shift $\varphi(\omega)$ of the oscillatory motion of the bead over a range of frequencies ω. This experimental setting can be simulated by an oscillating sphere that interacts with actin filaments by mutual repulsive forces. By solving the equation of motion for the sphere, we can obtain the amplitude $x_0(\omega)$ and the phase shift $\varphi(\omega)$ and thus compute storage and loss modulus:

$$G'(\omega) = \frac{F_0}{6\pi R\,|x_0(\omega)|} \cdot \cos\varphi(\omega) \tag{5}$$

$$G''(\omega) = \frac{F_0}{6\pi R\,|x_0(\omega)|} \cdot \sin\varphi(\omega)\,, \tag{6}$$

Up to now, only systems consisting of a small number of actin filaments (not more than 200, with a maximal length of 100 monomers) were simulated over a small range of frequencies (10–100 Hz). Therefore, the results cannot be compared with experimental results on a quantitative level. However, on a qualitative level we obtain an increase of the moduli with increasing frequency of oscillation. Comparison of an actin gel with an actin-myosin gel shows that the latter has a larger storage modulus and a smaller loss modulus. This is in good qualitative agreement with experimental results [25, 37].

3. Multiparticle Model of Cellular Motility

3.1. Model Description

The second model is aimed at the explanation of whole cell amoeboid shape changes on solid substrates from a physical point of view. As the main physical mechanism underlying cell motility, we propose a stochastic disturbance of an equilibrium between the internal hydrostatic pressure (as a global organizer that tends to increase the cell volume uniformly) and local stochastic contractive forces within the actin-myosin network (which tend to decrease the cell volume due to association with the envelope), but without any other regulatory elements. A more detailed description of the model can be found in [22].

The model consists of three kinds of particles: A-particles, which form either regions of high G-actin concentration with the capability of polymerization (presence of profilin, etc.) and stress fiber formation or micro-gel-like [33] regions of the actin-myosin cytoskeleton; M-particles, which form virtual elements of the membraneous cell envelope; and a single C-particle, which represents the central part of the cell consisting of nucleus and organelles and which is assumed not to play any active role in cellular motion. Every particle is represented by a point in space and an interaction radius (which has been set equally for all particles for simplicity). It is further characterized by its type and its state. Depending on these, a particle may interact with other particles of the same or of other types, leading to changes in its state or to motion induced by different kinds of experienced forces.

The fluid-like cell envelope is mimicked by a closed elastic chain of M-particles. No internal frictional or bending forces among the M-particles were incorporated in the model. Forces that can act on an M-particle are (i) spring forces due to association of the particle to its two neighbors, (ii) a random force leading to diffusion, (iii) a general frictional force representing the stochastic interactions of the particle with the virtual surrounding bulk medium, (iv) an excluded volume force acting between non-neighboring M-particles to prevent the formation of loops within the envelope chain, and (v) an outwards-oriented force that mimics the action of the internal hydrostatic pressure on the cell envelope.

The actin-myosin network is represented by A-particles that may be single or associated with each other, with M-particles, or with the C-particle. Association of A-particles is spatially anisotropic and results in the formation of a chain representing one of the long and thick actin fibers (stress fibers) that can be observed in living cells. *In vivo*, these bundles are known to be stabilized by cross-linker molecules like α-actinin [12]. These chains have a bending stiffness that tends to straighten them locally. A-particles of different chains may associate to form cross-links, thus leading to a fibrous network. This mimics the connection of stress fibers to each other within the cytoplasm by the diffuse portion of the cytoskeleton, thus leading to a stochastic mechanical continuum [15]. The existence of a coherent network is very important for transmission of local contractive forces to the cell envelope. These contractive forces act between neighbored A-particles and mimic the action of myosin II clusters.

In general, associative and contractive forces are of the spring type with a given equilibrium distance. The bending force acting on A-particles within a fiber is calculated from the sum of the two vectors that connect the particle with its two neighbors within the fiber. The hydrostatic force acting on every M-particle is assumed to depend on the deviation of the cell 'volume' (in the two-dimensional model the area of the model cell) from a predefined resting 'volume' in an exponentially decreasing manner. Stochastic contractive forces within the network are generated by randomly setting the equilibrium distance between two associated particles to zero. Repulsive forces are introduced between M-particles to avoid the formation of loops and further between A- and M-particles to keep the A-particles from leaving the model cell. Another kind of repulsive effect is introduced to prevent actin clumping by allowing movement of an actin particle to a new location only if the density of actin particles at the new location is below a certain threshold. Active transport of actin monomers to the cell periphery *in vivo* is mimicked in the model by a centripetal force acting on each A-particle being further away from the envelope than a certain threshold. Finally, a Gaussian random force acts on every particle.

Association, dissociation, and contractile conditions between two particles are simulated stochastically using rate constants and a constant time step. Movement of the particles due to all forces is computed using a Langevin-type equation approximating the overdamped limit (Eq. (7)):

$$\vec{v}_i = \frac{1}{\gamma}\left(\vec{F}_i + \frac{\mathrm{k_R}\vec{\xi}}{\sqrt{\Delta t}}\right) \tag{7}$$

$$\vec{r}_i(t + \Delta t) = \vec{r}_i(t) + \vec{v}_i\,\Delta t \tag{8}$$

In Eqs. (7, 8) $\vec{v}_i$ is the velocity of particle i, γ is the bulk friction coefficient, $\vec{F}_i$ is the sum of forces acting on particle i except the frictional force, Δt is the time interval, $\mathrm{k_R}\vec{\xi}$ is a white random noise with strength $\mathrm{k_R}$, and $\mathbf{r}_i$ is the position. The actual displacement of the particle is calculated from the velocity by a simple Euler scheme (Eq. (8)). The liquid portion of the cytoplasm is not modeled explicitly.

3.2. Results and Discussion of Computer Simulations

Beginning with the initial state of the model cell (represented by a starlike radial-symmetrically folded cell envelope containing the C-particle in the center and randomly distributed single A-particles, not shown), simulations of cell dynamics were carried out as follows. First, the A-particles were allowed to form a coherent network. Then, the pressure force and contraction probability were set > 0 to initiate dynamic shape changes. Simulations were performed for a fixed time interval. The cell envelope was folded in the initial state in a star-like manner in such a way that there was a nearly twofold length of the envelope with respect to the initial model cell inner area. This was done to mimic the surplus of membrane surface in living cells.

Figure 3 shows successive states of the model cell during a simulation with moderate contraction. After an initial increase, the cell 'volume' fluctuates around a mean value, which is a bit greater than the 'resting volume' and is determined by the relative strength of the hydrostatic pressure force and the contraction probability within the network. The model cell performs a random motion (not shown) and shows lively dynamics of its envelope, with transient protrusions and beginning bleb formation. Due to the inherent instability of the network, which is a result of the contractive forces and the stochasticity with respect to the connections between the A-particles, the network does not necessarily but possibly can collapse into one or several clusters and then no longer encloses the inner cell portion. This is a case of autonomous generation of polarity. In the simulation shown in Figure 3, the cell eventually reaches a bipolar state. Autonomous generation of polarity with monopolar or bipolar shapes has been experimentally observed with human keratinocytes by Alt et al. [2] and Libotte et al. [23]. Figure 4 shows an alternative representation of the same results: a three-dimensional plot of the model cell envelope *versus* time. The figure elucidates the transient protrusive behavior of the model cell, appearing in the plot as rows of bulges on the surface and even as a drifting protrusion in the lower left part. Such behavior has also been observed with motile keratinocytes on agar surfaces [2].

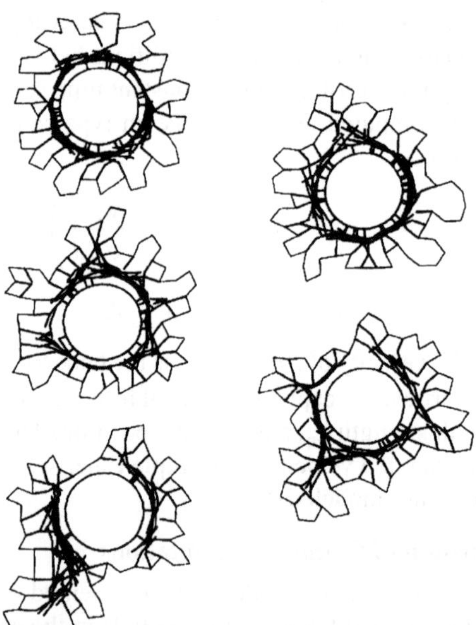

Figure 3. Successive states (from top to bottom) of the model cell during a simulation with moderate contractility. The lines indicate pairs of associated particles. See text for further explanation.

If contraction probability is increased, the model cell 'volume' decreases remarkably after an initial overshoot. However, it again stabilizes around a mean value. Under these conditions, intensive blebbing of the model cell envelope occurs (Figure 5). The roundish

Figure 4. Three-dimensional representation of shape dynamics of the model cell based on the simulation shown in Figure 3. The cell envelope is drawn *versus* time (from top to bottom) forming a surface.

shape of the blebs results from an equilibrium between the hydrostatic pressure force and the surface tension of the envelope [10]. The network always collapses into one coherent cluster, so that the model cell assumes a unipolar shape, finally. In the opposite case, that is, if contractibility is smaller than a certain threshold, the model cell 'volume' increases until the cell assumes a circular shape, determined by the hydrostatic pressure force and the surface tension within the envelope, as if the whole cell would behave like a single large bleb. Then, the network consists of many small transient clusters and is no longer coherent (and therefore unable to keep the cell volume in limits). The model cell also shows a rather apolar appearance.

The mechanical system underlying the dynamic shape changes in the present model results in more inhomogeneous and polar cellular states, the larger the ratio is between contractility and the pressure force.

On the way to a quantitative characterization of shape dynamics, we performed a two-dimensional analytical approach based on the calculation of the 'momental ellipse', which is the ellipse that has the same centroid and the same second-order moments as the original shape of the model cell [1]. From the momental ellipse, the shape parameters 'elongation' (E) and 'dispersion' (D) can be computed as

$$E = \log_2\left(\frac{R^+}{R^-}\right) \geq 0 \tag{9}$$

$$D = \log_2\left(\frac{\pi R^+ R^-}{A}\right) \geq 0 \tag{10}$$

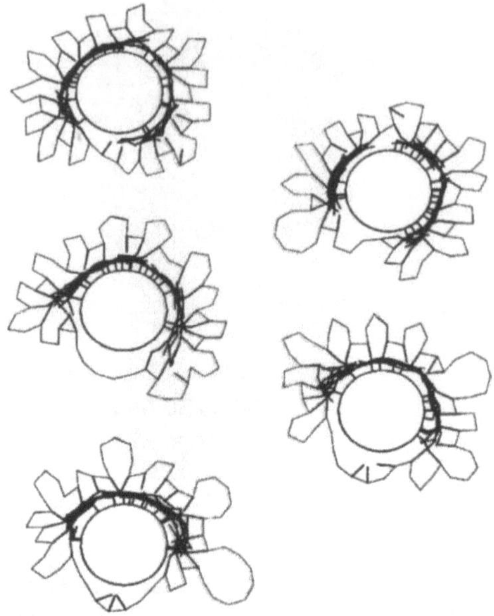

Figure 5. Successive states (from top to bottom) of the model cell during a simulation with high contractility. The lines indicate pairs of associated particles. See text for further explanation.

In Eqs. (9, 10) R^+ is the largest radius of the momental ellipse, R^- is its smallest radius, and A is its area. E is a measure of the deviation from radial symmetry, whereas D measures the dissection of the original shape. Figure 6 shows E *versus* time for the simulation shown in Figure 3. The value increases at the beginning and then fluctuates around 0.15 for some time. Finally, it increases to a value above 0.6 in coincidence with the shape transition of the model cell from more-or-less roundish to bipolar. The positions of two typical shapes of the model cell are also sketched in Figure 6. Figure 7 shows the trajectory of the model cell in the E-D-space, together with a representation of some typical shapes and their values of E and D (see the inset). The trajectory mostly stays within a region of values of E and D smaller than 0.2, with the exception of the mentioned escape to larger values of E. Then, the value of D is also larger. Libotte et al. (personal communication) determined elongation and dispersion for human keratinocytes over time on different kinds of surfaces. Taken together, the authors got values of 0.1–1 for E and of 0.01–0.02 for D (unpublished results), with the higher values for E being associated with more polar or even bipolar states of the cells.

The dynamic shape changes of the model cell are in good qualitative (and with respect to shape analysis, also quantitative) agreement with experimental observations on human keratinocytes, for which a pressure-based mechanism of motility has been proposed [2]. In our model system, protrusions form in those regions of the cell periphery where the network-envelope connection is stochastically weakened [8]. There, the cell

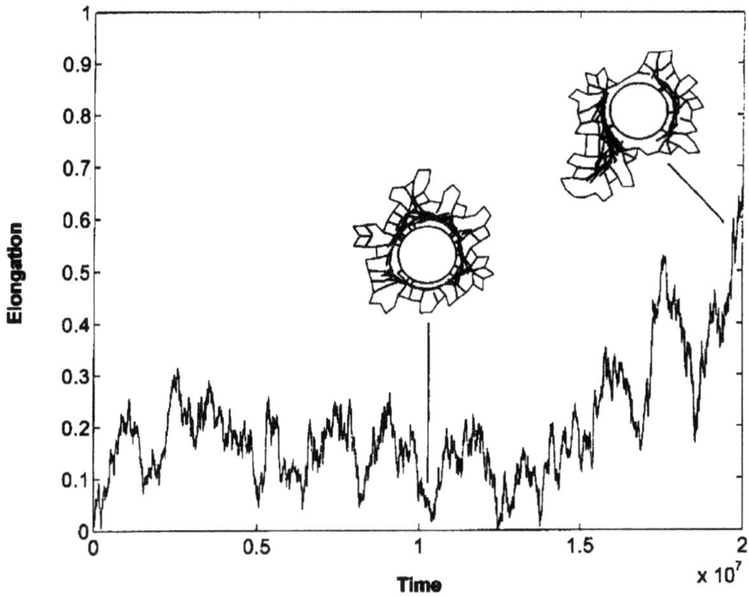

Figure 6. Plot of elongation of the model cell envelope *versus* time based on the simulation shown in Figure 3. The positions of two distinct shapes of the model cell are also indicated in the figure.

envelope is bulged out by internal hydrostatic pressure. The transient nature of the protrusions arises from the delayed reorganization of the network within the protruded area followed by a retraction of the envelope due to contraction of the network. In cases where the delay time is large, blebs will form.

4. Discussion

The phenomena described in this paper have been investigated by other groups using different models.

Farkas et al. [13] published a model for the motion of actin filaments on myosin-coated surfaces. The actin filaments are modeled by a series of mass points that are connected by weightless stiff rods. They are elastic and bend at the mass points. In the authors' numerical studies, a filament shows a sliding motion that follows a winding path.

A computer model for investigation of the shear modulus of a cross-linked network was developed by Frey et al. [16]. In contrast to the present model, they simulate a two-dimensional network consisting of (semi)flexible filaments that are connected by passive cross-linkers. The elastic response of the network under shear deformation is calculated.

These models use a more complex description of actin filaments, in contrast to the first model discussed here, where actin filaments are modeled as stiff rods. Therefore, an extension to our model will be the introduction of filament flexibility. This might enable us to reproduce, in simulations, the experimentally observed formation of actin rings and

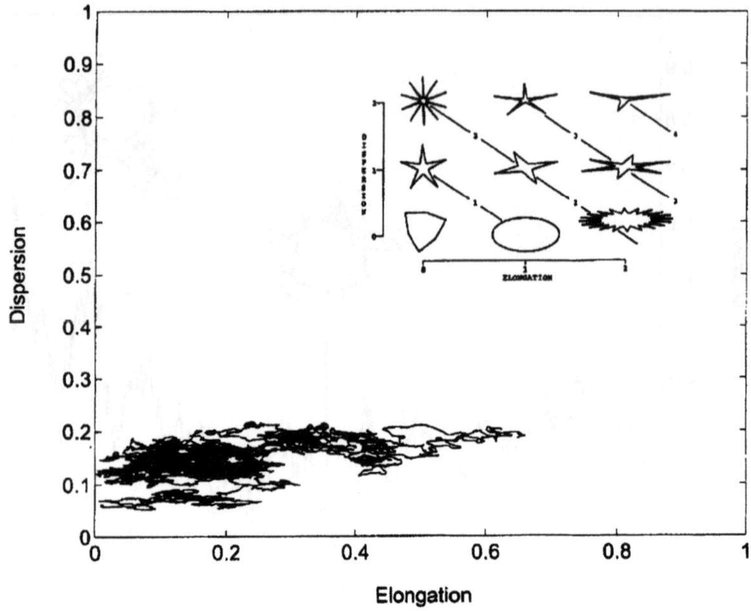

Figure 7. Phase-plot of dispersion *versus* elongation of the model cell envelope based on the simulations shown in Figure 3. The inset shows some typical cell shapes together with their positions in the dispersion/elongation phase-plot.

loops [32]. Pattern formation in a similar system consisting of microtubules and their associated motor proteins was investigated by Surrey et al [30]. They observed the self-organization of asters and vortices in experiment and computer simulation.

A model for cell motility similar to the one presented here was developed by Bottino and Fauci [5, 6, 7], based on the immersed boundary method [26], which considers the detailed fluid dynamics of the liquid cytoplasmic portion, whereas in the present model we follow the well-known Brownian Dynamics approach, in which the bulk fluid is incorporated *via* a general friction coefficient. Another similar particle-based model has been recently published by Baumgärtner and Sambeth [28, 29]. In their model, the protrusive forces are exclusively generated by actin polymerization near the cell membrane at the leading edge. This mechanism might play a dominant role in fish keratocytes [4, 35]. Here, we focus instead on the internal hydrostatic pressure as the main protrusive force, which might be true for cell types like fibroblasts or the keratinocytes of the human skin [3]. Indeed, one has observed a series of intermediate states between pressure-generated blebs and flat lamellipodia [19]. *In vivo*, one might face a mixture of different mechanisms, with one or the other mechanism being dominant for a certain cell type or a certain situation or cellular state. Yet another similar model for cell movement has been developed by Promayon et al. [27]. Their model is based on a new 'physical-object oriented' approach and can be found at the end of Part II.

The two models presented in this paper are related to each other, not only from a methodological point of view but also with respect to the modeled systems. The first model, on a mesoscopic scale, focuses on molecular interactions and spatial arrangements in actin-myosin solutions. The second model, on a macroscopic scale, describes the shape dynamics of tissue cells. Our knowledge of the detailed local and emergent dynamics of actin and myosin in aqueous solution, gained through simulations using the first model, was incorporated into the second model on the larger scale in a phenomenological way by the definition of rules mimicking the molecular behavior. Particularly, within the second model, we mimicked the phenomena of filament sliding, leading to contraction, and of parallelization of actin filaments, leading to more or less stiff bundles. Both phenomena appeared in the simulations based on the second model, by introducing virtual actin elements (A-particles) that, upon interaction, give rise to the generation of chains. Hence, there is a compression of information on the way from the first to the second model. On the other hand, mechanistic assumptions, incorporated phenomenologically into the second model, might be 'verified' in the future by a more detailed modeling on the lower level. This modeling of certain phenomena in the same context on different spatial scales can result in a fruitful synergistic effect on both levels.

References

[1] W. Alt. Statistics and dynamics of cellular shape changes. In M. A. Chaplain, G. D. Singh, and J. C. McLachlan, editors, *On growth and form*, pages 287–307. John Wiley & Son, Ltd., Chichester, 1999.

[2] W. Alt, O. Brosteanu, and B. Hinz. Patterns for spontaneous motility in videomicrographs of human epidermal keratinocytes (HEK). *Biochem. Cell Biol.*, 73:441–459, 1995.

[3] W. Alt and M. Dembo. Cytoplasm dynamics and cell motion: two-phase fluid models. *Math. Biosci.*, 156:207–228, 1999.

[4] G.G. Borisy and T.M. Svitkina. Actin machinery: pushing the envelope. *Curr. Opin. Cell Biol.*, 12:104–112, 2000.

[5] D.C. Bottino. Computer simulations of mechanochemical coupling in a deforming domain: application to cell motion. In Ph.K. Maini and H. Othmer, editors, *Mathematical Models for Biological Pattern Formation*, volume 121 of *IMA Volumes in Mathematics and its Applications, Frontiers in Applied Mathematics Series*, pages 295–314. Springer, New York, 2000.

[6] D.C. Bottino and L.J. Fauci. A computational model of ameboid deformation and locomotion. *Eur. Biophys J.*, 27:532–539, 1998.

[7] D.C. Bottino, et al. How nematode sperm crawl. *J. Cell Sci.*, 115:367–384, 2002.

[8] A. Boulbitch, A. Mogilner, T. Roberts, M. Steward, and G. Oster. Shape instability of a biomembrane driven by a local softening of the underlying actin cortex. *Phys. Rev. E*, 62:3974–3985, 2000.

[9] D. Bray. *Cell Movements*. Garland Publishing, 1992.

[10] C.C. Cunningham. Actin polymerization and intracellular solvent flow in cell surface blebbing. *J. Cell Biol.*, 129:1589–1599, 1995.

[11] M. Doi and S.F. Edwards. *The Theory of Polymer Dynamics*. Oxford University Press, 1986.

[12] M. Edlund, M.A. Lotano, and C.A. Otey. Dynamics of α-actinin in focal adhesions and stress fibers visualized with α-actinin-green fluorescent protein. *Cell Mot. Cytoskel.*, 48:190–200, 2001.

[13] Z. Farkas, I. Derenyi, and T. Vicsek. The dynamics of actin filaments in motility assays: a microscopic model and its numerical simulation. This issue.

[14] D. Felix. Modellierung und Simulation der dynamischen Eigenschaften von Aktin-Myosin-Polymernetzwerken. Diploma thesis, Mathematisch-Naturwissenschaftliche Fakultät der Rheinischen Friedrich-Wilhelms-Universität Bonn, 1999.

[15] G. Forgacs. Cooperative phenomena in physical networks. In D. Beysens and G. Forgacs, editors, *Dynamical networks in physics and biology*, pages 3–13. Springer, Berlin/Heidelberg/New York, 1998.

[16] E. Frey, K. Kroy, and J. Wilhelm. Physics of solutions and networks of semiflexible macromolecules and the control of cell function. In B. T. Stokke, editor, *The Wiley Polymer Networks Group Review Series, Volume 2*. Wiley, 1998.

[17] Y.C. Fung. *Biomechanics*. Springer Verlag, 1993.

[18] Y. Harada, A. Noguchi, A. Kishino, and T. Yanagida. Sliding movement of single actin filaments on one-headed myosin filaments. *Nature*, 326:805–808, 1987.

[19] A.K. Harris. Protrusive activity of the cell surface and the movements of tissue cells. In N. Akkas, editor, *Biomechanics of active movement and deformation of cells*, volume H 42 of *NATO ASI Series*, pages 249–294. Springer, Berlin/Heidelberg, 1990.

[20] P.A. Janmey, S. Hvidt, J. Käs, D. Lerche, A. Maggs, E. Sackmann, M. Schliwa, and T.P. Stossel. The mechanical properties of actin gels. *Journal of Biological Chemistry*, 269(51):32503–32513, December 1994.

[21] J. Käs, H. Strey, J.X. Tang, D. Finger, R. Ezzel, E. Sackmann, and P.A. Janmey. F-actin, a model polymer for semiflexible chains in dilute, semidilute, and liquid crystalline solutions. *Biophys. J.*, 70:609–625, 1996.

[22] J. Lenz. Many-particle simulation of ameboid motility. *Future Gener. Comput. Syst.*, 17:863–872, 2001.

[23] Th. Libotte, H.-W. Kaiser, W. Alt, and T. Bretschneider. Polarity, protrusion-retraction dynamics and their interplay during keratinocyte cell migration. *Exp. Cell Res.*, 270:129–137, 2001.

[24] H. Lodish, et al. *Molecular Cell Biology*. W.H. Freeman and Company, 2000.

[25] A. Palmer, T.G. Mason, J. Xu, S.C. Kuo, and D. Wirtz. Diffusing wave spectroscopy microrheology of actin filament networks. *Biophys. J.*, 76:1063–1071, 1999.

[26] Ch.S. Peskin. Numerical analysis of blood flow in the heart. *J. Comput. Phys.*, 25:220–252, 1977.

[27] E. Promayon, J.L. Martiel, and P. Tracqui. Physical-object oriented 3d simulations of cell deformations and migration. This issue.

[28] R. Sambeth and A. Baumgaertner. Autocatalytic polymerisation generates persistent random walk of crawling cells. *Phys. Rev. Lett.*, 86:5196–5199, 2001.

[29] R. Sambeth and A. Baumgaertner. Locomotion of a two-dimensional keratocyte model. *J. Biol. Systems*, 9:201–219, 2001.

[30] T. Surrey, F. Nédélec, S. Leibler, and E. Karsenti. Physical properties determining self-organization of motors and microtubules. *Science*, 292:1167–1171, 2001.

[31] K. Takiguchi. Heavy meromyosin induces sliding movements between antiparallel actin filaments. *Journal of Biochemistry*, 109:520–527, 1991.

[32] J.X. Tang, J.A. Käs, J.V. Shah, and P.A. Janmey. Counterion-induced actin ring formation. *Eur. Biophys. J.*, 30:477–484, 2001.

[33] M. Tempel, G. Isenberg, and E. Sackmann. Temperature-induced sol-gel transition and microgel formation in α-actinin crosslinked actin networks: a rheological study. *Phys. Rev. E*, 54:1802–1810, 1996.

[34] T.Q.P. Uyeda, S.J. Kron, and J.A. Spudich. Myosin step size. Estimation from slow sliding movement of actin over low densities of heavy meromyosin. *Journal of Molecular Biology*, 214:699–710, 1990.

[35] A.B. Verkhovsky, T.M. Svitkina, and G.G. Borisy. A network contraction model for cell translocation and retrograde flow. *Cell Behaviour: Control and Mechanism of Motility*, pages 207–222, 1999.

[36] J. Xu, W.H. Schwarz, J.A. Käs, T.P. Stossel, P.A. Janmey, and T.D. Pollard. Mechanical properties of actin filament networks depend on preparation, polymerization conditions, and storage of actin monomers. *Biophys. J.*, 74:2731–2740, 1998.

[37] F. Ziemann, J. Rädler, and E. Sackmann. Local measurements of viscoelastic moduli of entangled actin networks using an oscillating magnetic bead micro-rheometer. *Biophys. J.*, 66:2210–2216, 1994.

Polymer and Cell Dynamics – Multiscale Modeling
and Numerical Simulations, 91–112
ed. by W. Alt, M. Chaplain, M. Griebel und J. Lenz

Analysis of Cell Motility Combining Cytomechanical Model Simulations and an Optical Flow Method

Angélique Stéphanou[1], Xavier Ronot[1,2] and Philippe Tracqui[1]

1) Laboratoire TIMC-IMAG/Dyna Cell, Faculté de Médecine de Grenoble,
F-38706 La Tronche Cedex, France
2) E.P.H.E., Institut Albert Bonniot; F-38706 La Tronche, France

Summary. We developed a combined analysis of isolated cell dynamics based on both the extraction of relevant data obtained from time-lapse video-sequence analysis, such as velocities, periodic components, and preferential morphologies, and the simulation of cells' dynamical behavior from theoretical models. Two methods have been alternatively used for analyzing the videomicroscopy sequences. The first one is based on an optical flow approach and relies on the conservation of light intensity in order to estimate the parameters of an affine motion model of cell deformations. A refined analysis of localized cell movements such as lamellipodial extension and retraction has also been undertaken using a more classical segmentation of cell outlines. Various tools have been subsequently used to analyze the corresponding time-amplitude curves and contours maps, as well as time variation of the motion model eigenvalues, including Fourier analysis. In association with this quantification of cell shape dynamics, we consider a continuous cytomechanical model of cell deformations. The model relies on the description of mechanical and chemical interactions among the cytoplasm, the actin cytoskeleton, and the cell membrane. Different patterns of cell morphological changes have been simulated, considering different values of the model parameters. Specific illustration of these combined approaches is proposed for the study of L929 murin fibroblast deformations.

1. Introduction

The morphology of a cell in a tissue depends on its interactions with neighboring cells and with the extracellular matrix. Thus, changes in cell shape can obviously result from alteration of these cell interactions [15]. Most of the studies conducted on cell shape changes were interested in modifications induced by external stimuli of a different nature (chemical, mechanical, etc.). However, these already complex interactions and cellular responses cannot be completely understood if the spontaneous cell activity is not characterized. In this sense, the *in vitro* study of stationary isolated cells plated onto inert substratum in the absence of additional external stimuli can shed light on the self-organized character of cell deformations. This concept was already proposed in the early works of Goldacre [8] and

Weiss [22], the latter suggesting that a cell expends energy to put its surface layer "... in a travelling dynamic oscillation between higher and lower contractile states synchronized over a large fraction of the cell surface and sweeping over in the manner of a pulse". It is now well documented that shape changes are mainly due to the dynamic rearrangements of the cell actin cytoskeleton [17, 19], coupled with the dynamic polymerization process governing associated microtubule organization [21].

However, the mechanisms driving the formation of cellular protrusions are still controversial. Alternative views consider that cell protrusions can be generated by the formation of F-actin bundles which push forward the cytoplasmic membrane where the membrane is locally detached from the network [18, 12], or that the initiating process is the local polymerization of an actin gel in a free volume located between the plasma membrane and the cell actin cortex [4]. Besides the formulation of theoretical models that take into account such hypotheses [14, 16, 1, 13], it is of particular interest to study cell shape dynamics from the analysis of relevant spatiotemporal data obtained from time-lapse videomicroscopy. Determination of the modification with time of cell contour perimeters provides a first basis for analyzing dynamic changes of cell morphology. This has been done for different type of cells, including keratinocytes [2], leukocytes [1], as well as amoeba [9]. However, it is known that cell contour segmentation remains quite a time-consuming process that can rarely be conducted as a fully automated manner. An alternative approach, based on optical flow methods, has been recently proposed for analyzing globally the dynamic changes of cell morphology [7]. This approach relies on the formulation of a simple *a priori* motion model that takes into account most cell shape changes observed on isolated cells.

In this paper, both approaches are considered for studying the self-organized character of L929 murin fibroblast shape changes without net cell translocation. Indeed, cell morphology changes are commonly described as a random process, in which cell protrusion orientation and amplitude appear as stochastic processes that are temporally independent. We show here, in the case of L929 fibroblasts, that cell shape changes have a periodic component and that cell membrane protrusions are correlated and synchronized. Using Fourier analysis of time signals obtained by each approach, namely, cell contour time-amplitude maps and motion model eigenvalue time variations, we characterized preferential cell morphology and the cyclic aspect of pseudopods' extension and retraction. Since both approaches are quite different in nature, we also compare their relevance for analysis of cell deformation and quantification of protrusion velocity, as well as their relevance for extended analysis of cell locomotion.

2. Materials and Methods

2.1. Cell Culture

Cell culture experiments were performed on L929 murin fibroblasts (ATCC, CCL 1) that were cultured in antibiotic-free Minimum Essential Medium (MEM, Life Technologies), supplemented with 5% fetal calf serum (FCS, Life Technologies), and with 2 mM L-glutamine. The cells were cultured on glass for 48 hours.

2.2. Videomicroscopy

Image sequences were acquired using an imaging workstation composed of an inverted microscope (Axiovert 135M, Carl-Zeiss) equipped with a temperature- and CO_2-controlled chamber (37° C, 5% CO_2) (M-Incubator, Carl-Zeiss) and automatic shutters and video cameras (C2400-08 Hamamatsu and CoolView 2 Photonic Science) coupled with image software (SAMBA 2640TM, Unilog and Image Pro 4, Media Cybernetics). The magnification used to record the dynamics of spontaneous cell deformations was ×400, with a delay between two consecutive images of 2 min for short sequences (1.5-hour recording) and 3 min for long sequences (8-hour recording). These sampling times were short enough to be compatible with cell movement frequency.

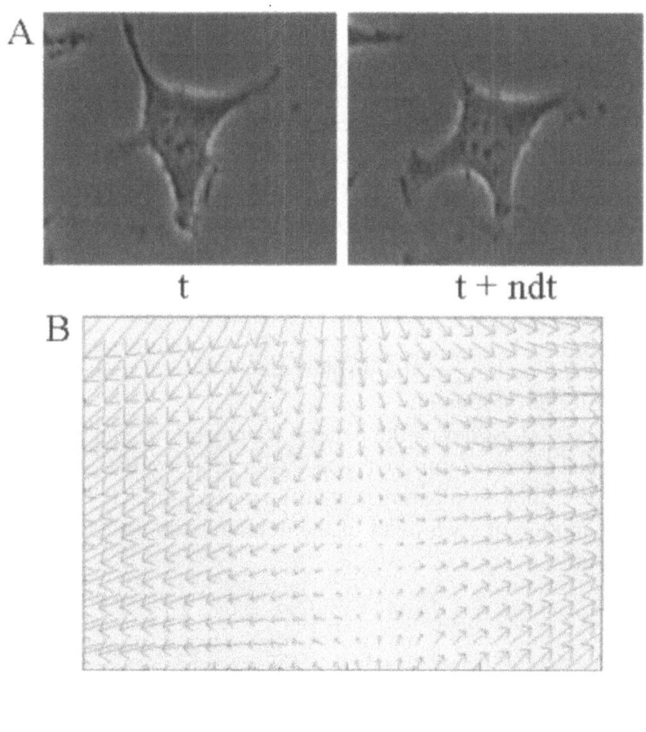

Figure 1. Successive images of the protrusive activity of an L929 fibroblast (A) and associated displacement field (B) determined when considering an affine motion model of deformation (reproduced at a larger scale). Time interval between the two images is 10 min. The displacement field reveals the cell protrusive dynamic during this time interval, namely, a diagonal extension of the cell protrusions synchronized with cell membrane retraction in the quasi-perpendicular direction. The motion field has been drawn over the entire region, not only for pixels corresponding to the cell, in order to provide a better graphical visualization of the cell deformation.

3. Analysis of Cell Dynamics from Time-lapse Sequences

3.1. Optical Flow Analysis

Cell shape change analysis from video sequences is especially challenging: cells are highly deformable objects and the contrast of videomicroscopy images is usually low. Therefore, in order to avoid the drawbacks of segmentation-based techniques, we developed a spatiotemporal analysis of cells' protrusive dynamics based on a direct exploitation of temporal and spatial variations of light intensity over the image domain R containing the cell [7]. At each pixel $p(x, y)$ of the domain, the spatial and temporal derivatives of the light intensity $I(p, t)$ were then evaluated and used for further estimation of the apparent velocity motion field, or optical flow, during the time interval Δt separating two successive images of the sequence. More precisely, the apparent displacement $w(p, t)$ during a time step at each pixel was determined assuming a conservation of the light intensity with time. Thus, $w(p, t)$ has to satisfy the following conservation equation:

$$DFD(w, p, t) = I(p + w(p, t), t + \Delta t) - I(p, t) = 0, \tag{1}$$

where $DFD(w, p, t)$ is called the displaced frame difference (DFD).
There is no analytical solution to this equation. Thus, one method consists in using a first-order Taylor development of $I(p + w(p, t), t + \Delta t)$ to get an equation (called the Motion Constraint Equation) that linearly relates the velocity at each pixel to the local spatial and temporal derivatives of $I(p, t)$:

$$\nabla I^T(p, t) w(p, t) + \frac{\partial I}{\partial t}(p, t) = 0. \tag{2}$$

This equation is valid as long as the displacements $w(p, t)$ remain sufficiently small to justify the first order-development.

While the above steps define the general framework of non-parametric optical flow methods, an original aspect of our approach was to propose that cell shape changes could be satisfactorily described by a parametric affine motion model. This assumption implies that the apparent motion of each pixel of the domain obeys at each time t the following parametric relation:

$$w(p) = \begin{pmatrix} a_0 \\ a_3 \end{pmatrix} + \begin{pmatrix} a_1 & a_2 \\ a_4 & a_5 \end{pmatrix} \begin{pmatrix} x \\ y \end{pmatrix}. \tag{3}$$

Time-dependent parameters a_0 and a_3 account for translation, whereas the parameters a_1, a_2, a_4, and a_5 determine more complex shape changes such as rotation, extension, or retraction. In agreement with phase plane analysis of two-dimensional linear differential systems, such behaviors are related to the complex or real nature of the eigenvalues of the square matrix $[a_1 \dots a_5]$ (Table 1). These parameters are the components of an unknown vector $\phi(a_0, a_1, a_2, a_3, a_4, a_5)$ determining $w(p) = w_\phi(p, t)$, which is estimated for each successive pair of video images (Fig. 1A) by minimizing the following least mean square expression:

$$E(\phi) = \sum_{p \in R} [\nabla I^T(p, t) w_\phi(p, t) + \frac{\partial I}{\partial t}(p, t)]^2. \tag{4}$$

Details of the estimation procedure can be found in [7]. The associated displacement field between time t and $t+\Delta t$ (Fig. 1B) is then calculated and plotted over the image domain by applying the affine relation above.

Table 1 Motion characterization from the eigenvalues

Eigenvalue	associated motion
λ_1 and λ_2 real; $\lambda_1\lambda_2 > 0$	Pure expansion
λ_1 and λ_2 real; $\lambda_1\lambda_2 < 0$	Combined expansion and retraction (in an orthogonal direction)
λ_1 and λ_2 pure imaginary	Pure rotation

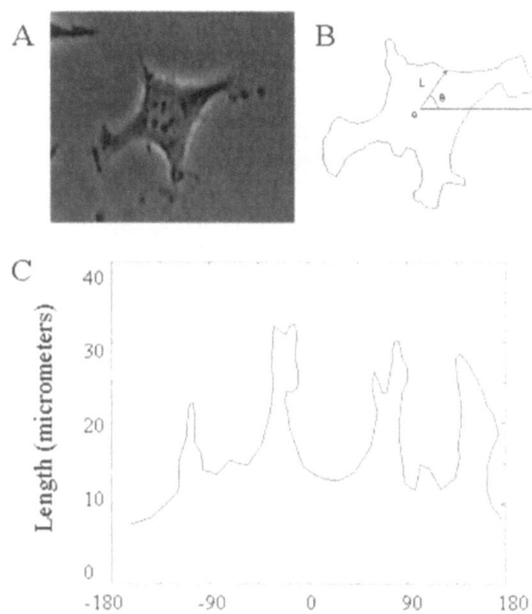

Figure 2. The boundary of an isolated L929 fibroblast (A) has been extracted by manual segmentation (B). A polar referential $[0,\ \theta,\ L(\theta)]$ is defined, with the cell nucleus barycenter as the origin 0. We then defined a polarity (or contour) map (C) as the polar reconstruction of the cell membrane position $L(\theta)$ plotted versus the angular direction θ.

3.2. Spatiotemporal Representation of Cell Deformations

The cell outlines are extracted manually (Figs. 2A,B) and reconstructed in polar coordinates (Fig. 2C). The cell nucleus barycenter is taken as the origin of a 2D polar coordinates system. The position of each point of the cell membrane with regard to this origin and according to an angular direction θ is defined at each time t by the estimated distance $L(\theta, t)$, with $0 < \theta < 2\pi$. From the videomicroscopy time-lapse sequences, one then gets

– (i) at any given sampling time $t_{n\Delta t}$, a contour or polarity map $C_n(\theta) = L(\theta, t)_{t=t_{n\Delta t}}$ represented as multi-valued graph – (ii) in a given angular window $[\theta_a, \theta_b]$ (see below) a time/amplitude map $A_\theta(t) = L(\theta, t)_{\theta \in [\theta_a, \theta_b]}$. Superimposition of the polarity maps at successive times $t = n\Delta t$ gives the 3D map reflecting the spatiotemporal shape changes of any given cell (Fig. 2).

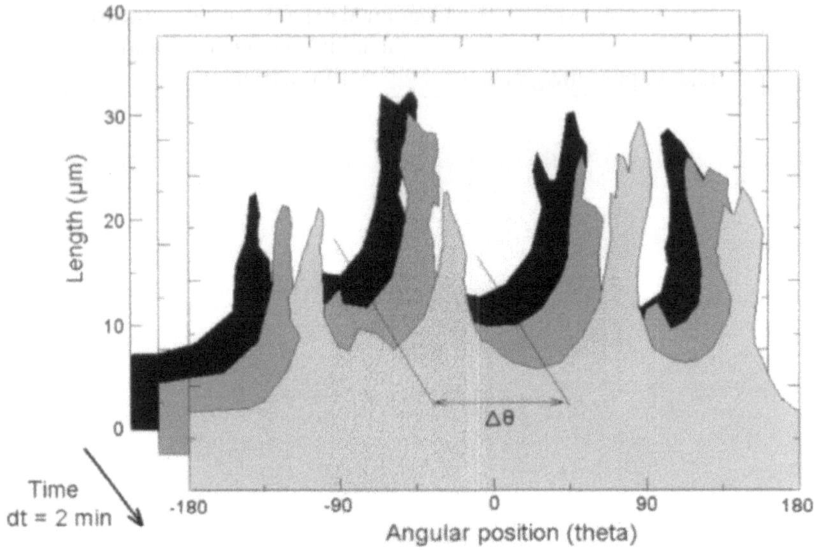

Figure 3. Spatiotemporal 3D reconstruction of the cell protrusion dynamic obtained by superimposing polarity maps at successive times. After appropriate thresholding, one can define small enough angular windows $\Delta\theta$ within which the oscillatory behavior of the cell protrusion takes place. A remarkable feature of isolated L929 cells is the persistence with time of each window orientation (see Fig. 4). It is thus possible to extract for each protrusion time/amplitude curves (see Fig. 5) corresponding to the evolution with time of the maximum protrusion length within the angular window.

3.3. Local Analysis of Membrane Extension/Retraction Dynamics

Different scales of membrane deformations are revealed by the polarity maps, with large amplitude of deformation occurring selectively in specific angular windows (Fig. 3). We therefore performed an analysis of individual cell protrusion dynamics by defining specific angular windows of width $[\theta_{a_i}, \theta_{b_i}]$ and by retaining in these windows the largest value of $A_\theta(t)$. The successive values of $A_\theta(t)(\theta)_{\theta \in [\theta_{a_i}, \theta_{b_i}]}$ give an approximative representation of the extension/retraction dynamics of each protrusion in a given mean orientation, as long as this orientation is preserved, which is indeed the case in the experiments (Fig. 4).

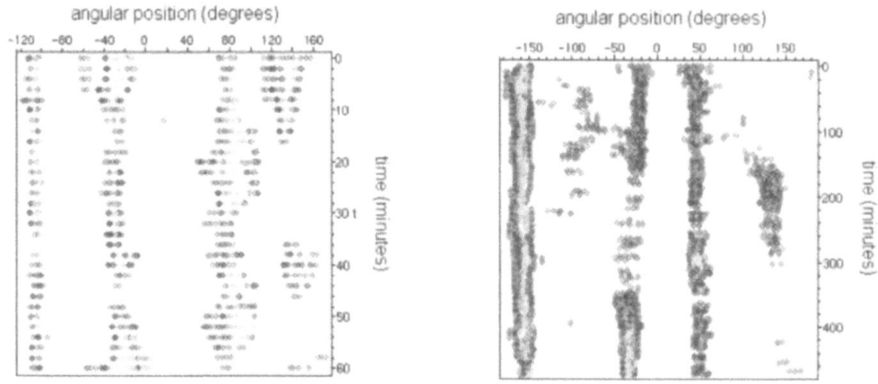

Figure 4. 2D projections of the 3D spatiotemporal polarity maps (Fig. 3) above a length threshold that suppresses roughly the mean cell envelope. Only significant membrane protrusions thus remain in the graph. The map on the left shows the evolution of the membrane deformations recorded during approximately one hour in one experiment, while the map on the right is obtained for another fibroblast in an experiment conducted for 8 hours. These two cases are quite representative of the considered population of L929 cells, which can exhibit persistence of protrusive directions on both short and long time scales.

3.4. Fourier Analysis

Period detection on time/amplitude curves $A_\theta(t)$ has been performed through Fourier analysis. The discrete Fourier transform is given by:

$$A_\theta^n = \sum_{n=0}^{N} x_k \exp\left(\frac{2i\pi kn}{N}\right), \tag{5}$$

with

N number of samples of the curve,
x_k amplitude for the sample k, and
A_θ^n amplitude for the frequency

$$\nu_n = \frac{n}{T} = \frac{n}{N\Delta_t}, \tag{6}$$

with Δ_t length of sampling interval.

Evaluation of the energy is given by $|A_\theta^n|$. The limited nucleus movements have been taken into account by recording its barycenter displacement with time. This motion corresponds to a low frequency signal of about 0.4 mHz, which determines the frequency used to filter the time/amplitude curves before their Fourier analysis.

4. Theoretical Modeling of Cell Membrane Deformation

4.1. Formulation of a Cytomechanical Model

We consider the polymerization and reorganization of actin between its globular form (G-actin) and filamentous form (F-actin), as well as the associated cytoplasmic convection of F-actin near the cell membrane, as the leading dynamical factors responsible for cell shape oscillatory changes. Indeed, this polymerization/depolymerization process is predominant near the cell membrane, exerting on it a local stress that depends on the F-actin concentration. On the other hand, membrane displacement is limited by passive elastic stress exerted by the cytoskeleton and transmitted to the membrane by the cascade of proteins insuring F-actin/integrins connections. The sub-membrane periphery of the cell, the cell cortex, can thus be viewed as a dense viscoelastic scaffold. Furthermore, the actin-myosin interaction gives rise to a contractile network embedded in the viscous cytoplasm. The local membrane curvature induces on the cell cortex an additional stress whose direction depends on the curvature orientation.

All these phenomena were originally considered in the cytomechanical model developed by Alt and Tranquillo [3]. We use this model as a basis for a further analysis of the observed spontaneous oscillatory dynamics of L929 fibroblasts. The model considers the 2D annular domain bounded by the cell cortex on one side, the plasma membrane on the other side, this latter defining a free-moving boundary. We note $L(x,t)$ the width of the annular domain along any radial direction defined by the angle x ($0 \leq x \leq 2\pi$). In order to simplify the model formulation, we further consider as an approximation that $L(x,t)$ remains small enough with respect to the mean radius R_0 of the cell body (experimentally the membrane extension is measured from the cell nucleus barycenter; thus, $L_{\text{exp}}(\theta,t)$ corresponds to $L_{\text{simul}}(x,t) + R_0$). This is a clear limitation of the model since large membrane extensions are observed with the fibroblasts. However, our aim is to investigate whether this simplified description can explain qualitatively and partly quantitatively the observed oscillatory pulsating star-shape morphologies of the fibroblasts. A positive answer would support a unified description of cell dynamical morphologies, from rounded (keratinocytes, etc.) to stellar (fibroblasts, etc.) shapes, based on a common modeling of actin network polymerization dynamics and cell cortex mechanical non-equilibrium.

As a further simplification, we have kept the approximations proposed by Alt and Tranquillo [3], namely, that within the cortex, both the variations of F-actin concentration and the tangential velocity of F-actin are small in the radial direction. This approximation can be viewed as a representation of F-actin bundles emerging radially from the cell actin cortex into the cell membrane protrusion.

This 1D approximation considers the cell cortex as a shell-like network of F-actin with small thickness. The formulation of the 1D cytomechanical model then consists of three equations, the first one being a conservation of F-actin concentration, the two others defining the force balance equilibrium in the radial and tangential directions, respectively. The three associated variables are the local annular width $L(x,t)$, the F-actin concentration $a(x,t)$, and the tangential velocity $v(x,t)$. The radial velocity of the network is given by the time derivative $\frac{\partial L}{\partial t}$.

The conservation equation for the F-actin along the width $L(x,t)$ is expressed by:

$$\frac{\partial}{\partial t}(La) + \frac{\partial}{\partial x}(Lav) = \eta L(a_0 - a), \tag{7}$$

where η characterizes the net polymerization rate. The threshold between polymerization and depolymerization is given by the critical concentration value a_0.

The cell membrane position depends on the balance of the forces applied on the membrane. We take into account the protrusive force generated by internal hydrostatic pressure P_1, counter-balanced by an elastic resistive stress $\gamma_1 La$ exerted by the actomyosin network on the cortex and assumed to be proportional to the local F-actin density. We also consider the curvature stress T_{c_1} depending on the local cortical actin network-membrane complex elasticity $\tau_1(x)$. In the radial direction, the force balance equation is then

$$\Phi_1 a\frac{\partial L}{\partial t} = P_1 - \gamma_1 La + T_{c_1} \qquad \text{with} \qquad T_{c_1} = \frac{\partial}{\partial x}\left(a\tau_1(x)\frac{\partial L}{\partial x}\right), \tag{8}$$

where Φ_1 is a friction coefficient corresponding to the drag of the cell membrane motion on the glass substratum.
The equation for the mechanical equilibrium on the actomyosin network in the tangential direction is given by

$$\frac{\partial}{\partial x}[\sigma_v + \sigma(a) - T_{c_0}] = \Phi_0 av, \tag{9}$$

where Φ_0 is a friction coefficient of the same origin as the previous one, σ_v is the viscous stress component of the cytoplasm-actomyosin network complex, $\sigma(a)$ is the contractile component of the actomyosin network, and T_{c_0} is a curvature-induced stress depending on the local actin cortex elasticity $\tau_0(x)$. There, respective expressions are given by

$$\sigma_v = \mu_0 a\frac{\partial v}{\partial x}, \tag{10}$$

$$\sigma(a) = \psi_0 a^2 e^{-a/a_{sat}}, \text{ and} \tag{11}$$

$$T_{c_0} = \frac{\partial}{\partial x}\left(a\tau_0(x)\frac{\partial L}{\partial x}\right), \tag{12}$$

with μ_0 as the viscosity coefficient for the actin cortex. The parameter ψ_0 is the coefficient defining the intensity of the contractile activity of the actomyosin network: it depends on the F-actin saturation concentration a_{sat}, beyond which the network contractility starts to decrease exponentially due to an effect of compaction of the network. As in the model of Alt and Tranquillo [3], we assume that the functions $\tau_0(x)$ and $\tau_1(x)$ are both constant.

The above 1D partial differential equations are non-dimensionalized on the circle $[0, 2\pi]$. We define the following dimensionless variables:

$$\tilde{t} = \frac{t}{T} = t\eta \qquad \tilde{a} = \frac{a}{a_0} \qquad \tilde{L} = \frac{L}{R_0}. \tag{13}$$

$$\tilde{P} = \frac{P_1}{\Phi_1 a_0 L_0} \quad \tilde{\gamma} = \frac{\gamma_1 T}{\Phi_1} \quad \tilde{\mu} = \frac{\mu_0}{\Phi_0} \quad \tilde{\tau} = \frac{\tau_0 T^2}{\Phi_0} \quad \delta = \frac{\tau_1 \Phi_0}{\tau_0 \Phi_1} \tag{14}$$

$$\tilde{\psi} = \frac{\psi_0 a_0 T}{\Phi_0 L_0} \quad \tilde{a}_{sat} = \frac{a_{sat}}{a_0}. \tag{15}$$

Dropping the tildes for notational simplicity, we obtain the following dimensionless system of equations:

$$\frac{\partial(La)}{\partial t} + \frac{\partial}{\partial x}(Lav) = L(1-a) \tag{16}$$

$$a\frac{\partial L}{\partial t} = P - \gamma La + \delta T_c \tag{17}$$

$$\frac{\partial}{\partial x}[\mu a \frac{\partial v}{\partial x} + \sigma(a) - T_c] = av \tag{18}$$

$$\text{with } T_c = \frac{\partial}{\partial x}\left(a\tau\frac{\partial L}{\partial x}\right) \quad \text{and} \quad \sigma(a) = \psi a^2 e^{-a/a_{sat}}. \tag{19}$$

4.2. Conditions for Oscillatory Dynamics of Cell Membrane Protrusions

A simple way to look for possible self-sustained oscillations of the plasma membrane is to find model parameters which correspond to a destabilization of the uniform steady state through a Hopf bifurcation. Furthermore, according to the experimental results, we are looking for spatially non-uniform solutions in order to represent the appearance of extending/retracting protrusions along different directions. These directions will be associated with specific unstable modes of the cytomechanical model solutions.

Linearizing the nonlinear equations around the uniform steady state

$$a_s = 1 \qquad L_s = \frac{P}{\gamma} \qquad v_s = 0 \tag{20}$$

with solutions of the form

$$(a, L, v) \sim e^{\lambda t + imx}, \tag{21}$$

leads to the following linearized system

$$\begin{pmatrix} \lambda + \gamma + m^2\delta\tau & P & 0 \\ \lambda & \frac{P}{\gamma}(1+\lambda) & im\frac{P}{\gamma} \\ -im^3\tau & -im\sigma & 1+\mu m^2 \end{pmatrix} \begin{pmatrix} L - L_s \\ a - a_s \\ v - v_s \end{pmatrix} = 0, \tag{22}$$

where $\sigma = \frac{d\sigma}{da}_{|a_s}$. The eigenvalues λ of the matrix M are given by $det(M) = 0$, which leads to the following dispersion equation:

$$\lambda^2 + \lambda(-\gamma + A + B) + AB + C = 0, \tag{23}$$

with

$$A = (\gamma + m^2\delta\tau) \quad B = \left(1 - \frac{m^2\sigma}{1+\mu m^2}\right) \quad C = \frac{m^4\tau P}{1+\mu m^2}. \tag{24}$$

The roots of the dispersion equation are

$$\lambda(m) = \frac{1}{2}\left[-1 + \frac{m^2\sigma}{1+\mu m^2} - m^2\delta\tau\right]$$

$$\pm\frac{1}{2}\left[\sqrt{\left(2\gamma - 1 + \frac{m^2\sigma}{1+\mu m^2} + m^2\delta\tau\right)^2 - 4\gamma(\gamma + m^2\delta\tau) - 4\frac{m^4 P\tau}{1+\mu m^2}}\right]. \tag{25}$$

When the expression under the square root is negative, $\lambda(m)$ is complex. A Hopf bifurcation then occurs for the first integer value of the mode m, such that the largest real part $Re[\lambda(m)] = 0$, with

$$Re[\lambda(m)] = \frac{1}{2}\left(-1 + \frac{m^2\sigma}{1+\mu m^2} - m^2\delta\tau\right). \tag{26}$$

The real value m_0 of m such that

$$Max\,[Re[\lambda(m)]] = 0 \tag{27}$$

is given by:

$$m_0 = \sqrt{\frac{1}{\mu}\left(\sqrt{\frac{\sigma}{\delta\tau}} - 1\right)}. \tag{28}$$

Let us remark that the coefficients P and γ, which respectively represent the protrusion and the retraction of the lamellipod, have no influence on the determination of the deformation modes. The key parameters are the cytoplasm viscosity μ, the contractility σ, and the friction parameter $1/\delta\tau = \Phi_1/T^2$.

As intuitively expected, larger values of τ and μ correspond to smaller values of m_0, i.e., to a small number of potential membrane protrusions. Conversely, large values of parameter σ, which monitors the contractility of the F-actin network, can lead to large values of m_0 and thus to a larger number of cell membrane protrusions.

By evaluating $Re[\lambda(m_0)]$, one obtains the critical bifurcation value

$$\sigma_* = \delta\tau + 2\sqrt{\mu\delta\tau} + \mu. \tag{29}$$

Further substitution in the expression of m_0 gives

$$\mu\delta\tau = \frac{1}{{m_0}^4}. \tag{30}$$

The bifurcation parameters for a given mode are then chosen by fixing two of the following parameters: μ, τ, δ. We usually set cytoplasm viscosity μ and the cell cortex elasticity τ. We then evaluate δ using the relation $\delta = ({m_0}^4\mu\tau)^{-1}$ and calculate σ given by Eq. (29). Since m has to be an integer, the closest integer value to m_0 defines the first unstable mode m_1.

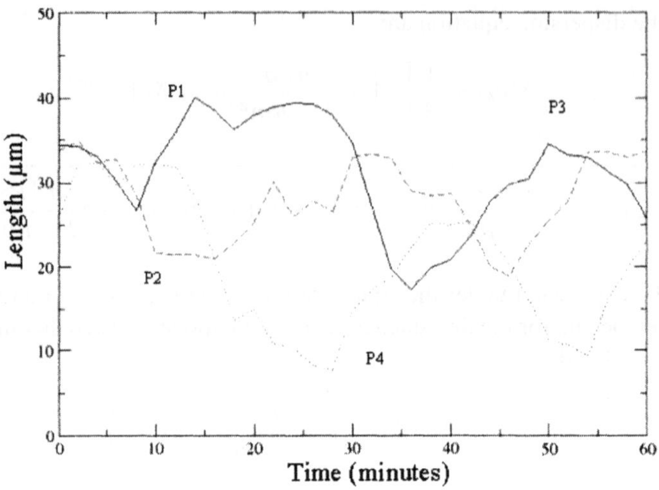

Figure 5. Evolution with time of the maximum length, within an angular window of 60 degrees width, of each protrusion of an L929 fibroblast. Four protrusions appear on this time scale, along the four main directions given by $\theta_1 = -110$ degrees (P1 curve), $\theta_2 = -24$ degrees (P2 curve), $\theta_3 = 75$ degrees (P3 curve), and $\theta_4 = 148$ degrees (P4 curve), respectively.

5. Results

5.1. Characterization of Spontaneous Dynamical Behavior of Fibroblasts

5.1.1. Fibroblasts Morphologies Adherent cells such as fibroblasts adopt a starry configuration when cultured on glass. Cell protrusions are long, thin membrane extensions, with flattened terminations that can increase cell substratum adhesion. Short, very thin extensions of filopod type can also be observed. We will not consider these deformations in our subsequent theoretical analysis since their duration does not exceed a few minutes. When isolated, as in sub-confluent cultures, L929 fibroblasts exhibit various quasi-symmetrical stellar morphologies with two to five protrusions. Figure 4 shows typical morphologies with four main membrane protrusions, usually oriented along two more or less perpendicular axes. Spatiotemporal localization of the cell protrusions (Fig. 4) show that, within an angular window of approximately 40 degrees width, the different directions of protrusion can remain quasi-stationary on time scales as large as 8 h (Fig. 4, right).

5.1.2. Identification of a Pulsating Movement Oscillatory extension/retraction motion of each cell protrusion can be analyzed through the time/amplitude representation. The four time/amplitude curves in Figure 5 are associated with typical oscillatory morphology of a cell exhibiting four protrusions. Further smoothing of each curve by a spline function enhances the main cell membrane displacements (Fig. 6). First, a relatively steady pulsating movement is observed for each protrusion around a mean amplitude value, different for each of them and ranging from 10 μm in the θ_2 direction (θ_2 = -24 degrees) to 30

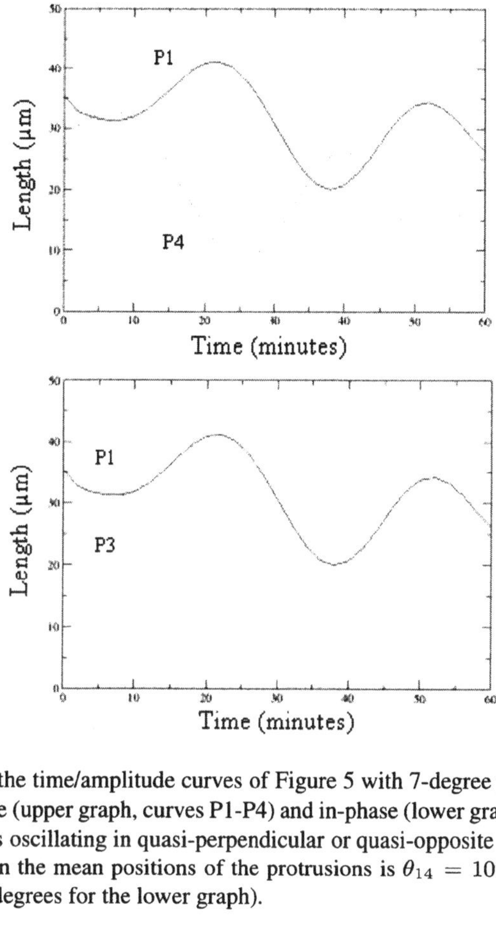

Figure 6. Smoothing the time/amplitude curves of Figure 5 with 7-degree polynoms reveals more clearly the out-of-phase (upper graph, curves P1-P4) and in-phase (lower graph, curves P1-P3) character of the protrusions oscillating in quasi-perpendicular or quasi-opposite directions, respectively (angular width between the mean positions of the protrusions is $\theta_{14} = 102$ degrees for the upper graph and $\theta_{13} = 185$ degrees for the lower graph).

μm in the θ_4 direction (θ_4 = 148) degrees. Despite this variability between the protrusion length, time/amplitude representation allows us to characterize the synchronization of cell protrusions occurring in the different directions. We clearly see from Figure 6 the in-phase and out-of-phase protrusions' oscillatory dynamic appearing in opposite (θ_1 = -110 degrees, θ_3 = 75 degrees, Fig. 6, top) and perpendicular directions (θ_2 = -24 degrees, θ_4 = 148 degrees, Fig. 6, bottom), respectively. Similar results have been obtained for several cells of our image sequence. For clarity, the above figures display a restricted range of 1 hour, but complete Fourier analysis of time/amplitude curves recorded for 8 hours confirms this result.

As introduced in Section 3.1, a global analysis of image motion, avoiding cell boundary segmentation, is provided by our optical flow approach. In that case, we showed that the eigenvalues associated with the affine cell motion model are reliable descriptors of the cell morphological changes [7]. We can thus transform the spatiotemporal analysis of cell shape changes into the analysis of the time signals given by the variation of these motion descriptors along the video-image sequence. Using standard Fourier analysis, we thus obtain from the 8-hour video-tracking of a fibroblast with four protrusions the eigenvalue variations displayed in Figure 7. The graph corresponds to the real parts of the eigenvalues of the affine model matrix (see Section 3.1) corresponding to the protrusive areas of the cell. The eigenvalues oscillate between positive (protrusion extension) and negative (protrusion retraction) values. Moreover, most of the time they have opposite signs, which indicates a synchronized out-of-phase oscillatory pulsation between the two protrusive zones (see Table 1). Thus, the optical flow analysis not only confirms, as expected, the periodic pulsatile behavior characterized from time amplitude maps (Fig. 6). It further provides an easier and more automatic analysis of such cell morphological changes over video sequences composed of a quite large number of images (typically 160 images for an 8-hour recording period with sampling acquisition time of 3 min.).

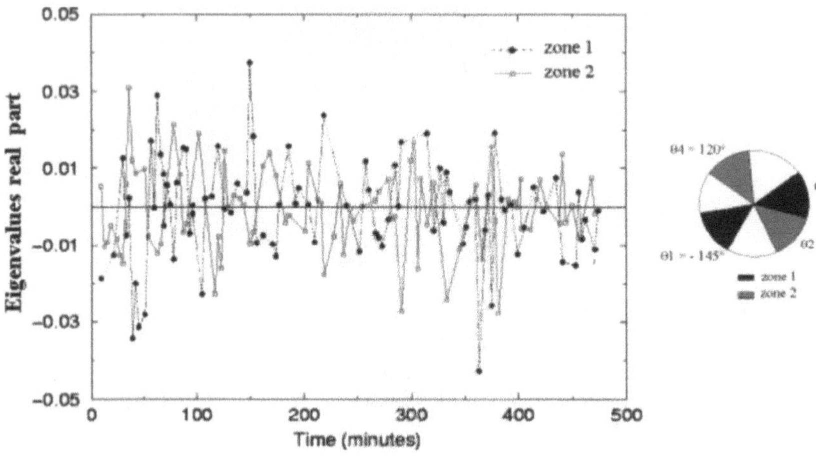

Figure 7. Evolution with time of the real part of the affine motion model eigenvalues (Section 3.1). Each of the associated eigenvectors remains in the angular zone 1 and zone 2, respectively oriented along a direction included in one of the two zones defined on the circle on the right. Each zone corresponds to the mean orientation of each protrusion ±15 degrees.

5.1.3. Quasi-periodicity of the Cell Shape Changes The time/amplitude curves presented in Figure 6 suggest a periodicity of about 30 minutes. We thus look for the periodicity of cell protrusion dynamics by calculating the Fourier transform of the 8-hour time/amplitude signals recorded for each protrusion. On this time scale, a non negligible local motion of the cell nucleus is observed and associated with low frequency movements. After drift correction and signal filtering to eliminate this nucleus displacement, Fourier transform of the time/amplitude curves exhibits a main peak of frequency in the range 0.6–0.7 mHz (Fig. 8), corresponding to a periodicity ranging from 24–28 min, depending on the cells. Similar periodicity analysis has been performed from synchronized

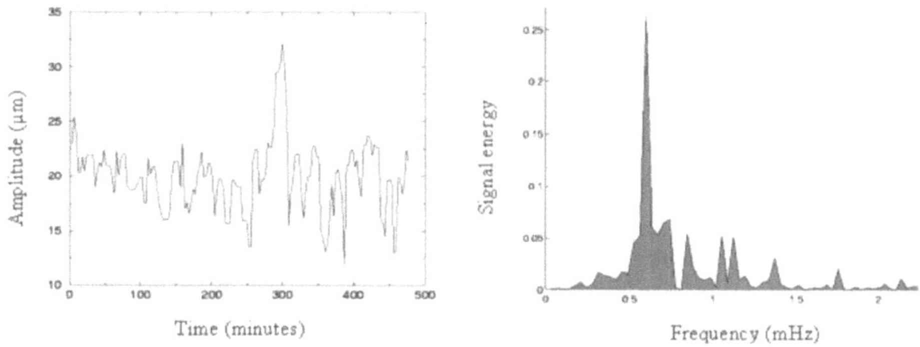

Figure 8. Left: time/amplitude curve of an oscillatory protrusion recorded for 8 hours for the fibroblast analyzed in Figure 4 (mean orientation $\theta = 40$ degrees). Right: associated fast Fourier transform and frequency spectrum. The highest peak corresponds to a frequency of 0.7 mHz, i.e., to a period of approximately 24 min.

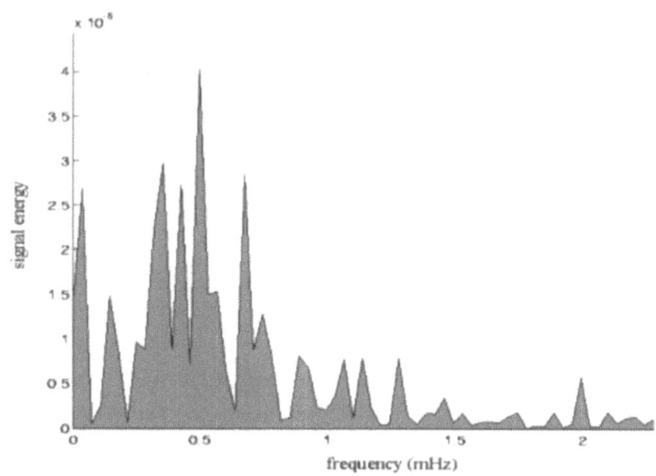

Figure 9. Fast Fourier transform of the variation with time of first eigenvalue real part (zone 1, Fig. 7) determined for the fibroblast previously analyzed in Figure 8. The frequency spectrum indicates a main peak around 0.5 mHz, a value very close to the 0.7 mHz obtained by the Fourier analysis of the protrusion time/amplitude curve of Figure 8.

oscillations of the cell motion model eigenvalues (Fig. 9). We notice the presence of low frequencies which could be attributed to the movement of the nucleus, as it has been evoked previously. Main peaks of frequency are detected around 0.7 mHz, which is quite coherent with the results obtained from the segmentation approach and associated analysis of time/amplitude curves.

5.2. Numerical Simulation of Cell Membrane Deformations

The numerical resolution of the model is based on the finite difference scheme of Crank-Nicholson. The partial differential equations system (Eqs. (16)–(19)) is solved on the unit circle. The associated matrix inversion was solved using the Thomas algorithm [20] adapted to periodic boundary conditions. The elliptic equation (balance of mechanical stress acting on the actomyosin network) is solved by the successive over-relaxation method SOR. The initial conditions are random perturbation of F-actin concentrations around the homogeneous steady state.

5.2.1. Simulations of Oscillating States The stability analysis shows that several cell protrusions can be generated. However, it has been experimentally observed that isolated cells preferentially develop four protrusions. Considering the stability analysis for the choice $m = 4$, and imposing values for the cytoplasm viscosity $\mu = 2$ and membrane elasticity $\tau = 1.5$, Equations (30) and (29) give $\delta = 0.001$ and $\sigma_* = 2.1$, respectively. This choice of parameters corresponds to the Hopf bifurcation point.
The parameter σ corresponding to the contractile intensity of the actomyosin complex, has been chosen as the bifurcation parameter. In order to destabilize the homogeneous state, the value $\sigma = 5.5$ has been chosen. Parameters P and γ have been fixed to the value 0.5.

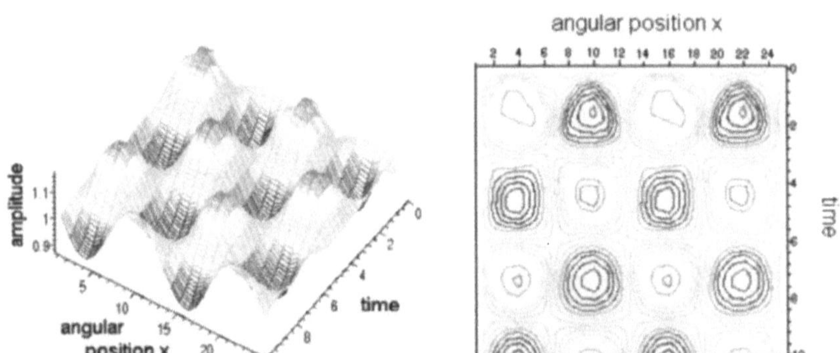

Figure 10. 3D spatiotemporal graph and associated 2D iso-level contours of simulated amplitude of cell membrane deformations $L(x,t)$ obtained with the following set of model parameters (Eqs. (16)–(19)): $\mu = 2$, $\tau = 1.5$, $\delta = 0.001$, $P = \gamma = 0.5$, $\psi = 5.5$, $a_{sat} = 2$. The asymptotic state is a standing wave corresponding to alternated periodic oscillations of the protrusion length along two perpendicular directions. (Note: the time associated with the simulations t_{simul} is given in normalized time units.)

Figure 10 presents the asymptotic, spatially symmetric state showing the periodic oscillatory variations of the membrane extension $L(x,t)$ relative to the cell cortex surface. This periodic pulsating behavior of the cell protrusion appears along steady perpendicular protrusive directions, in agreement with experimental observations [7]. Figure 11 presents simultaneous plots of the three model variables' $L(x,t)$, $a(x,t)$, and $v(x,t)$ quantifying

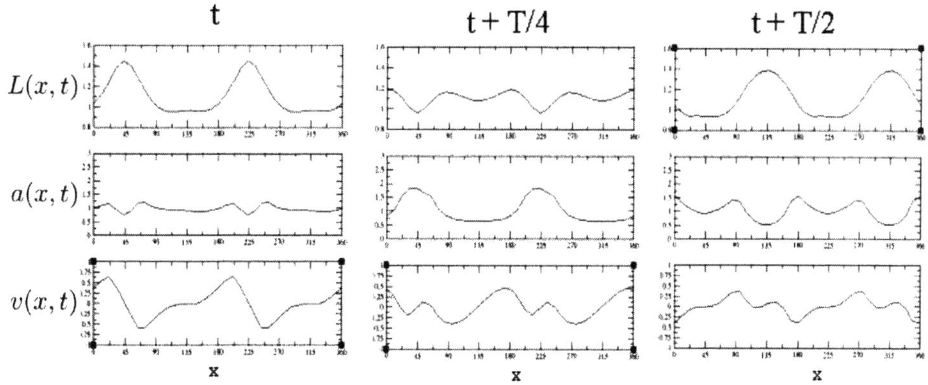

Figure 11. Simulated polarity maps of the three model variables $L(x,t)$, $a(x,t)$, and $v(x,t)$ drawn at three consecutive times t (left), $t+T/4$ (middle), and $t+T/2$ (right). Parameter values are as in Figure 10 with a larger number of points n to define the cell boundary ($n = 35$). Maxima of protrusion length $L(x,t)$ coincide with minima of actin concentration $a(x,t)$, while the tangential velocity $v(x,t)$ is zero.

position of the cell membrane from the surface of the cell body, actin concentration, and velocity, respectively, along the cell outline and for three successive times t, $t+T/4$, and $t+T/2$, where T is the oscillation period. We observe that maxima of protrusion lengths occur for associated minima of the actin density. Actually, when the actin density is high, cell cortex contraction and associated membrane retraction are strong. Conversely, for low actin concentrations, the cortex experiences weaker mechanical stresses and the membrane protrudes more freely under the effect of the intracellular pressure.

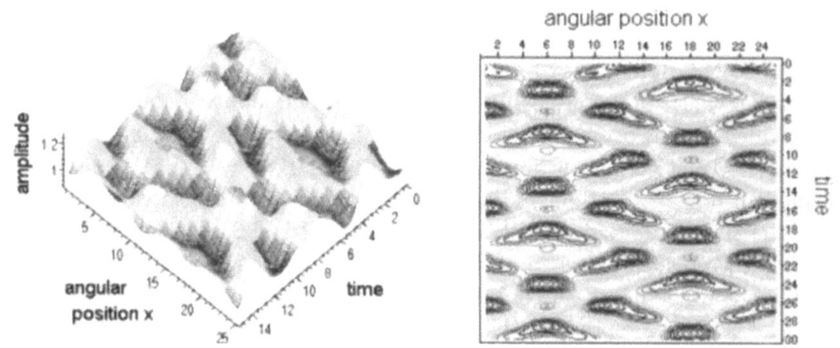

Figure 12. 3D spatiotemporal oscillations of simulated cell membrane deformations obtained with a less viscous ($\mu = 1$), less contractile ($\psi = 3$) and softer ($\tau = 1$) cell cortex (compare with Fig. 10). A more complex asymptotic regime of mixed-mode oscillating protrusions is generated.

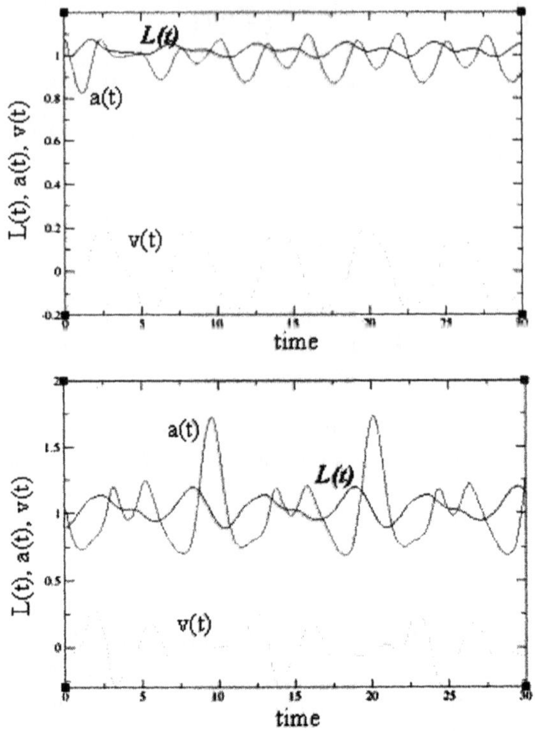

Figure 13. Simulated time/amplitude curves of the three model variables $L(x,t)$, $a(x,t)$, and $v(x,t)$. Top graph: periodic oscillatory behavior corresponding to global cell dynamic presented in Figure 10 (for the angular position $x = 22$, i.e., 317 degrees). Bottom graph: periodic oscillatory behavior corresponding to cell shape changes of Figure 12 (for the angular position $x = 18$, i.e., 259 degrees).

More complex oscillations of the cell membrane could be generated when considering decreasing values of the viscosity, elasticity, and contractility of the cell cortex. The model behavior has thus been simulated for the values $\mu = 1$, $\tau = 1$, and $\psi = 3$. The other parameters values are as in Figure 10 ($\delta = 0.001$, $P = \gamma = 0.5$), and the leading unstable mode is now $m = 6$. For similar initial conditions, the corresponding asymptotic state (Fig. 12) shows oscillations alternating between 2 and 3 protrusions periodically occurring at the same locations.

These standing wave behaviors are only part of a larger diversity of spatiotemporal states that the model is able to reproduce. Thus, by decreasing both τ and μ while increasing γ, i.e., the amplitude of the resistive stress exerted on the cell cortex, rotating waves of protrusions can be simulated. This dynamical behavior can be directly related to experimentally observed cell shape changes already reported for different cell lines [10, 2].

5.2.2. Estimation of the Polymerization Rate from the Periodicity The temporal evolution of the model variables $L(x,t)$, $a(x,t)$ and $v(x,t)$ associated with the two previous simulations (Figs. 10 and 12) are plotted in Figure 13 for a given value of the angular position x. The simulation time is $t_{simul} = 30$. According to our time scaling (Section 4.1), the corresponding real time t_{real} is derived from the relation $t_{real} = t_{simul}/\eta$. Considering on the one hand the estimated periodicity of cell movements of about 24 to 28 minutes and on the other hand a 10 time units period determined from the second simulated curve $L(x,t)$ (Fig. 13, bottom graph), we thus get a predicted rate of $\eta = 0.42\ min^{-1}$. This value can be compared to the value $\eta = 2.0\ min^{-1}$ used in [3] and to the F-actin polymerization rate of 3 min^{-1} reported by Korn et al. [11]. However, this is just a rough estimation since obviously the simulated period depends on other model parameter values such as, most importantly, the viscosity coefficient μ.

6. Discussion

It is commonly admitted that cell membranes fluctuate without any privileged direction in space and without any particular coherence with time. However, different results support the self-organized character of these cell shape changes. Ehrengruber et al. [6] show that neutrophils undergo periodic cytoskeletal rearrangements that lead to cycles of shape change, ultimately resulting in cell translocation. Rotating and pulsating standing waves were also characterized in the case of keratinocytes [2].

Our study shows that isolated L929 murine fibroblasts cultured *in vitro* exhibit an auto-organized behavior. The characterization of spontaneous cell deformations, i.e., without external stimuli except cell adhesion on the glass substratum, is realized on the basis of image analysis of videomicroscopy sequences. A standard method of analysis is cell boundary segmentation, in order to get a spatiotemporal representation, or contour maps, for the dynamics of cell deformations. This method, as exemplified in this paper, allows us to obtain very detailed information on the cell dynamics (amplitudes of extension/retraction, velocity of the protrusions, measures of periodicities for the overall dynamics). However, this approach is quite difficult and fastidious, as automatic detection of the cell outline remains inefficient in most cases because of the weak video images contrast between cell membrane and glass substratum.

We proposed an alternative method based on optical flow analysis. As it was shown, this method allows a rapid and efficient characterization of the nature of the cell dynamics (pulsation, oscillation, migration, etc.) and provides a direct measure of the possible periodicity of the cell deformations. We thus characterized quadrupolar preferential cell morphologies associated with a periodic and synchronized pulsating movement between two perpendicular protrusive directions, where pseudopod retraction along one axis corresponds to the extension in the perpendicular axis.

Comparing these results with those obtained for keratinocytes [2], we observed one interesting similarity, i.e., the fact that an extension in one direction is accompanied by simultaneous retraction in another direction. Beyond the chemical characterization of the recurring pseudopod protrusion and retraction, which appear to be driven by actin polymerization/depolymerization cycles (in neutrophils, sinusoidal oscillations of F-actin, with a period of 8–10 s, were reported [6]), cyclic alteration of the cell body shape can be

viewed as the result of an isovolumetric or isomass feedback control. As proposed by Dunn and Zicha [5], the spread area of a cell could be actively regulated by feedback mechanisms that adjust the rate of protrusion in order to compensate spontaneous fluctuations in retraction. The explanation proposed by Dunn and Zicha [5] is that retraction involves disassembly of cellular material which is then rapidly transported to other parts of the cell to generate protrusive structures.
The time for recycling of the material, which is most probably actin, thus depends on the cell morphologies. In the case of keratinocytes' rounded shapes, circulation of the material through the rotating wave is easy; thus, the periodicity reported by Alt et al. [2] is around 2 minutes, whereas for fibroblasts that exhibit long polarized structure, displacement of the material from the tip of one protrusion to another is a much longer process, taking, according to our measure of periodicity, around 30 minutes. Despite the strong morphologic differences between the two types of cells, it is possible to suppose a common origin for the underlying processes involved in their respective self-organization.

Following the work of Alt and Tranquillo [3], we investigated concomitantly the dynamical properties of a cytomechanical model which integrates various hypotheses regarding the biochemical and biomechanical processes underlying cell deformations. Model parameters refer to intrinsic cell properties (elasticity of the cell sub-membranous cortex, contractility of the actomyosin network, etc.). The underlying idea is to provide a way to reinforce the link and the feedbacks between experimental and theoretical approaches through the determination of pertinent dynamical criteria used as a basis for biological understanding and model validation. In this paper, we exemplified some of these links, both qualitatively in terms of cell morphologies and quantitatively through determination of periodic cell protrusive activity or protrusion velocity field characterization. However, such connections remain difficult because of the large number of parameters defining a given experimental situation and cell type. Nevertheless, integration of both combined experimental and theoretical analysis of cell deformation should provide a way to standardize the determination of cytomechanical model parameters of different cell lines. A consequence will be the setup of a database gathering both static (structural) and dynamic features of various cell lines with associated experimentally measured parameters on the one hand and modeling identified parameters on the other hand.

Acknowledgments

A.S. is funded by the Région Rhône-Alpes. This work is part of the project "Métrologie du vivant" supported by a grant from the Région Rhône-Alpes.

References

[1] W. Alt. Mathematical models and analysing methods for the lamellipodial activity of leukocytes. *NATO ASI (Adv. Sci. Inst.), Ser. H, Cell Biol.*, 42:407–422, 1990.

[2] W. Alt, O. Brosteanu, B. Hinz, and H.W. Kaiser. Patterns of spontaneous motility in videomicrographs of human epidermal keratinocytes. *Biochem. Cell. Biol.*, 73:441–459, 1995.

[3] W. Alt and R.T. Tranquillo. Basic morphogenetic system modeling shape changes of migrating cells: How to explain fluctuating lamellipodial dynamics. *J. Biol. Syst.*, 3:905–916, 1995.

[4] J. Condeelis. Life at the leading edge: the formation of cell protrusions. *Annu. Rev. Cell Biol.*, 9:411–444, 1993.

[5] G. Dunn and D. Zicha. Dynamics of fibroblasts spreading. *J. Cell Sci.*, 108:1239–1249, 1995.

[6] M.U. Ehrengruber, T.D. Coates, and D.A. Deranleau. Shape oscillations: a fundamental response of human neutrophils stimulated by chemotactic peptides. *FEBS Lett.*, 259:229–232, 1995.

[7] F. Germain, A. Doisy, X. Ronot, and P. Tracqui. Characterisation of cell deformation and migration using a parametric estimation of image motion. *IEEE Trans. Biomed. Eng.*, 46:584–600, 1999.

[8] R.J. Goldacre. The role of the cell membrane in the locomotion of amoebae, and the source of the motive force and its control by feedback. *Exp. Cell Res. Suppl.*, 8:11–16, 1961.

[9] T. Killich, P.J. Plath, E.C. Hass, X. Wei, H. Bultmann, L. Rensing, and M.G. Vicker. Cell movement and shape are non-random and determined by intracellular, oscillatory rotating waves in dictyostelium amoebae. *Biosystems*, 33:75–87, 1994.

[10] T. Killich, P.J. Plath, X. Wei, H. Bultmann, L. Rensing, and M.G. Vicker. The locomotion, shape and pseudopodial dynamics of unstimulated Dictyostelium cells are not random. *J. Cell Sci.*, 106:1005–1013, 1993.

[11] E.D. Korn, M. Carlier, and D. Pantaloni. Actin polymerisation and ATP hydrolysis. *Science*, 238:638–644, 1987.

[12] J. Lee, A. Ishihara, and K. Jacobson. How do cells move along surfaces ? *Trends Cell Biol.*, 3:366–370, 1993.

[13] A. Mogilner and G. Oster. The physics of lamellipodial protrusion. *Eur. Biophys. J.*, 25:47–53, 1996.

[14] G. Oster and A.S. Perelson. Cell spreading and motility: a model lamellipod. *J. Math. Biol.*, 21:383–388, 1985.

[15] P.P. Parnigotto, M.T. Conconi, V. Bassani, S. Pastore, E. Contiero, and R. Cortivo. Interaction between keratinocytes and fibroblastes cultured in vitro: morphology, morphometry and growth. *Ital. J. Anat. Embryol.*, 98:31–39, 1993.

[16] C.S. Peskin, G.M. Odell, and G.F. Oster. Cellular motions and thermal fluctuations: the brownian ratchet. *Biophys. J.*, 65:316–324, 1993.

[17] D.A. Schafer, M.D. Welch, L.M. Machesky, P.C. Bridgman, S.M. Meyer, and J.A. Cooper. Visualization and molecular analysis of actin assembly in living cells. *J. Cell Biol.*, 143:1919–1930, 1998.

[18] J. Small, A. Rohlfs, and M. Herzog. Actin and cell movement. *Cell Behaviour: Adhesion and Motility, Eds G. Jones, C. Wigley and R. Warn. The Society of Experimental Biology*, pages 57–71, 1993.

[19] J.V. Small, I. Kaverina, O. Krylyshkina, and K. Rottner. Cytoskeleton cross-talk during cell motility. *FEBS Lett.*, 452:96–99, 1999.

[20] J.C. Strikverda. Finite difference schemes and partial differential equations. *Wadsworth and Brooks/Cole, Mathematics Series*, 1989.

[21] C.M. Waterman-Storer and E.D. Salmon. Positive feedback interaction between microtubule and actin dynamics during cell motility. *Curr. Opin. Cell Biol.*, 11:61–67, 1999.

[22] P. Weiss. Guiding principles in cell locomotion and cell aggregation. *Exp. Cell Res. Suppl.*, 8:260–281, 1961.

Polymer and Cell Dynamics – Multiscale Modeling
and Numerical Simulations, 113–124
ed. by W. Alt, M. Chaplain, M. Griebel und J. Lenz

Micropipet Aspiration of the Human Neutrophil

Jeanie L. Drury and Micah Dembo

Department of Biomedical Engineering, Boston University, 44 Cummington Street, Massachussets, MA 02215-2407, USA

Summary. Micropipet aspiration has been widely employed to study the mechanical behavior of passive neutrophils. The observed dynamics have been rationalized in terms of mechanical models that describe these cell as "slippery droplets" of viscous fluid enclosed by a surface or cortical tension. Here, a low Reynolds number hydrodynamics code based on the finite element method was utilized to analyze several such fluid models (e.g., Newtonian and shear thinning). We investigated the behavior of these models as a function of surface tension, droplet radius, viscosity, aspiration pressure, and pipet radius. In addition, for the simplest case of a Newtonian fluid droplet, we tabulated a dimensionless factor, M, which was utilized to approximate the apparent viscosity of the neutrophil. This analysis resulted in a useful method by which rheological parameters could be determined and subsequent models tested. Overall, of the models tested, the mechanical behavior of the passive neutrophil during micropipet aspiration can be best represented by a fluid droplet consisting of a shear thinning *bulk* viscosity, a shear thinning *dilatational surface* viscosity, and an area-dependent surface tension.

1. Introduction

The primary functions of neutrophils (neutrophilic leukocytes) are to defend the body against foreign materials (e.g., bacteria and viruses) and remove dead cells and tissues. Thus, these cells have evolved to optimally accomplish such tasks. For example, neutrophils have developed methods to sense foreign bodies and then actively move out of the circulation. Additionally, they have large volumes that enable them to engulf such materials. Remarkably, despite this volume (and, thus, large radius when compared to typical capillaries), they are able to circulate, in a *passive* state, to all regions of the body. They do this by deforming from their normally spherical shape into an elongated "cigar-like" shape and then recovering as they passively traverse the systemic and pulmonary circulatory systems. The properties of passive neutrophils that allow this circulation and affect the flow of blood through the microcirculation have been of great interest.

In an attempt to probe the mechanical nature of the passive neutrophil, various micromanipulation techniques have been applied. Of these, the most common is that of micropipet aspiration. Using micropipet aspiration, several common details about the mechanical nature of these cells have been determined. In general, the cytoplasm of the passive neutrophil behaves like a highly viscous fluid with very little elasticity [9, 10, 17].

This cell has a highly wrinkled plasma membrane which unfolds and smooths to accommodate the resulting increase in surface area [10, 22]. Furthermore, the neutrophil has an effective cortical tension [10, 18] which increases with the area expansion of the cell [1, 18]. Finally, it has been confirmed that neutrophils also have an effective surface viscosity [8, 10, 28]. Thus, when estimating mechanical properties and rheological moduli, these cells have generally been regarded as "slippery fluid droplets" with a cortical tension.

However, these cells are far from homogeneous. They contain a nucleus, numerous organelles and granules, a polymeric cytoskeleton (composed primarily of actin, microtubules, and intermediate filaments), and cytosol. The individual contributions of these structures to the overall mechanical properties of the cell are largely unknown. Much research has been dedicated to determining the properties of these structures both under conditions of isolation and *in situ*. Conceivably, understanding the structural and molecular basis of the mechanical properties may be the key to understanding the passive behavior and properties of white blood cells.

It has long been hypothesized that, of all the cytoskeletal polymers, actin is the primary structural protein that determines the leukocyte's geometry, mechanical properties, and ability to generate the forces necessary for chemotaxis and phagocytosis [14, 19, 20, 23]. *In vitro* experiments on purified actin filament gels (using a cone and plate viscometer) have indicated that actin filament preparations behave as a shear-rate-dependent fluid with a viscosity that ranged from 225 to 3200 centipoise, depending on the strain rate and the purity of the actin [2, 15, 27, 29]. When cross-linkers (e.g., α-actinin) were included in the gels, the viscosity was found to still be shear rate dependent and to increase by nearly 40-fold [21]. The addition of cross-linking brings the viscosity of actin gels into the range of the estimated apparent viscosity of the neutrophil ($\sim$ 1000 poise) [10, 17]. Furthermore, the dependence of the viscosity on the rate of shear was found to be a power-law relationship in which the viscosity decreased as the rate of shear increased. Such a dependence has also been noted for the passive neutrophil during aspiration experiments [8, 26].

In addition to the polymer cytoskeleton, the cytoplasm is 75–80% water (cytosol) by weight [13]. The viscosity of the cytosol is on the order of 0.98 centipoise at 37 °C [16]. This is essentially equivalent to the viscosity of bulk water and 5 orders magnitude lower than the accepted viscosity value of passive neutrophils. While the polymer cytoskeleton and cytosol are the major constituents of the interior of the neutrophil, there is also a nucleus and numerous membrane-bound organelles. However, the effects of the lobed nucleus of the neutrophil on the cell's ability to deform has been shown to be negligible [5, 25, 28].

Despite this heterogeneity, treating the passive neutrophil as a slippery homogenous droplet has been amazingly successful in explaining the general behavior of the passive neutrophil during aspiration and has yielded first-order approximations of many rheological moduli. Early models of the neutrophil all considered the neutrophil to be a fluid droplet enclosed by a constant surface tension. The only differences were in the characterization of the fluid. Common models included a Newtonian fluid [7, 8, 10, 11, 17, 18, 24], a Maxwell fluid [4–6], and a shear-thinning or power-law fluid [8, 26].

In general, these models have been somewhat successful at describing specific aspects of neutrophil behavior under limited conditions. However, they have all failed miserably at describing and predicting the behavior of these cells over large ranges of applied pressures and/or pipet sizes. Recently, a model that incorporated both a shear-thinning cytoplasm and cortical dissipation was able to explain the time scale of aspiration over large ranges of pressure and across pipet sizes [8]. Following is a discussion of how this final model was attained. Overall, a low Reynolds number hydrodynamics code based on the finite element method was implemented to analyze these models and determine the rheological constants. Concurrent with these calculations, aspiration experiments were conducted to provide the data that were necessary to determine parameters and to test the models.

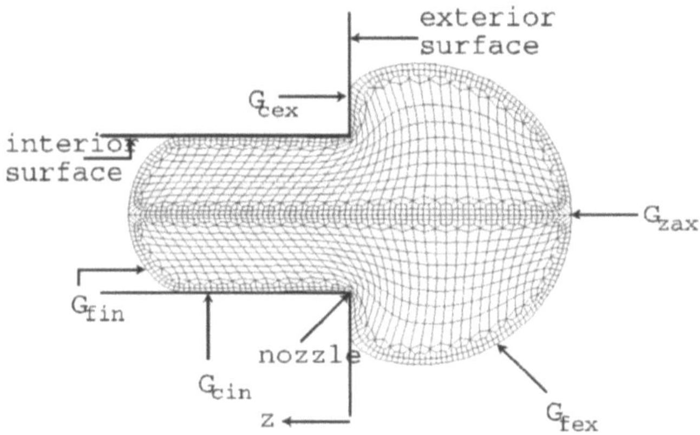

Figure 1. Geometry of a *pipet* and *droplet* after partial aspiration. The "interior", "exterior", and "nozzle" of the pipet are indicated. Also labeled are Γ_{fin}, Γ_{fex}, Γ_{cin}, and Γ_{cex}, which are the free-interior, free-exterior, constrained-interior, and constrained-exterior boundaries of the droplet, respectively. There is a fifth, *logical* boundary, Γ_{zax} which is the droplet axis of symmetry.

2. Methods

2.1. Continuum Description of the Neutrophil

Our theoretical approach to the mechanics of neutrophil aspiration was a generalization of earlier analyses [7, 8]. Briefly, cylindrical coordinates were introduced and utilized to define the *pipet* (cylindrical cavity with interior radius, R_p) and the cell (a slippery and incompressible *droplet*, which is initially a sphere of radius, R_c). Within the droplet, the velocity and pressure of the cytoplasm were assumed to satisfy the Stokes equations and were subjected to several boundary conditions: Γ_{fin}, Γ_{fex}, Γ_{cin}, and Γ_{cex} (free-interior, free-exterior, constrained-interior, and constrained-exterior surfaces, respectively) (see Fig. 1). The free surfaces were able to move both normally and tangentially as a result of the balance between pressure and surface stresses. In contrast, the constrained boundaries were able to slide tangentially because of the slippery droplet surface, but could not move normally because they were in contact with the hard surface of the pipet. All of these manifolds could grow or shrink in extent and shape as functions of time, and in some configurations, certain manifolds (e.g., Γ_{cex}) did not exist at all. There was also a fifth boundary, Γ_{zax}, which existed at the axis of rotation. This was a purely *logical* boundary that was important numerically but that had no physical significance.

In the simplest case, the viscosity, μ, is a constant (Newtonian). However, experimental data has suggested that μ is instead a decreasing function of the applied pressure. Thus, to incorporate both possibilities, the following constitutive law was developed:

$$\mu = \mu_0 \left(1 + \frac{s}{s_c}\right)^{-b}, \tag{1}$$

where s is the *shear rate* (an objective and coordinate invariant measure of the local instantaneous rate of fluid deformation). This functional form of μ yields a kind of generalized *power-law* fluid in which μ_0, s_c, and b are material constants. The Newtonian limit of Eq. (1) exists when $b \to 0$ or $s/s_c \ll 1$.

The total surface stress (γ_{total}) at any location on the boundary can be given by the sum of the *static* tension (γ) and a *viscous* tension ($\eta \nabla_\tau \cdot \mathbf{v}$):

$$\gamma_{total} = \gamma + \eta \nabla_\tau \cdot \mathbf{v}. \tag{2}$$

In accordance with established experimental measurements of neutrophils under various conditions of static deformation [1, 10, 18], $\gamma = \gamma_0 \left(\frac{A}{A_0}\right)^3$, depending on the relative cell surface area $\frac{A}{A_0}$. In Eq. (2), the coefficient, η, is the *dilatational surface viscosity*. Because the physical processes underlying shear thinning were expected to be similar in both the cortex and bulk cytoplasm, we assumed that μ and η had the same dependence on s. Thus, $\eta = \eta_0 \left(1 + \frac{s}{s_c}\right)^{-b}$.

Finally, the plasma membrane of the neutrophil was modeled by introducing a surface field, ρ, (cm^2 of actual membrane bilayer per cm^2 of "projected" mesh area). As aspiration proceeded, the membrane was transported laterally *via* the surface flow of the cytoplasm. The amount of wrinkled membrane also increased or decreased as the resolved mesh surface expanded and contracted. Finally, the plasma membrane could slip with respect to the cytoplasm under the influence of its own internal tensions. All this was taken into account by allowing ρ to evolve according to the following transport equation: $d_t\rho = \nabla_\tau \cdot (D_m \nabla_\tau \rho) - \rho \nabla_\tau \cdot \mathbf{v}$, with excess area diffusion coefficient D_m, which is chosen as zero in the simulations below.

2.2. Numerical

The computational solutions of the governing equations just described were generated using a low Reynolds number hydrodynamics transport code based on the Galerkin finite element method. A detailed description of this method as it was applied to micropipet aspiration has appeared elsewhere [7, 8].

2.3. Experimental

Measurements of cell aspiration were conducted under eight combinations of pipet size and aspiration pressure (see Table 1). The basic micropipet aspiration system was similar to that used previously by Evans and Yeung [10] and described in detail elsewhere [8]. Briefly, micropipets were pulled with a vertical pipet puller (Model 700C, David Kopf Instruments) and chosen such that their internal diameter was between 4 and 6.5 μm. Cells were obtained from a single donor *via* a finger prick and selected for study if they were in a quiescent state and had a diameter between 8.5 and 9 μm. Following the initial characterization, the suction pressure in the pipet was suddenly ramped (maximum rate of $100,000$ dynes/cm^2/sec) to the desired final value with a Harvard Apparatus PHD 2000 syringe pump (Holliston, Massachusetts) and measured with a continuous manometer system (transducers by Validyne). The ramping rate and final pressure were computer controlled. Overall, experiments were completed within 40 minutes of the finger prick. After this time, the cells were discarded and a new sample was prepared.

Hoffman modulation contrast [12]; (see also `http://micro.magnet.fsu.edu/primer/techniques/hoffman.html`) was utilized to visualize the cells. A CCD72 video camera (DAGE-MTI) was used to capture the Hoffman contrast image of the cell and pipet. This image was then displayed on a monitor and recorded on video tape. A line marker (three video lines wide) was positioned along the center of the long axis of the pipet, and an intensity profile along this marker was collected every $1/30$ s. These intensity values were then added to obtain a single intensity profile along the center axis of the pipet and the aspirating cell. Fringes in this intensity profile were used to determine the position of the pipet nozzle as well as the positions of the leading and trailing edges of the cell relative to the nozzle (designated as $L_p(t)$ and $Z_{\min}(t)$, respectively).

For each experimental condition listed in Table 1, the data for all cells were combined so as to determine $\bar{t}(L_p)$, (the time required for an average cell to reach a given value of L_p). For each experimental condition, the final column in Table 1 gives the population mean ($\pm$ SD) of $\dot{L}_p$ at the point where $\sim 50\%$ of the cell volume had been aspirated (denoted by $\dot{L}_{p50}$). The ratio $\bar{s} \equiv \dot{L}_{p50}/R_p$ provided a useful empirical measure of the average shear rate of the cytoplasmic flow field at each experimental condition. Overall, this approach resulted in a single, average aspiration curve for each experimental condition.

3. Results

3.1. Parameter Estimation

We initially considered the case of a Newtonian fluid droplet with constant surface tension as a benchmark model. This was because of its simplicity and because, as a first-order approximation, it reproduced many of the behaviors of the passive neutrophil during aspiration. Applying this scheme, we developed a method to estimate γ and μ for the passive neutrophil.

Bearing in mind that cell aspiration occurs at a very low Reynolds number (thus, kinetic energy is negligible), the work due to pressure ($\Delta P\dot{Q}$) is completely balanced by the rate of dissipation into heat ($\int_\Omega \mu\Phi d\Omega$) plus the rate of storage as mechanical potential

ΔP (dynes/cm^2)	R_p (μm)	$\dot{L}_{p50} \pm SD$ (μm/sec)
470	2.2	0.0277 ± 0.0074
687	2.2	0.0570 ± 0.025
1430	2.2	0.137 ± 0.024
3172	2.2	0.702 ± 0.24
14204	2.2	9.173 ± 4.70
164	3.2	0.109 ± 0.014
1346	3.2	2.023 ± 1.08
5124	3.2	22.37 ± 12.02

Table 1. Experimental conditions: Here, ΔP is the average applied pressure, R_p is the pipet radius, and $\dot{L}_{p50} \pm SD$ is the average rate of entry at the point where the cell was 50% aspirated, plus or minus the standard deviation.

($\gamma\dot{A}$) as follows:

$$\Delta P\dot{Q} = \int_{\Omega} \mu\Phi d\Omega + \gamma\dot{A}. \tag{3a}$$

In Eq. (3a), the only quantity not readily available or measurable is the heat dissipation integral. Thus, to use Eq. (3a), values of this integral had to either be numerically tabulated or analytically estimated. To this end, we numerically tabulated a non-dimensional dissipation factor, M, at various stages and conditions of aspiration, according to the following equation (see Appendix A [7]):

$$M(V_{asp}, \Delta P, R_p) = \frac{R_p^3 \int_{\Omega} \Phi d\Omega}{\dot{Q}^2}. \tag{3b}$$

By combining Eqs. (3a) and (3b) and approximating $\dot{A}$ and $\dot{Q}$ in terms of experimentally measurable parameters ($\dot{A} \approx 2\dot{Q}\,(1/R_p - 1/R_{\max})$ and $\dot{Q} \approx \pi R_p^2 \dot{L}_p$), we obtained the following analytical formula for the viscosity:

$$\mu \approx \frac{\Delta P R_p - 2\gamma\,(1 - R_p/R_{\max})}{\pi M \dot{L}_p}. \tag{4}$$

It is important to note that Eq. (4) is valid only for Newtonian droplets or for droplets that can be approximated by a Newtonian fluid at a given instant in time.

We applied this approximation to simultaneously solve for both μ and γ at 50% aspiration (using the two lowest pressure datasets for $R_p = 2.2$ μm; see Table 1). The resulting value of γ_{50} (0.06 dynes/cm; this value is in agreement with previous experimental observations [1, 18] that γ is area dependent) was then used to solve for μ at all five pressures for the chosen pipet. Finally, Eq. (1) was fit to these results to determine the following parameter values: $\mu_0 = 6000$ poise, $b = 0.5$, and $s_c = 0.05$ sec (see Table 2 for the full set of parameters for each model).

Parameter	N	ST	SVS
R_c (μm)	4.25	4.25	4.25
μ_0 (poise)	6000	6000	1100
s_c (1/sec)	∞	0.05	0.05
b	0	0.5	0.5
γ_0 (dynes/cm)	0.03	0.03	0.03
α	3	3	3
ρ_0 (cm/cm)	2.1	2.1	2.1
D_m (cm^2/sec)	0	0	0
η_0 (poise $\cdot$ cm)	0	0	100

Table 2. Model parameters: Here, N refers to the Newtonian droplet model, ST refers to the shear-thinning droplet model, and SVS refers to the shear-thinning droplet model with surface viscosity. In addition, each parameter is defined as follows: R_c is the average cell radius, μ_0 is the low-shear *bulk* viscosity, s_c is the characteristic shear rate for shear thinning, b is the shear-thinning exponent, γ_0 is the resting surface tension, α is the exponent of the surface tension constitutive law, ρ_0 is the initial excess area density, D_m is the excess area diffusion coefficient, and η_0 is the low-shear *dilatational surface* viscosity.

3.2. Test of Model Validity

Using the parameters in Table 2, a full set of dimensional calculations (for all eight experimental conditions) was completed for both the Newtonian and shear-thinning models. The L_p versus $time$ results of those calculations are shown in Figure 2. Here, the Newtonian model predicted intermediate entry rates that were very close to those of the data for low pressures (Fig. 2a and 2b). However, it grossly underestimated the high-pressure and large pipet data. In addition, the Newtonian model underestimated the magnitude of the initial or final rapid rates of entry, which are characteristic of neutrophil aspiration. These results indicated that as a first order approximation, the Newtonian model could reproduce some of the behaviors of neutrophil aspiration (namely, the intermediate entry rate), within a given set of conditions. However, the Newtonian model had almost no predictive capabilities.

As demonstrated in Figures 2a–e, use of the shear thinning model resulted in the proper intermediate entry rates over the entire range of pressures for the smaller pipet. Unfortunately, the large pipet entry rates (Figs. 2f–h) were still underestimated. Similar to the Newtonian model, the shear-thinning model underestimated the magnitude of the initial or final rapid entry rates. Thus, while the addition of shear thinning produced slightly better results than the pure Newtonian model for data from a single pipet (used to determine the parameters), it also had little to no predictive value.

In an attempt to capture the apparent differences in entry rates across pipet sizes, a cortical dissipation or surface viscosity (previously suggested by Evans and Yeung [10] to explain this discrepancy) was added to the shear-thinning model. An energy analysis, similar to that above, was completed to estimate a new set of parameters for the modified

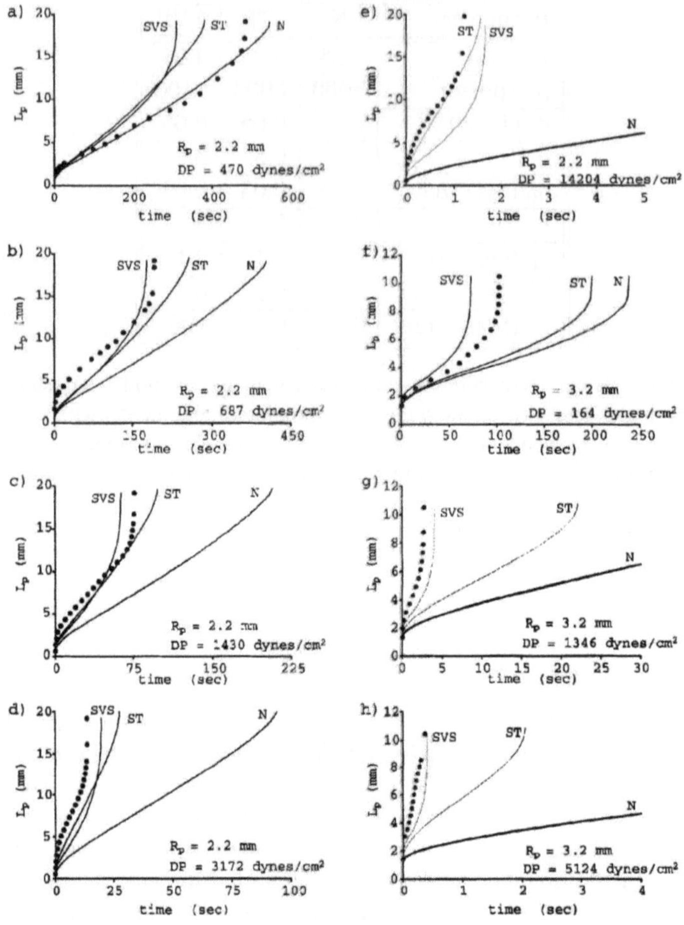

Figure 2. L_p versus *time* data (closed circles) and model fits (solid lines). In each case, N designates the fit of a Newtonian droplet, ST designates the fit of a shear-thinning droplet, and SVS designates the fit of a shear-thinning droplet with surface viscosity. In (e), (g), and (h), the Newtonian droplet curves are truncated due to the extreme overestimation of the time for complete aspiration.

shear-thinning model. This analysis resulted in the following analytical formula:

$$\Delta P = \mu\left(\pi M_\mu \frac{\dot{L}_p}{R_p}\right) + \eta\left(M_\eta \frac{\dot{L}_p}{4R_c^2}\right)\left(\frac{2}{R_p} - \frac{2}{R_{\max}}\right)^2 + \frac{\gamma}{R_p}\left(\frac{2}{R_p} - \frac{2}{R_{\max}}\right). \quad (5)$$

Assuming that $M_\mu = M_\eta = M$, an attempt was made to simultaneously estimate γ, μ, and η. Unfortunately, this assumption was invalid, and both M_μ and M_η need to be numerically tabulated to use Eq. (5). Regardless, we were able to solve this model using the following parameters, which were fit to the data by an iterative process: $\mu_0 = 1100$

poise, and $\eta_0 = 100$ poise·cm (all other parameters were the same as those used in the shear-thinning model; see Table 2). Once again, calculations were completed for all eight experimental conditions. As seen in Figure 2, this resulted in curves that fit the data across both pipet sizes and pressures. Furthermore, adding surface viscosity to the shear-thinning model resulted in the prediction of the final rapid entry rate. Yet, the initial rate of entry was still not predicted.

In addition to examining the L_p versus $time$ aspiration curves, the shape of the neutrophil during the aspiration process was also an important means by which to test these models. Figure 3 contains images of nearly completely aspirated neutrophils, overlaid with the outline of the droplets at the same stage of aspiration. When examining this data, there were two very striking features of the shape of both the Newtonian and shear-thinning models. First, both droplets were greatly flattened against the exterior of the pipet when compared to the aspirating neutrophil (Figs. 3a and 3b). This flattening was not seen when surface viscosity was added to the shear-thinning model (Fig. 3c). In fact, those droplets appeared to be almost too rounded. The second feature of both the shear-thinning and Newtonian droplets was the unround and flattened shape of the unaspirated portion of the droplet (Figs. 3a and 3b), which eventually led to the development of a reentrant jet (droplet appeared to be folding in on itself) during the final moments of aspiration. This flattening and reentrant jet formation were never observed during aspiration of neutrophils and were not present when surface viscosity was added to the model (Fig. 3c). If anything, the addition of cortical dissipation resulted in droplets that were too round when compared to the neutrophil. This was most likely due to a slight overestimate of the surface viscosity.

4. Conclusions

Overall, computational fluid dynamics has been a very powerful tool. It has allowed us to construct new methods to estimate rheological moduli (instead of fitting them with an iterative process), thus saving time and removing some of the subjectivity. In addition, it has given us the ability to analyze and modify several models of the passive neutrophil, thus resulting in an improved understanding of its mechanical behavior and properties during micropipet aspiration.

As shown above and in more detail in [8], a model that incorporated both a shear-thinning cytoplasm and cortical dissipation (surface viscosity) was able to explain the time scale of aspiration over large ranges of pressure and across pipet sizes. This model was also able to describe the final rapid entry behavior. However, it was unable to account for the initial rapid entry behavior of these cells.

In an effort to capture this behavior, we have proposed a model in which the droplet is composed of two intermixing fluids (see also [3]). One of the fluids is a low-viscosity solvent (like the cytosol), while the other is a highly viscous fluid *network* (like the actin cytoskeleton). This network has a given concentration within the droplet. It is able to contract or swell and frictionally interacts with the solvent in the form of a hydraulic drag force. Like actin gels and recent neutrophil observations, the network is assumed to

a)
b)
c)

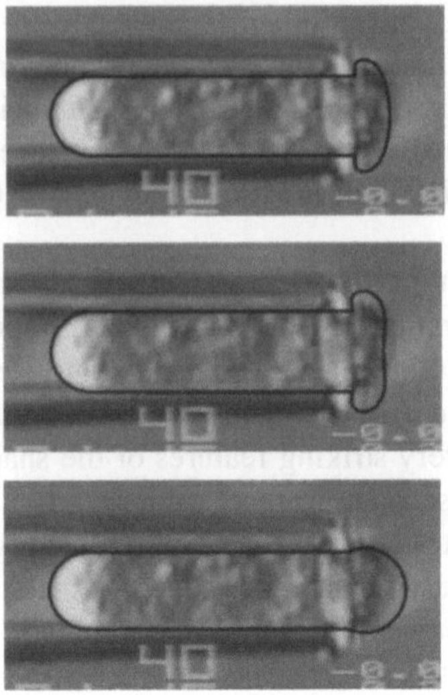

Figure 3. Geometry of neutrophil and droplet near the completion of aspiration. In all cases, the neutrophil and droplet were aspirated under the conditions $R_p = 2.2\ \mu$m and $\Delta P = 3172$ dynes/cm^2. To compare the geometries, the droplet outline was superimposed over the neutrophil image. In (a), the droplet is Newtonian, in (b), the droplet is shear thinning, and in (c), the shear-thinning droplet includes dilatational surface viscosity.

have a shear-thinning viscosity. As in previous models, we consider this *interpenetrating* droplet to be slippery and to have both an area-dependent surface tension and a shear-thinning dilatational surface viscosity. A thorough analysis of this model is still required to determine its validity and ability to predict the initial rapid entry behavior. In addition, to make an unbiased estimate of all the parameter, tables for the dissipation factors, M_μ and M_η, must be completed.

Acknowledgments

The authors wish to thank the Boston University Scientific Computing Center for use of their facilities. Special thanks to Dr. Evan Evans and his lab group at the University of British Columbia for their advice and assistance with the aspiration experiments. Work was supported by a Whitaker Special Opportunities Award in Cellular Mechanics.

References

[1] Albarran, B., H.P. Ting-Beall, and R.M.Hochmuth (2000) Effect of surface area change on cortical tension in passive neutrophils. *Biophysical Journal* **78** (Part 2 of 2), 187A

[2] Buxbaum, R.E., T. Dennerll, S. Weiss, and S.R. Heidemann (1987) F-actin and microtubule suspension as indeterminate fluids. *Science* **235** 4795, 1511–1514

[3] Dembo, M. (1989) Field theories of the cytoplasma. *Comments on Theoretical Biology*, **1**, 159–177

[4] Dong, C., R. Skalak, K.-L.P. Sung, G.W. Schmid-Schönbein and S. Chien (1988) Passive deformation analysis of human leukocytes. *Journal of Biomechanical Engineering* **110**, 27–36

[5] Dong, C., R. Skalak and K.-L.P. Sung (1991) Cytoplasmic rheology of passive neutrophils. *Biorheology* **28**, 557–567

[6] Dong, C., and R. Skalak (1992) Leukocyte deformability: finite element modeling of large viscoelastic deformation. *Journal of Theoretical Biology* **158**, 173–193

[7] Drury J.L. and M. Dembo (1999) Hydrodynamics of micropipet aspiration. *Biophysical Journal* **76**, 110–128

[8] Drury J.L. and M. Dembo (2001) Aspiration of human neutrophils: Effects of shear thinning and cortical dissipation. *Biophysical Journal* (in press)

[9] Evans, E. and B. Kukan (1984) Passive material behavior of granulocytes based on large deformation and recovery after deformation tests. *Blood* **64**, 1028–1035

[10] Evans, E. and A. Yeung (1989) Apparent viscosity and cortical tension of blood granulocytes determined by micropipet aspiration. *Biophysical Journal* **50**, 151–160

[11] Hochmuth, R.M., H.P. Ting-Beall, B.B. Beaty, D. Needham and R. Tran-Son-Tay (1993) Viscosity of passive human neutrophils undergoing small deformations. *Biophysical Journal* **64**, 1596–1601

[12] Hoffman, R. and L. Gross (1975) The modulation contrast microscope. *Nature* **254**, 586–588

[13] Hou, L., F. Lanni, and K. Luby-Phelps (1990) Tracer diffusion in F-actin and ficoll mixtures: toward a model for cytoplasm. *Biophysical Journal* **58**, 31–43

[14] Howard, T.H. and R.G. Watts (1994) Actin polymerization and leukocyte function. *Current Opinion in Hematology* **1**, 61–68

[15] Janmey, P.A. (1991) Mechanical properties of cytoskeletal polymers. *Current Opinion in Cell Biology* **2**, 4–11

[16] Luby-Phelps, K., S. Mujumda, R.B. Mujumdar, L.A. Ernst, W. Galbraith, and A. S. Waggoner (1993) A novel fluorescence ratiometric method confirms the low solvent viscosity of the cytoplasm. *Biophysical Journal* **65**, 236–242

[17] Needham, D. and R.M. Hochmuth (1990) Rapid flow of passive neutrophils into a 4 μm pipet and measurements of cytoplasmic viscosity. *Journal of Biomechanical Engineering* **112**, 269–276

[18] Needham, D. and R.M. Hochmuth. 1992. A sensitive measure of surface stress in the resting neutrophil. *Biophysical Journal* **61**, 1664–1670

[19] Omann, G.M., R.A. Allen, G.M. Bokoch, R.G. Painter, A.E. Traynor, and L.A. Sklar (1987) Signal transduction and cytoskeletal activation in the neutrophil. *Physiology Review* **67**, 285–322

[20] Pettit, E.J. and F.S. Fay (1998) Cytosolic free calcium and the cytoskeleton in the control of leukocyte chemotaxis. *Physiology Review* **78**, 949–967

[21] Sato, M., W.H. Schwartz, and T.D. Pollard (1987) Dependence of the mechanical properties of actin/α-actinin gels on deformation rate. *Nature* **325**, 828–830

[22] Schmid-Schönbein, G.W., Y.Y. Shih, and S. Chien (1980) Morphometry of human leukocytes. *Blood* **56**, 866–875

[23] Sha'afi, R.I. and T.F.P. Molski (1988) Activation of the neutrophil. *Progressive Allergy* **42**, 1–64

[24] Tran-Son-Tay, R., D. Needham, A. Yeung, and R.M. Hochmuth (1991) Time-dependent recovery of passive neutrophils after large deformation. *Biophysical Journal* **60**, 856–866

[25] Tran-Son-Tay, R., H.P. Ting-Beall, D.V. Zhelev, and R.M. Hochmuth (1994) Viscous behavior of leukocytes. In: *Cell Mechanics and Cellular Engineering* (V.C. Mow, F. Guilak, R. Tran-Son-Tay, and R.M. Hochmuth, eds) Springer-Verlag, New York, NY, 22–32

[26] Tsai, M.A. , R.S. Frank, and R.E. Waugh (1993) Passive mechanical behavior of human neutrophils: Power-law fluid. *Biophysical Journal* **65**, 2078–2088

[27] Xu, J., W.H. Schwarz, J.A. Käs, T.P. Stossel, P.A. Janmey, and T.D. Pollard (1998) Mechanical properties of actin filament networks depend on preparation, polymerizations conditions, and storage of actin monomers. *Biophysical Journal* **74**, 2731–2740

[28] Yeung, A. and E. Evans (1989) Cortical shell-liquid core model for passive flow of liquid-like spherical cells into micropipets. *Biophysical Journal* **56**, 139–149

[29] Zaner, K.S. and T.P. Stossel (1982) Some perspectives on the viscosity of actin filaments. *The Journal of Cell Biology* **93**, 987–991

Polymer and Cell Dynamics – Multiscale Modeling
and Numerical Simulations, 125–138
ed. by W. Alt, M. Chaplain, M. Griebel und J. Lenz

Physical-object-oriented 3D Simulations of Cell Deformations and Migration

Emmanuel Promayon, Jean-Louis Martiel and Philippe Tracqui

Lab. TIMC-IMAG, Faculté de Médecine de Grenoble, F-38706 La Tronche Cedex, France

Summary. Many aspects of individual and collective cell motion are still poorly understood, and physical models dealing with both aspects appear clearly as valuable tools for understanding cell biomechanical behavior. We theoretically analyze and simulate individual cell mechanical properties and the related chemotactic behavior of single cells collectively involved in an aggregation process. Cell objects are defined as three-dimensional elastic bodies moving on a viscous medium and submitted both to internal cohesive and external attractive forces (gravity and chemoattraction). We first investigate individual cell mechanical response to externally applied forces and compare the simulated cell object deformations with experimental data obtained with optical tweezers. We then examine the simulated cell population reorganization submitted to a two-dimensional chemoattractant gradient over a solid plane simulating a glass coverslip. Simulations are carried out for different values of cell chemotactic response. We especially simulate cell sorting between pre-spore and pre-stalk cells during *Dictyostelium discoideum* aggregation process by considering two-cell populations exhibiting differential cell-cell interactions. We conclude that our physical-object-oriented (POO) approach for modeling individual cell cytomechanics also satisfactorily reproduces the two-dimensional aggregation at a cell population level. Because individual cell mechanical behavior can be compared to a wide range of cell microrheology experiments, the cell object parameters can be estimated. This framework should thus be adequate for biologically realistic multiscale analyses leading to a better understanding of how, through modulation of mechanical factors, individual cell behavior controls collective cell reorganization.

1. Introduction

Spatial aggregation and reorganization of cell populations is crucial in physiological and pathological developmental processes, including embryogenesis, tissue healing, or cancerogenesis. Adhesive interactions between adjacent cells (cell-cell adhesion) and between a cell and the extracellular matrix (cell-matrix adhesion) play a major role in the maintenance or reconstruction of multicellular structures. Cell motility, possibly controlled by extracellular factors such as chemoattractants, also requires that such adhesions can be modulated and eventually broken in order to ensure a net spatial translocation of the cell.

Numerous theoretical models have been proposed for analyzing how different spatial organization of cell populations could emerge from such complex interactions. One can distinguish the following approaches:

- continuous approaches based on a global description of the spatiotemporal evolution of cell density and on the determination of instability conditions leading to an heterogeneous distribution of the cells [15, 24, 30] and

- discrete approaches, based on the description of the individual properties and behavior of cells and looking for the emergence of macroscopic cellular organization and rearrangements resulting from individual interactions, as observed in morphogenetic processes [12, 27], inculding cell sorting [9]. In cellular automata models, cell-cell interaction rules and neighborhood conditions are specified [3, 8], but generally this is done without considering the intrinsic mechanical properties of individual cells, such as elasticity and architecture.

The other widely used discrete approach is based on a polygonal tessellation of the plane. For example, Duvdevani-Bar et al. [6] proposed a strictly topographic cellular organization in which each cell is reduced to the vertex of a given triangulated graph that defines the neighborhood relationships between cells. An abstract law is implemented as a recombination process over the neighborhood relationships and yields to a mechanical-like behavior. Further model enhancement simulates biological processes such as cell proliferation or motility. Honda et al. [11] had previously used a similar approach to predict the morphological response of a tissue to a given cell division rate. They first proposed to determine the best topographic approximation of a given fully connected tissue with a Voronoi diagram, which divides the space into cell objects according to shorter distance relationships. A cell is assimilated to a Voronoi region and a recombination process is defined in order to reorganize the diagram after the division of the given cell. Even if such a process captures some of a mechanical behavior, it is a very limited model for cellular organizations and cellular biomechanics. Extending this geometrical formalization, Dugnolle et al. [7] simulated the mechanical cell monolayer rearrangement occurring during *in vitro* wound-healing experiments by considering cells as elastic proliferating bodies moving on a viscous surface and submitted to adhesion forces.

This work proposes a totally different approach to modeling collective cellular processes: it considers the mechanical interactions of individual cells having defined mechanical identity, cell-cell interactions and random motion. Our approach is based on the consideration of "cell objects", proposed as model of real cells and defined as 3D elastic physical objects with specific geometry and internal ("subcellular") structure. At first glance, this approach shares some common features with the finite element method (FEM), since both methods rely on the geometric discretization of the objects on a mesh and the evaluation of physical variables (forces, strain, etc.) at each node of this mesh. However, in the physical-object-oriented (POO) approach, this geometric mesh is realized through a bottom-up modeling procedure: the defined geometric structure aims to mimic the physical structure, as in cell tensegrity models [28, 29]. In contrast, the mesh in the FEM provides only a discretization of the material continuum with arbitrary geometric elements (triangle, tetrahedron, etc.). This specificity of the POO approach will appear more explicitly in the examples presented in this paper.

More precisely, we illustrate the potentiality of such a POO approach in the formulation of a multiscale model by addressing the following questions:

- Is the cell object relevant enough to describe some of the mechanical properties of individual living cells reported in cellular microrheological experiments?
- Is it possible to account for emergent cell collective behavior, such as cell sorting and aggregation, by considering, for example, the response of such mechanical cell objects to extracellular gradients?

The first section of this paper deals with the formulation of the 3D physical cell object. The second section explores some of its mechanical properties. We especially compare the results of our POO computational approach to experimental red blood cell deformations realized with optical tweezers. The third section focuses on spatiotemporal aspects of cell population organization, sorting and migration, with special consideration of the *Dictyostelium discoideum* aggregation process. The *Dictyostelium discoideum* slime mold is indeed widely used as an experimental model for studying complex self-organizing processes in biology [1, 17]. There are a lot of experimental data and models on this organism related to cAMP pulses generation [14], to spiral cAMP waves in aggregating population [23], or cell movement and aggregation [16].

2. Description of the Computational Model

2.1. Objectives of the Physical-object-oriented (POO) Approach

To be able to model individual cell behavior, we considered each cell as a physical entity. Each cell object has its own history, properties, and actions. To be able to manipulate such entities, a computer model and associated algorithms are developed within an object-oriented modeling framework. This approach provides an efficient and precise structure for the methodological organization of concepts and their relationships. The object-oriented framework allows us to describe properties and actions for a whole category of objects, called a class. The class describes the state and behavior of a cell object. Each cell object is thus a particular instance of this template and has an individual behavior. The behavior laws described in the class could depend on intrinsic properties and interactions. They can also change with time, either explicitly or implicitly, through time-varying variables such as extracellular factors or concentrations. Intrinsic properties are mainly mechanical properties such as elasticity, contractility, and incompressibility [19]. They could also include biochemical or genetic processes. Furthermore, cell object interactions with the environment, through biased migration (chemotaxis, haptotaxis, etc.), or cell-cell interactions, e.g., collision and adhesion, can be defined in this proposed framework, as exemplified below.

2.2. The POO Approach

Mass-spring networks are often used to model 3D physical object dynamics. However, the main drawbacks of these models lie in the control and assessment of parameters. We initially developed a different approach, in the context of human breathing [19]. This discrete model is based on computer graphics modeling [4, 21]. Natural motions and realistic-looking flexible and elastic objects are efficiently modeled and simulated by means of physically based computer graphics models. These models use a small amount of data (object geometry and relations between objects). From this, an animation motor (using forces, energies, or direct displacements) integrates movement and deformation laws to compute the evolution of shapes and positions. Generally, constraints are added to control movements and deformations or to model complex physical properties.

We modeled the human abdomen using an elastic and incompressible 3D object and the human diaphragm using an elastic and contractile membrane. These objects were linked together and attached to rigid objects modeling the rib cage. An explicit algorithm was developed to compute the dynamic evolution of these objects as well as to insure the necessary constraints. Incompressibility and connections were considered as constraints.

This model, allowing for rigid, incompressible elastic or contractile objects, is the groundwork, extended here to model cell deformation and migration. Other properties such as incompressibility or cell interaction have also been taken into account, as detailed below.

2.2.1. Geometry. Each cell object requires a 3D geometry. We do not yet consider cell division, thus the cell topology remains unchanged. To simplify the calculations, we only consider the cell contour and define a cell object by a 3D closed surface. This contour is represented by nodes linked in a triangular mesh. Triangular meshes are simple and considerably simplify the incompressibility constraint and collision resolutions.

2.2.2. Forces. To generate forces and motions, a mass is assigned to each contour node. Forces are exerted on the nodes to generate displacements and deformations. Three kinds of forces can be used in our model: force fields (e.g., gravitation force), locally applied forces (e.g., user manipulation), and linear actuator forces (LAFs). We introduce a LAF when a target position P_{target} is known for a given node of position P. To minimize the distance $|PP_{target}|$, a spring is created between P and P_{target}. It generates a force that linearly attracts P towards P_{target}. The expression of the LAF is $F = k_{elas}\,(P - P_{target})$, where k_{elas} is the elastic modulus of the spring. LAF can be seen as a potential force that tends to minimize a distance. LAFs permit us to model any kind of forces that could be defined by target positions. P_{target} can depend on geometry or on constraints and can change dynamically.

2.2.3. A specific case: elasticity. Calibration of spring-mass networks is known to be difficult. Therefore, to model elasticity, we define a local shape memory [18]. The elastic property of a cell object is simply defined as its ability to recover its original shape after mechanical deformation. To model this property, we construct a local shape coordinate system where each contour node position is defined relative to its neighboring nodes by three parameters (α, β and γ) (Fig. 1). The initial value of this three-parameter set defines the rest shape. During the simulation, according to the neighboring node positions, a target position is computed for each node under consideration. This target position satisfies the local shape and defines a LAF. Therefore, a cell object is defined by a contour where the shape memory force is applied to all its nodes. A unique elasticity modulus k_{elas} is used by the corresponding LAF.

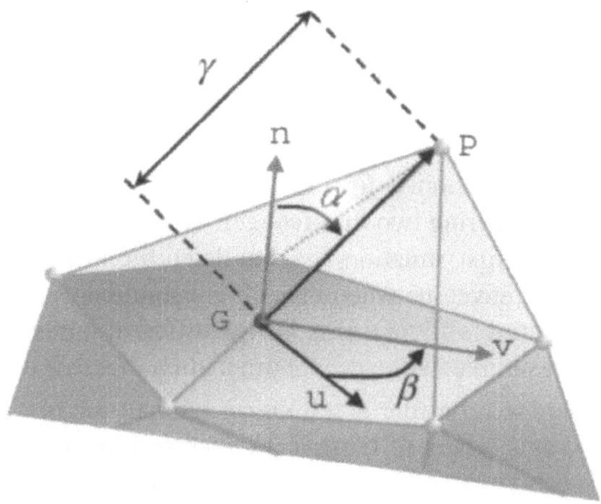

Figure 1. Local shape memory force to model elasticity. The local shape is defined by two angles α and β and a distance γ. A node position P is defined by α, β, and γ and the current position of its neighbors. G is the center of masses of the neighbors. **n** is an approximate normal from which **v** is deduced, **u** is the cross product between **n** and **GP**. At each iteration this allows us to set a target position for each node.

2.2.4. Constraints. Considering forces is often not enough to realistically model complex behavior of physical objects. Constraints are thus added to maintain some additional conditions such as boundary conditions (non-penetrating area) or incompressibility. Our algorithm considers constraints as non-quantified force components. Thereby, it is possible to handle the total incompressibility of a contour [18]. This allows the model to verify the incompressibility constraint exactly and efficiently for all the cell objects individually.

2.2.5. Numerical integration. The following algorithm is used to compute the dynamical behavior of a cell object from time t to time $t + dt$:

- For each node: compute the sum of applied forces.
- Use a discrete integration scheme for equation of motion (e.g., Newton-Cotes) to compute the new velocity and position of the nodes, without considering any constraints.
- Take local and global constraints into account and impose given displacements to the nodes (these displacements have to be computed at each time step).
- Adjust the velocity to include the imposed displacements.

3. POO Simulation of Individual Cell Mechanical Behavior

An increasing amount of experimental work is devoted to the analysis of mechanical properties of cells, which play a central role in a wide range of biological processes, from cell migration to the modulation of the cell response to extracellular mechanical stimuli. In addition to micropipette aspiration techniques (cf. Drury and Dembo, this volume, Part II), the recent development of experimental techniques such as microplates [22] or optical tweezers [10, 20] makes possible the manipulation of isolated living cells and the application of specific mechanical forces or deformations. Such experiments can moreover lead to quantitative results, usually in terms of linear stress-strain relationships, from which cell mechanical moduli can be inferred.

Our purpose in this section is to illustrate the application of our POO approach in cytomechanics by considering two different cell object prototypes, characterized by a specific architecture. In a first situation, we simulated the influence of transverse links within an elastic discrete envelope when the cell is submitted to uni-axial compression. This situation is a caricature of living cells having different intracellular organization of their cytoskeleton, for example, with specific orientation of stress fibers.

As a second case, we consider a cell object without internal cross-links: this cell object can be viewed as a model of living cells such as human erythrocytes, where the cell membrane is entirely responsible for the elastic cell deformation, the inner cytoplasm being only viscous. In order to test the validity of our modeling approach with regard to real experiments, we simulated optical tweezers manipulations in which calibrated forces, between 0 and 56 pN, have been diametrically applied to human erythrocytes [10]. We specifically considered the suspension of erythrocytes in hypotonic buffer, which transforms the initial discoid cell morphology into a spherical or nearly spherical shape.

3.1. Influence of the Cell Object "Cytoskeleton": *in silico* Compression Loading

In this first situation, we consider a cell objet topology defined by 66 nodes. Three different cell prototypes are considered (Fig. 2). The cell0 architecture is assumed to be only an external envelope, with no internal links. In cell1, we considered a reinforcement of the cell architecture with horizontal elastic cross-links modeling some cytoskeleton fibers. Finally, cell2 architecture includes internal diagonal links connecting the apical and basal cell surfaces. In each case, a vertical force F_c is locally applied at the top of the cell, on four or five upper nodes. The five lowest nodes of the basal cell surface are fixed (zero displacement) and simulate the partial cell attachment to a rigid substratum.

Figure 2 presents the initial 3D shape of each cell object (first raw) and the equilibrium state they reached (second raw) when a given vertical loading force F_c is applied for a fixed value k_{elas} of the elasticity modulus.

Figure 3 shows the evolution with time of the height $z(t)$ defining the coordinate of the node located on the top of each cell object. The first part of the curve corresponds to the cell uni-axial compression, and the second phase corresponds to the elastic relaxation simulated after removal of the loading force.

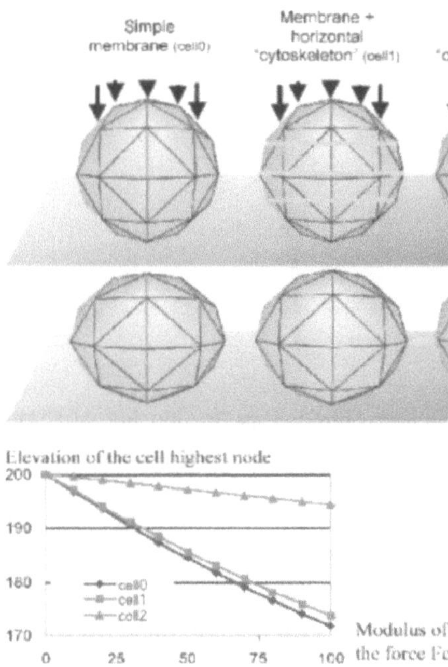

Figure 2. Influence of the cell object "cytoskeleton" on the cell mechanical response to vertical load. Three cell types are considered: no "cytoskeleton" (cell0), horizontal links (cell1), diagonal links (cell2). The first row presents the initial 3D shape of each cell prototype. For a given fixed value k_{elas} of the elasticity modulus of each cell object, the second row in the figure indicates the equilibrium state which is reached when a vertical loading force F_c is applied on the 5 (cell0 and cell1) or 4 (cell2) nodes marked by arrows. The insert at the bottom of the figures indicates the vertical position at equilibrium (normalized units) of the highest node of the cell object as F_c is increased (horizontal axis, normalized units).

3.2. *In silico* Optical Tweezers Experiments

We simulated here the experimental protocol used by Hénon et al. [10] to deform nearly spherical erythrocytes (red blood cells, RBCs) using optical tweezers. The force F is exerted on two silica microbeads that are stuck on the RBC membrane in diametrically opposite positions. By slowly incrementing the distance between the two trapped beads, an increasing stress is applied to the RBC membrane. Proper force calibration then gives the experimental decrease of the RBC diameter (measured in a plane perpendicular to the applied force) as a function of the increasing value of F. Experimentally, the variations are linear up to 10pN, which gives an elastic shear modulus close to 2μN/m. However, for a range of force between 10 and 35 pN, the curve is almost flat with no significant decrease of the RBC diameter [10].

In order to simulate this experiment with our POO approach, a force F_s is locally exerted on two opposite nodes of the cell object contour, pulling them apart (Fig. 4). The

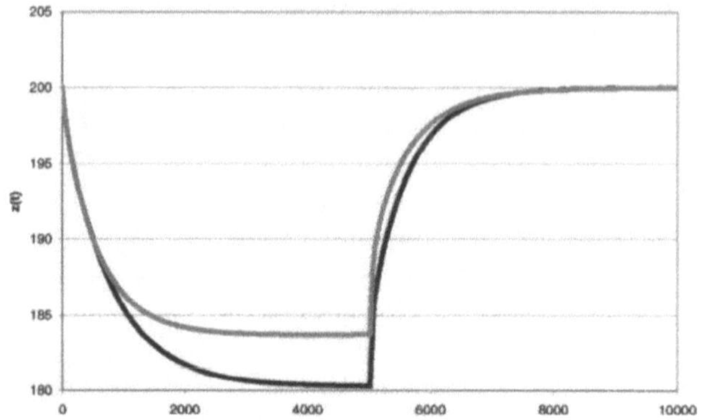

Figure 3. Associated compression and relaxation curves of cell0 and cell1. The figure shows the overall evolution with time of the height $z(t)$ defining the coordinate of the node located on the top of each cell. The first part of the curve corresponds to the compression phase, and the second part corresponds to the nonlinear elastic relaxation simulated after removal of the loading force F_c.

simulation parameters are the elasticity of the cell, i.e., k_{elas}, and the modulus of F_s. Figure 4 shows that k_{elas} can be adjusted such that the modeled cell and the real RBC exhibit the same mechanical response.

4. Cell Aggregation and Sorting

4.1. Model Definition

The *Dictyostelium discoideum* slime mold aggregation is a complex process. Upon starvation, isolated cells undergo a differentiation program, which allows cell aggregation into a slug. The aggregation process is under the control of cyclic cAMP signals synthesized by isolated cells in the population [1, 17]. Recent model studies have shown the importance of cell to cell interactions in the process of mound formation and cell sorting [2, 24, 26]. *Dictyostelium* cells represent an important cellular model to explore self-organizing structures by short-range interactions between cells [5].

Within the mathematical framework presented in Section 2, we simulated cell migration in a population of cell objects subjected to a central force field; each of these mathematical cell objects is equivalent to a biological isolated cell. The differentiated cell type would be coded by different properties of the corresponding model cell object. The central force field is the equivalent of the cAMP gradient in *D. discoideum* aggregation process; however, other situations may be considered, such as force transmission through mechanical constraints in tissues or cell and extracellular matrix interactions.

In the present model version, we assume that a clump of cell objects is held fixed at a central position (Fig. 5). This initial aggregate exerts an attractive force field towards isolated cell objects in the neighborhood. The force field is mediated through diffusion of a chemoattractant molecule, produced by the clump. This diffusion, which is assumed to be linear, generates a gradient of chemoattractant concentration: this induces the attraction

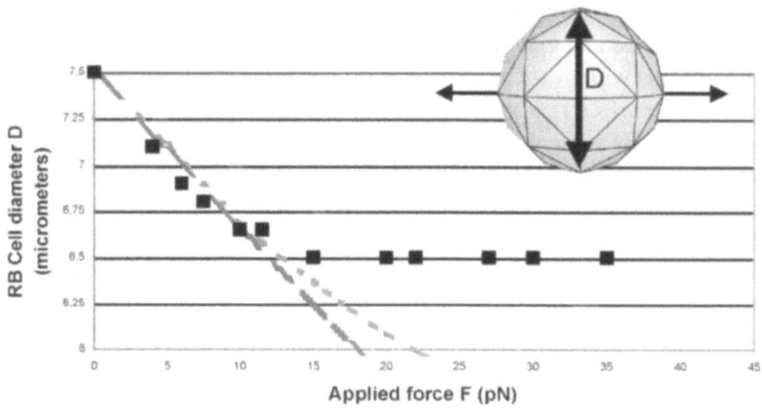

Figure 4. *In silico* optical tweezers experiments. A spherical cell object is proposed as a model of nearly spherical red blood cells (RBC) suspended in a hypotonic solution. Optical tweezers double trap is simulated by exerting a force F_s on two opposite nodes of the cell object contour (upper insert). The variation with load of the cell object diameter D(F_s) in a plane perpendicular to the direction of load is simulated and compared to real data published by Hénon et al. [10]. The simulation parameters are the elasticity of the cell, i.e., k_{elas} and the modulus of F_s. With appropriate scaling of the force, one can adjust the parameter k_{elas} such that the experimental mechanical response of RBC can be nicely fitted into the linear elastic regime. Dashed line represents the linear fit to the experimental data (elastic shear modulus $\sim$ 1,9 μN/m). Dotted line represents the simulated mechanical response of the RBC object. One can notice a departure from linear elasticity at larger traction forces.

force exerted on cell objects, with the source of chemoattractant defining a target position for the cells. The resulting LAF "elastic" modulus is called $k_{attractant}$. This force is applied to every cell node.

In order to mimic realistic situations, we also considered the possibility of a random movement of the cell objects. Brownian movement is modeled using a constant direction during a given latency time called t_{random}. When the latency time is elapsed, a new random direction and latency time are generated. We set: $t_{random} = -p * log(1 - r)$, where p is a constant of persistency and r is a randomly generated number between 0 and 1. The persistency parameter could be different for each cell object or type of cell object. The random direction is computed using θ_{random}, a randomly generated angle between 0 and 2π. For the simulations presented here, the random direction occurs on a plane, but it could easily be computed in a 3D space.

Finally, we also assume that the cell object population is a mixture of two cell object types, denoted, respectively, A and B, which behave differently in response to the force field. Type A and B correspond to the differentiated *D. discoideum* pre-spore or pre-stalk cell types. Movement rules mimic the fact that (pre-spore/pre-stalk) cell types of *D. discoideum* exhibit different responses to cAMP gradients, either in the aggregation phase or during the cell movement within the mound. During cell object displacement, collisions are expected and give rise to three possible pair formations.

Cell collisions will create aggregates during simulation. It should be noted that the destruction of aggregates is also possible if the bonds between cell objects are weak, which occurs typically when the cell objects are of different types. During simulations, if two cell objects c_1 and c_2 collide, a dynamic link is built between them. If c_1 and c_2 are not already part of any aggregate, a new aggregate is created. If c_1 is in a different aggregate than c_2, the two aggregates are merged. During simulation, if an aggregate contains no more than one cell object, it is destroyed. The collision between some nodes of these cell objects makes a dynamic link between two cell objects. The closest two facets are then chosen and a LAF that links the nodes of these two facets is created. To generate these forces, each node of the chosen facet in c_1 uses one of the three nodes of the chosen facet in c_2 as target position (and *vice versa*). A differentiation table gives the stiffness to be used for this LAF, depending on the type of c_1 and c_2.

4.2. Cell Aggregation Simulation

We start with a population of 36 cell objects lying on an infinite plane representing the substratum. More precisely, each cell object is an icosahedron (12 nodes, 20 facets) that is totally incompressible. Half of the cell objects correspond to a cell type A. The two types differ in their ability to respond to cell-cell interactions (see previous section).

In Figure 5, we present three steps of the aggregation process simulation. Initially, the 36 cell objects were scattered uniformly (panel a). As time elapses, cells begin to move randomly and in the direction of the central force (panel b). For a long time, aggregation is almost complete (panel c). Accompanying cell movement towards the center, random collision between cell objects builds up sorting. Consequently, the final cell aggregate is organized into cellular clumps, each of them being associated with one of the two cell object types A and B. This exemplifies that cell object sorting occurred during the aggregation process. The effect of random movement is particularly important when the central force is weak. In these conditions, only aggregates with a small number of cell objects are observed. None of these aggregates would grow and give rise to a unique clump of cell object (results not shown).

5. Discussion

We have presented an object-oriented approach to model both individual and collective mechanical and motility cell behavior. This approach is based on the definition of a cell object, whose architecture, mechanical properties, cell-cell and cell-medium interactions can be precisely defined according to given working hypotheses. It is indeed known that the mechanical behaviors of living cells are hardly explainable by a viscous fluid balloon response only. Even if it remains a matter of controversy [13], the concept of cell tensegrity architecture proposed by D.E. Ingber has provided a theoretical background for different models of living cells [25, 28, 29]. We develop here an alternative modeling approach whose relevance is illustrated by considering two different scales of investigations.

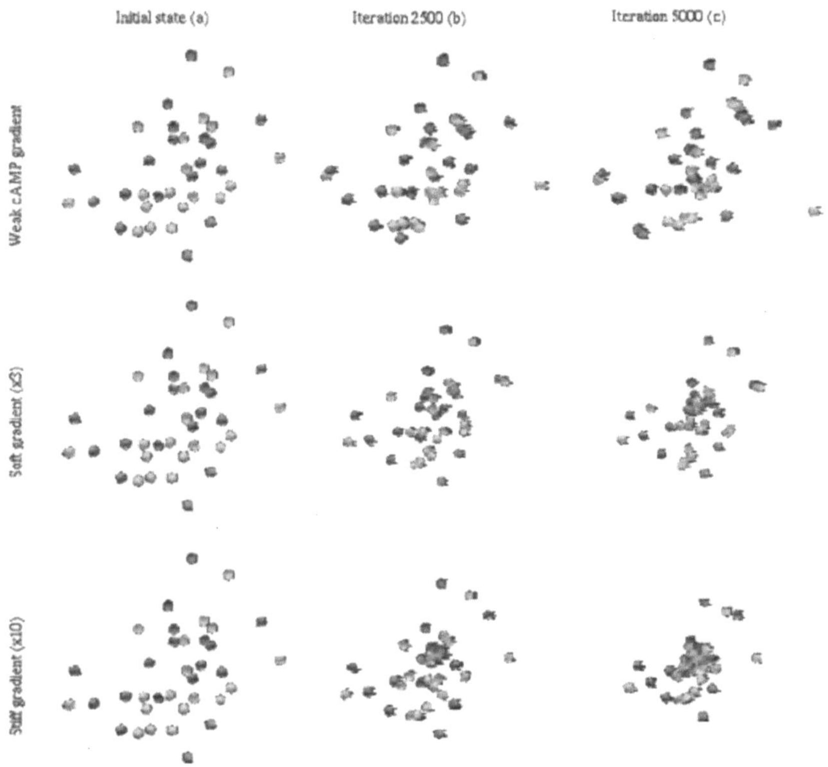

Figure 5. Simulation of the aggregation of a 36 cell objects population. These cell objects are subjected to a central force field whose effect is to gather the cell objects into an unique aggregate at the center position. Additionally, cell objects can move randomly. We also assumed that two cell object types A and B are equally present in the population. (A/A) and (B/B) cell object collisions give rise to stable bonds whereas (A/B) ones have no effect on aggregation.

The first one deals with the emergence of individual cell mechanical response from specific subcellular architectural elements. Such considerations can be closely related to specific organization of the cytoskeleton in living cells. Basic cell response to uni-axial loading has been illustrated in this paper. We furthermore proposed a more refined analysis of an incompressible cell object only modeled by a finite number of nodes on its external "membrane". We compared the simulated responses to real experiments undertaken on human red blood cells and showed that our cell object satisfactorily reproduces the experimental data in the linear deformation regime.

We have integrated the individual mechanical cell object properties in the simulation of collective and coordinated motion of two different populations of cell objects submitted to an external attraction field. Thus, the collective cell object re-organization and sorting result directly from the cell-cell mechanical and biochemical interactions, superimposed to "phenotypic" individual cell properties like random motion and differential sensitivity to the external gradient. Simulations showed the potentiality and the simplicity of our POO approach to account for collective dynamical living cell behavior. When compared to *D. discoideum* behavior, simulation results prove that very simple interaction

[6] Duvdevani-Bar, R. and Segel, L.A (1992) On topological simulations in developmental biology. *J Theor Biol* **166**, 33–50

[7] Dugnolle, P., Garbay, C. and Tracqui, P (1998) A mechanical model to simulate cell reorganisation during in vitro wound healing. *Proc. of the European Simulation Multiconference '98, Manchester, "Simulation Tools in Biology"*, 343–347

[8] Ermentrout, G.B. and Edelstein-Keshet, L (1993) Related Articles Cellular automata approaches to biological modeling *J Theor Biol* **160**(1), 97–133

[9] Graner, F. and Sawada, Y (1993) Can surface adhesion drive cell rearrangement? Part II: a geometrical approach. *J Theor Biol* **164**, 477–506

[10] Hénon, S., Lenormand, G., Richert, A. and Gallet, F (1999) A new determination of the shear modulus of the human erythrocyte membrane using optical tweezers. *Biophys J* **76**, 1145–1151

[11] Honda, H., Yamanaka, H. and Dan-Sohkawa, M (1984) A computer simulation of geometrical configurations during cell division *J Theor Biol* **106**, 423–435

[12] Honda, H., Tanemura, M. and Yoshida, A (2000) Differentiation of wing epidermal scale cells in a butterfly under the lateral inhibition model-appearance of large cells in a polygonal pattern. *Acta Biotheor* **48**(2), 121–36

[13] Ingber, D. E., Heidemann, S. R., Lamoureux, P. and Buxbaum, R. E (2000) Opposing views on tensegrity as a structural framework for understanding cell mechanics. *J Appl Physiol* **89**(4), 1663–1678

[14] Martiel, J.L. and Goldbeter, A (1987) A model based on receptor desensitization for cyclic AMP signaling in Dictyostelium cells. *Biophys J* **52**, 807–828

[15] Murray, J. D., Maini,P. K. and Tranquillo, R. T (1988) Mechanochemical models for generating biological pattern and form in development. *Physics Rep* **171**, 60–84

[16] Palsson, E. and Othmer, G (2000) A model for individual and collective cell movement in *Dictyostelium discoideum*. *Proc Natl Acad Sci USA* **97**(19), 10448–10453

[17] Parent, C.A. and Devreotes, P.N (1996) Molecular genetics of signal transduction in Dictyostelium. *Annu Rev Biochem* **65**(2), 411–40

[18] Promayon, E., Baconnier, P. and Puech, C (1996) Physically-based deformations constrained in displacements and volume. *Computer Graphic Forum* **15**(3), 155–164

[19] Promayon, E., Baconnier, P. and Puech, C (1997) Physically-based model for simulating the human trunk respiration movements. *Lecture notes in Computer Science* Springer Verlag **1205**, 379–388

[20] Sleep, J., Wilson, D., Simmons, R. and Gratzer, W. (1999) Elasticity of the red cell membrane and its relation to hemolytic disorders: an optical tweezers study. *Biophys J* **77**(6), 3085–95

[21] Terzopoulos, D., Platt, J., Barr, A. and Fleischer, K (1987) Elastically Deformable Models *Proc. Computer Graphics (ACM SIGGRAPH'87)* (M.C. Stone Ed.) **21 (4)**, 205–214

[22] Thoumine, O. and Meister, J.J (2000) Dynamics of adhesive rupture between fibroblasts and fibronectin: microplate manipulations and deterministic model. *Eur Biophys J* **29**(6), 409–19

[23] Tyson, J.J. and Murray, J.D (1989) Cyclic AMP waves during aggregation of Dictyostelium amoebae. *Development* **106**, 421–426

[24] Vasiev, B. and Weijer, C.J (1999) Modeling chemotactic cell sorting during *Dictyostelium discoideum* mound formation. *Biophys J* **76**(2), 595–605

[25] Volok, K.Y., Vilnay, O. and Belsky, M (2000) Tensegrity architecture explains linear stiffening and predicts softening of living cells. *J Biomech* **33**(12), 1543–1549

[26] Weijer, C. J (1999) Morphogenetic cell movement in Dictyostelium. *Semin Cell Dev Biol* **10**(6), 609–19

[27] Weliky, M., Minsuk, S., Keller, R. and Oster, G (1991) Notochord morphogenesis in Xenopus laevis: simulation of cell behavior underlying tissue convergence and extension. *Development* **113**(4), 1231–1244

[28] Wendling, S., Oddou, C. and Isabey, D (1999) Stiffening response of a cellular tensegrity model. *J Theor Biol* **196**, 309–325

[29] Wendling, S., Planus, E., Laurent, V., Barbe, L., Mary, A., Oddou, C. and Isabey, D (2000) Role of cellular tone and microenvironmental conditions on cytoskeleton stiffness assessed by tensegrity model." *Europ J Appl Physics* **9**, 51–62

[30] Yamaguchi, T., Yamamoto, Y. and Liu, H (2000) Computational mechanical model studies on the spontaneous emergent morphogenesis of the cultured endothelial cells. *J Biomech* **33**(1), 115–26

Part III
Multicellular Dynamics

"The form-shaping episodes of embryogenesis depend on the coordinated migrations and changes in form of entire populations of cells. These movements are controlled ultimately by genetic mechanisms: by the executive expression of genes in response to spatial or temporal cues. The physical transformation of genetic blueprints into eyes, legs, and wings, however, takes place on the factory floor of the cytoskeleton."

[Denis Bray (1982) Cell Movements. Garland Publ., New York and London, p. 357]

Introduction to Part III

Having already considered molecular dynamics in Part I and then polymer dynamics and cellular motility in Part II, we finally arrive at our last topic, which deals with motion and interaction at the level of many cells, all the articles in this part of the book being concerned with modeling multicellular systems.

Until a few years ago, a section such as this would undoubtedly simply have been filled with models based around a deterministic, continuum 'mean-field' framework, i.e., systems of coupled, nonlinear (probably), parabolic partial differential equations. However, over the last few years there has been a genuine appreciation within the mathematical biology modeling community of the need for a more accurate description of how individual 'biological entities' (e.g., cells) respond to external signaling cues. There is also a similar recognition of how to incorporate signaling cues from intracellular processes occurring at 'lower scales'. Generally speaking, there is now a greater awareness of the need to develop accurate 'multiscale' models of biological systems. Indeed, one of the aims of this book is to indicate how one may, for example, link models at the molecular, polymer and, finally, cell levels.

This recent focus on multiscale modeling has, in turn, led to the development of novel modeling techniques that are capable of combining and, examining events at an individual level with events at a continuum level, e.g., individual (discrete), interacting, migrating cells responding to and influencing the (continuum) flow or concentration fields of reacting and diffusing chemicals. It is worthwhile to note that the resulting novel modeling techniques are not merely 'phenomenological' and that the model results are not merely 'pretty pictures' that only in the final analysis can be compared with 'biological reality'. As will be seen from the articles in this Part III, all the models are firmly grounded in biophysical reality and, as the articles of Fogelson et al., Alt et al. and Rice & Barocas in particular show, can be firmly placed within an appropriately chosen and rigorous mathematical setting. A brief overview of each article now follows.

The paper by **Fogelson, Yu & Kuharsky** deals with two important aspects of blood clotting – 3D platelet aggregation and coagulation. In order to investigate the dynamics of 3D platelet aggregation, it is necessary to track the motion and behavior of individual particles (platelets) as they interact with each other, blood vessel walls, the suspending fluid (blood), and certain activating chemicals within a 3D fluid dynamic setting. This involves a very elegant 'mixture' of an Eulerian modeling framework (all fluid quantities) and a Lagrangian modeling framework (all particulate quantities such as platelets or blood vessel walls). The modeling therefore involves a blend of 3D computational fluid dynamics (Navier-Stokes equations) and the immersed boundary method. In the second part of the article, a different approach is taken to study the phenomenon of coagulation from a biochemical viewpoint. The emphasis is reversed somewhat with a simplified fluid dynamic approach coupled with a detailed modeling description of the signal transduction pathways (enzyme reactions) taking place during coagulation. The result is a sophisticated system of ordinary differential equations. Given the complexity of the reaction kinetics of the system, the model is shown to be of predictive value in the understanding of the complicated coagulation dynamics.

In some sense, a related hydrodynamical problem, but with much simpler 'particle' kinetics for immobile microbial cells in a 'biofilm' growing at the boundary of a flow chamber, is treated in the contribution by **Bungartz, Kühn, Mehl, Würtz**. Such biofilms have been modeled either as conglomerates of ball-shaped individuals that grow and divide depending on nutrients diffusing out of the fluid stream or almost equivalent cellular automata (CA) with corresponding dynamical rules on the same spatial microscale (see the review by Drasdo mentioned below). The authors here use the CA approach for simplicity. However, in any case, the (pseudo-steady) Navier-Stokes flow equations have to be solved in the exterior of the domain 'occupied' by microbial cells. Its complex, time-varying boundary requires the use of efficient numerical solution methods (adaptive multi-grids) and suitable parallelization of the discrete CA simulation steps, which alternate with the discretized partial differential equation (PDE) solution step for flow, transport, and diffusion. The case study presented here provides an easily generalizable basic framework for efficient numerical solutions of such hybrid continuum-discrete evolution models with complex geometries on a microscale level, and it offers the option to be embedded into appropriate macroscale simulation procedures that would allow long-term data comparison and control in standardized experimental assays.

The review paper by **Drasdo** provides an excellent and truly comprehensive survey of several key individual-based modeling approaches and simulation tools that have arisen recently and are now being used to great effect in modeling the dynamics of multicellular biological systems. The first class presented comprises various 'lattice' models (including CA) with discrete (stochastic) interaction, motion, and transition rules on spatial grids that are differently scaled or shaped relative to the size and type of the individual biological cells being represented. Next, as a less artificial class, we have the 'off-lattice' models with smooth (stochastic) motion of a collection of cells represented by balls or by deformable ellipsoids or polyhedra. The review compares several model descriptions through physical interaction laws or geometric side conditions and the resulting system of generalized Newton's equations for individual cell motion (with more or less frictional damping, depending on spatiotemporal scaling). Finally, some other alternative modeling approaches are mentioned, such as topological systems for foam dynamics or the classical Lindenmeyer systems. Many illustrative examples are offered pointing to applications such as tumor growth, embryonic cell sorting, and mophogenesis of intestinal crypts and other endo- or epithelial growth and differentiation dynamics. The review concludes with a valuable synopsis and a comprehensive reference list.

Following on from Drasdo's review we have, a series of articles dealing specifically with individual-based models for multicellular systems. The first of this group of articles, namely, that by **Czirók, Deutsch & Wurzel**, considers the phenomenon of aggregation of individual cells into larger groups ('cohorts') generated through the active migration and interaction of the individuals. In this instance, cells are modeled as locally interacting, self-propelled and -oriented entities moving in a plane. Initially, a simple lattice-based model is introduced where particles are located at some spatial location and move with some given velocity oriented in a particular direction. Particles interact locally with each other through a simple rule that states that at each time step a given particle assumes the average direction of motion of the particles in some defined local neighborhood. A cellular automaton description of interacting, migrating cells is then considered, with local interactions being governed by an interaction energy field whose 'strength' can be varied. The numerical simulations of these models show very interesting dynamical behavior (aggregation, swarming, flocking, and cooperative behavior) arising from simple interaction rules and depending on order parameters such as cell density or noise amplitude. The techniques introduced here have enormous potential applications in modeling many different aspects of cellular interactions.

In the article by **Alt, Bretschneider & Müller**, the very general concept of a 'bime', or **b**iological **i**nteracting and **m**oving **e**ntity, is introduced. This concept formalizes how individuals move and interact with each other while communicating *via* an underlying dynamic interaction field, which itself can be modified by the 'bimes' and in turn then influences the dyamics of the 'bimes'. In this article, the whole framework of 'bimes', interaction dynamics, and interaction fields is placed on a rigorous (general) mathematical foundation. Examples in one, two and three dimensions are given as to how to apply the general theory to a specific biological setting. The examples range widely from two individual cells interacting on a viscoelastic substratum *via* tensiotaxis to amoebae aggregation (*Dictyostelium discoideum*) *via* chemotaxis to the dynamics of flocks of birds. Other potential and important applications of this technique include wound healing, cancer cell invasion of tissue, and other important cell migration problems observed in many aspects of developmental biology (embryology). For the example of swarm dynamics, the article mentions a method of how to derive approximative stochastic continuum fluid equations from the interactive multiparticle ODE system in a suitable scaling limit for nearest-neighbor distance.

The contribution by **Rice & Barocas** is also concerned with the afore-mentioned problem of cell motion and interaction in a viscoelastic medium but is motivated by a specific experimental assay: contractile tissue cells such as fibroblasts placed in a collagen gel (modeling an extracellular fibrous tissue matrix) are able to deform the gel under certain (e.g., isometric) conditions. From measurements of gel deformation or boundary tension, one wants to estimate cell traction parameters, which describe the (averaged) amount of traction force exerted by each of the cells. As one step towards a general method to calibrate different measurement and model devices, the authors present a comparison of 2D numerical and analytic solutions for a two-phase flow model of anisotropic viscous gel deformation. In contrast to a previously treated nodel version, here the cells are distributed within the gel as discrete points with traction forces along individually given line directions. Elementary sample simulations reveal that the discretized numerical (Galerkin finite element) solution does not reproduce well the velocity singularities around the cells and therefore should be replaced with a hybrid solution method which locally implements special singular (Green's-type) basis functions.

Finally, the last two articles of **Part III** bring us back in some sense to the first two contributions by combining continuum models with discrete models for collective cell motion. The first article of **Anderson** presents a novel modeling technique that is highly applicable to modeling multicellular systems responding to multiple external signaling cues. In the first instance a 'traditional' (parabolic) 'master' partial differential is formulated to decribe the spatiotemporal dynamics of cell populations that are considered to migrate in response to the external stimuli of reacting and diffusing chemicals in the system. The migratory response of the cells contains two main components – a random or unbiased migration term (analogous to modlecular diffusion) and a directional or biased term that is governed by gradients of the external chemical stimuli. By using a numerical discretization of the PDE, a biased random walk equation governing the motion of an individual cell can be derived. This discrete stochastic equation is then used to evolve the individual cells in time. In addition to these migration rules, other individudal processes associated with cells (e.g., proliferation) can be incorporated into the simulation process using cellular-automaton-like rules. In the final article, **Anderson & Pitcairn** apply this technique to model three important biological systems: tumor-induced angiogenesis, cancer cell invasion of healthy tissue, and cell aggregation (*Dictyostelium discoideum*).

Polymer and Cell Dynamics – Multiscale Modeling
and Numerical Simulations, 145–154
ed. by W. Alt, M. Chaplain, M. Griebel und J. Lenz

Computational Modeling of Blood Clotting: Coagulation and Three-dimensional Platelet Aggregation

Aaron Fogelson[1], Haoyu Yu[2] and Andrew Kuharsky[3]

1) Department of Mathematics, University of Utah, 155 South 1400 East, JWB 233, Salt Lake City, Utah 84112, USA
2) Physiome Sciences, 307 College Road, Princeton, New Jersey 08540, USA
3) Quantlab Financial, 4200 Blvd, Houston, Texas 77006, USA

Summary. Platelet aggregation and coagulation are the two main components of blood clotting, both of the normal hemostasis that maintains vascular integrity and of the pathological thrombosis that is the cause of most heart attacks and strokes. Two approaches to modeling aspects of blood clotting are presented. The first emphasizes the mechanical interactions among platelets, fluid, and vessel, in platelet aggregation. It uses the immersed boundary method as a modeling/computational tool on which to base three-dimensional simulations of platelet aggregation. The second emphasizes coagulation biochemistry and uses a much simplified description of flow and transport. The dynamics of coagulation are studied and novel predictions about the regulation of the system are made.

1. Introduction

The blood-clotting process is comprised of two major interacting parts, platelet aggregation and coagulation. Platelet aggregation involves the cohesion of groups of activated platelets. Platelets are activated by contact with molecules embedded in the injured vessel wall or released into the plasma. Cohesion between two activated platelets is accomplished by formation of bridges made up of plasma molecules (fibrinogen and von Willibrand factor) bound to specific glycoprotein receptors (GPIIb/IIIa receptors) on the two platelets' surfaces. Coagulation is a set of tightly regulated enzymatic reactions that lead to the formation of a fibrin polymer gel in and around the platelets in an aggregate in order to mechanically stabilize the aggregate. There are a number of excellent reviews of platelet aggregation [1, 14] and blood coagulation [5, 6, 10].

In this paper we will give an overview of two aspects of our work on modeling blood clotting. The first is three-dimensional microscopic-scale modeling of platelet aggregation, models that track the interactions of a group of discrete platelets with one another, the blood vessel wall, and the fluid continuum in which they are immersed. The second topic concerns models of coagulation in which the biochemistry is treated in considerable detail while flow, transport, and platelet deposition are modeled in a simple manner. We have also developed continuum models of aggregation [2, 3, 15], but we will not discuss these here.

2. Microscopic Aggregation Model

In these models, we track the motion and behavior of a collection of individual platelets as they interact with one another, the vessel walls, the suspending fluid, and activation-inducing chemicals. Model platelets typically enter into the domain of interest (a segment of a vessel or laboratory flow chamber) in a nonactivated state. They can be activated by exposure to a sufficiently high concentration of activating chemical or by direct contact with the 'injured' reactive portion of the vessel wall. While any platelet can stick to the injury, only activated platelets can stick to each other. The mechanical interactions among platelets, and among platelets, fluid, and vessel walls, are captured using the immersed boundary (IB) method [11, 12, 13]. With this approach each platelet and each vessel wall is modeled as an elastic object immersed in a viscous, homogeneous, incompressible fluid. In the IB method, fluid quantities are described in Eulerian terms, while quantities having to do with the immersed objects are described in Lagrangian terms. The important fluid quantities are the velocity $\mathbf{u}(\mathbf{x}, t)$, pressure $p(\mathbf{x}, t)$, and force density $\mathbf{f}(\mathbf{x}, t)$. Here, $\mathbf{x}$ denotes a point in space. Each immersed object, platelet, or vessel wall, is described by a Lagrangian variable; $\mathbf{X}_i(\alpha, t)$ is the location in space at time t of the point labeled by α on the ith object. The equations of motion that describe the mechanical interactions listed above are

$$\rho(\mathbf{u}_t + \mathbf{u} \cdot \nabla \mathbf{u}) = -\nabla p + \mu \Delta \mathbf{u} + \mathbf{f}; \qquad \nabla \cdot \mathbf{u} = 0\,, \tag{1}$$

$$\mathbf{f}(\mathbf{x}, t) = \sum_i \left[\int_{\Gamma_i} \mathbf{F}_i(\mathbf{X}_i(\cdot, t)) \delta(\mathbf{x} - \mathbf{X}_i(\alpha, t)) d\alpha \right], \tag{2}$$

$$\mathbf{F}_i(\mathbf{X}_i(\cdot, t)) = \tilde{\mathbf{F}}_i(\mathbf{X}_i(\alpha, t), \mathbf{X}_{i,\alpha}(\alpha, t), \mathbf{X}_{i,\alpha\alpha}(\alpha, t), t) + \sum_l \mathbf{F}_{il}\,, \tag{3}$$

and

$$\frac{\partial \mathbf{X}_i}{\partial t}(\alpha, t) = \int_\Omega \mathbf{u}(\mathbf{x}, t) \delta(\mathbf{x} - \mathbf{X}_i(\alpha, t)) d\mathbf{x}\,. \tag{4}$$

Eqs. (1) are the Navier-Stokes equations which governs the motion of the fluid driven by a force-density distribution $\mathbf{f}$. This force-density is obtained by transmission of the forces generated by the IB objects to the surrounding fluid, an operation expressed by Eq. (2). In this equation, Γ_i denotes the surface of the ith IB object, $\mathbf{F}_i(\mathbf{X}_i(\alpha, t))$ is the elastic force at $\mathbf{X}_i(\alpha, t)$ generated by deformation of this surface as well as by stretching of any cohesive bonds which connects the ith IB object to other IB objects, and $\delta(\mathbf{x})$ is the Dirac delta function. Eq. (3) defines $\mathbf{F}_i$ as a sum of a term, $\tilde{\mathbf{F}}_i$, due to elastic deformation of the ith IB object and of terms, $\mathbf{F}_{il}$, one for each cohesive bond l connecting a point on object i to a point on another object. Eq. (4) is the expression of the no-slip condition for a viscous fluid evaluated at the point $\mathbf{X}_i(\alpha, t)$ of object i. Here, the no-slip condition gives the equation of motion for the points on an IB object, i.e., that each such point moves at the local fluid velocity.

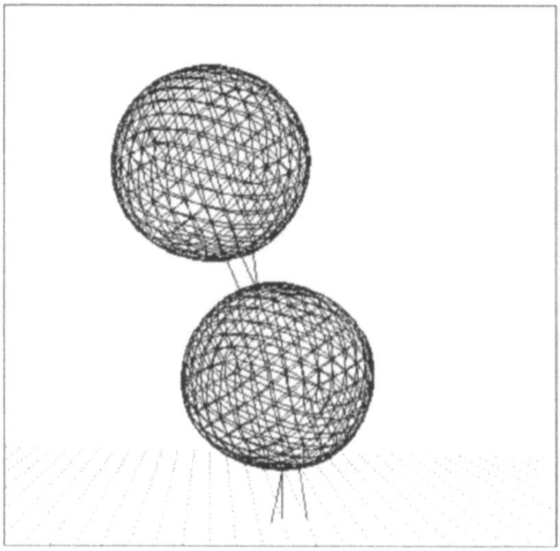

Figure 1. IB discretization of platelet surfaces.

For computations, the Navier-Stokes equations are solved using a Chorin-type projection method on a uniform rectangular computational grid. Each IB object is represented by a discrete set of points that generally do not coincide with points of the fluid grid. The points representing each platelet are the vertices of a triangulation of its surface (see Fig. 1). The vertices of each triangle are connected to one another by elastic springs with specified spring properties; these springs give the platelet its elastic characteristics. The δ-function in Eqs. (2, 4) is replaced by a desingularized approximation, with the result that for a given IB point, its force is spread to a $4 \times 4 \times 4$ portion of the fluid grid, and its velocity is obtained by interpolating from the same set of grid points. The solver used for the Navier-Stokes equations 'sees' the platelets and vessel walls *only* through their contributions to the force density $\mathbf{f}$, not as a change in domain geometry. It is the spreading of the IB forces to a small volume of the fluid around each IB point that prevents fluid leakage across the IB surface, even though only a discrete set of points is used to represent the surface.

Transport of the activating chemical is described by an advection diffusion equation:

$$c_t + \mathbf{u} \cdot \nabla c = D \Delta c + s\,, \qquad (5)$$

in which $c(\mathbf{x}, t)$ is the concentration of chemical at $(\mathbf{x}, t)$ and s is a source term accounting for secretion of the chemical by platelets as they are activated. The discrete solution of Eq. (5) is accomplished using LeVeque's algorithm [9] for the advection term and a multicolored, multigrid method for the implicit time-stepping of the diffusion term. The final features of the aggregation model and code are logical conditions to determine when a platelet becomes activated or adheres to a veseel wall, and when interplatelet bonds

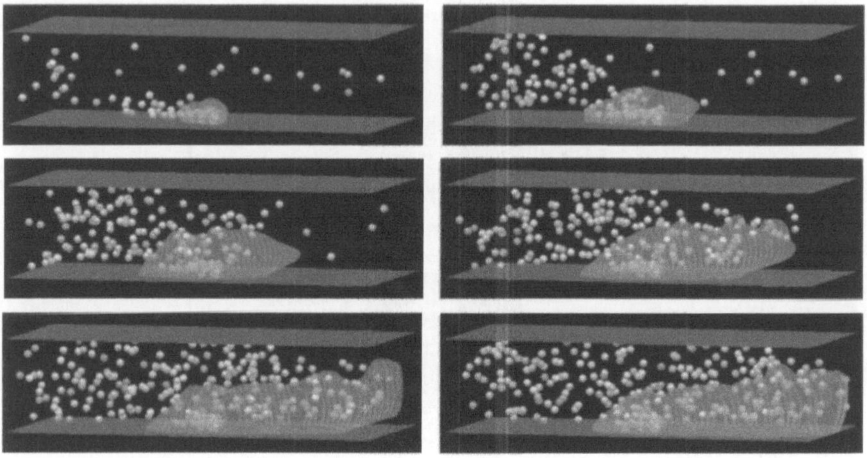

Figure 2. Development of platelet aggregate: Platelets (balls) enter the domain on the left and move to the right with the pressure-driven fluid. Platelets contacting the injury on the central part of the bottom wall adhere to the wall and secrete activating chemical. Other platelets activated by this chemical can form cohesive bonds with each other and the wall-adherent platelets. Bonds can break if subject to high stresses. The isosurface inside of which the activating chemical's concentration is sufficient to activate platelets is shown.

form and break. Figure 2 shows the development of a small aggregate. Many more details about the aggregation models and computations can be found in [16].

3. Coagulation Model

The most important parts of the coagulation reaction network are illustrated in Figure 3. Factors VII, IX, X, and prothrombin are inactive enzyme-precursors (zymogens), and Factors V and VIII are inactive cofactors. The corresponding active molecules are Factors VIIa, IXa, Xa, thrombin, Va, and VIIIa. Tissue factor (TF) is also an active cofactor present in the subendothelial matrix that is exposed to the passing blood when injury disrupts the normal endothelial cell (EC) lining of the vessel. The initiating enzyme is the TF:VIIa complex formed on the subendothelium. The coagulation response is propagated and amplified through a succession of enzyme reactions that culminate in the production of the enzyme thrombin. Thrombin plays numerous roles in coagulation as well as platelet aggregation. It cleaves the plasma protein fibrinogen into fibrin monomers. The monomers polymerize and cross-link (the latter is promoted by a thrombin-activated enzyme Factor XIIIa) to form a gel in and around the aggregating platelets. Thrombin is also a potent platelet-activating chemical. In addition, it promotes its own production by activating the cofactors VIIIa and Va. These molecules bind with enzymes IXa and Xa on the surface of an activated platelet to form the important and highly efficient enzyme complexes VIIIa:IXa (tenase) and Va:Xa (prothrombinase), which activate Factor X to

Xa and prothrombin to thrombin, respectively. Factor Xa free in plasma can also catalyze the activation of prothrombin to thrombin, but the efficiency of the platelet-bound Va:Xa complex is *5 orders of magnitude* greater than this solution-phase enzyme. Similar increases in efficiency are found for the platelet-bound VIIIa:IXa enzyme complex and the subendothelium-bound TF:VIIa enzyme complex compared to the corresponding solution-phase enzyme without its cofactor. Several chemical inhibitors are believed to play major roles in regulating the coagulation reactions: The plasma protein TFPI in complex with Factor Xa inhibits the action of the TF:VIIa complex. ATIII binds with Factors IXa, Xa, and thrombin to competitively inhibit them. APC inactivates the cofactors VIIIa and Va of the tenase and prothrombinase complexes, largely destroying the effectiveness of these enzymes. It is interesting to note that APC, which is the activated form of the plasma protein C, is upregulated by a complex on the healthy endothelium cells (not shown) adjacent to the injury. This complex involves thrombin, so thrombin upregulates an inhibitor of its own production, in addition to its positive feedback roles mentioned before.

One important feature of the coagulation system is stepwise enzyme activation in which an enzyme activated in one step is the catalyst for the next reaction. Coagulation also involves feed-forward and feed-back loops in which an enzyme produced in one step promotes or inhibits earlier or later reactions. Some of the coagulation reactions happen in the plasma, but several apparently critical reactions involve *membrane-bound* enzyme complexes. Despite the extreme importance of coagulation in both normal and pathological clotting, we have only general ideas about what the different features of the coagulation reaction network may be for; the multitude of steps for signal amplification, the loops for tight regulation, the involvement of membrane-bound complexes for spatial localization, and our understanding of the *dynamics* of the coagulation system remains poor. The complexity of the reaction system and the multitude of chemical and physical factors involved in coagulation make it very difficult to synthesize a coherent picture of coagulation dynamics using traditional laboratory approaches and make mathematical modeling a potentially very fruitful *complementary* tool for studying coagulation.

Some of the many open questions about the workings of the coagulation system are:

- If enzyme Factor VIIa were always available in the plasma, why is coagulation not happening all the time? If VIIa is not present in the plasma, how does an injury that exposes TF, but not VIIa, initiate the coagulation process?
- Why are there dual pathways to Factor Xa activation, a direct one by TF:VIIa and an indirect one involving TF:VIIa activation of IXa, and then the formation of the Xa-activating complex VIIIa:IXa on the activated platelet surface? The direct pathway is known to be kinetically sufficient, yet deficiency in either of the proteins necessary for the indirect pathway, Factors VIII and IX, results in the bleeding problems of hemophilia.
- How is the production of thrombin affected by changes in the density of surface binding sites for the enzymes? TF is the binding site for VIIa on the subendothelium, and there are specific binding sites on activated platelet surfaces for Va, VIIIa, IXa, and Xa.

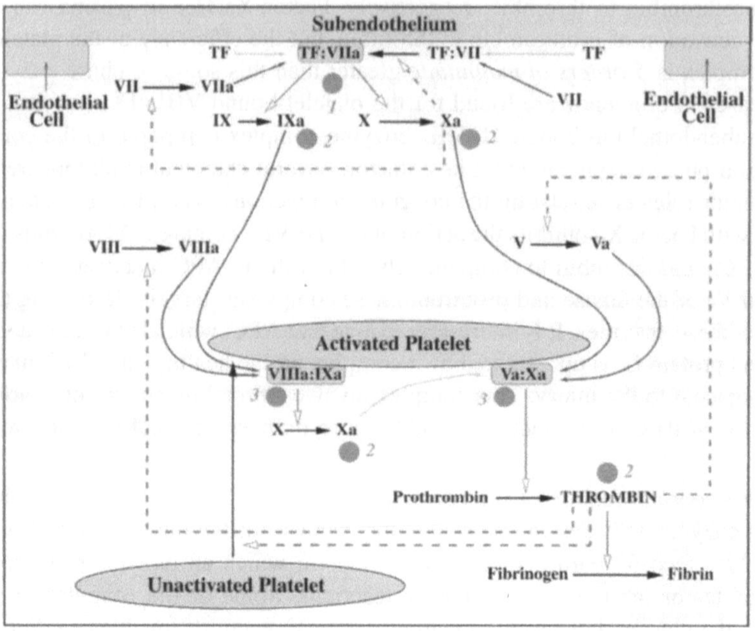

Figure 3. Coagulation reaction network: Surface-bound molecular complexes (TF:VIIa on the subendothelium, and VIIIa:IXa and Va:Xa on the surface of an activated platelet) are indicated by small rectangles. Thin dark or light solid arrows indicate movement of a chemical species through the plasma or on a surface, respectively. Thick lines with filled arrowheads indicate a molecular or cellular activation process. Solid lines with open arrowheads depict the action of the enzymes that catalyze the activation reactions. Dashed lines with open arrowheads indicate strong enzymatic feedback activity. Numbered filled circles indicate major inhibitors and their targets: 1 is TFPI, which complexes with Xa and inhibits TF:VIIa; 2 is ATIII, which inhibits IXa, Xa, and thrombin; 3 is APC, which inactivates the complexes VIIIa:IXa and Va:Xa. Not shown is the postulated role of adherent platelets in physically covering TF molecules on the subendothelium.

- How do the putative major chemical inhibitors affect the system? Are they enough to account for the observed behavior of the system? Might non-chemical inhibitors play a role?

To begin to address these and other questions we formulated a model (see [4, 7, 8] for details) of coagulation that takes into account plasma-phase and phospholipid-membrane-bound enzymes and zymogens, coagulation inhibitors, and activated and unactivated platelets. The model includes both plasma-phase and membrane-phase reactions. We assume that all reactions of interest occur in a thin 'boundary layer' shell above a small (e.g., $10\,\mu m \times 10\,\mu m$) area of subendothelium. We assume that all species (plasma-phase and membrane-phase) in this shell are well-mixed, so that all of the reactions can be described using (59) ordinary differential equations in time.

Initially the reaction shell is filled only with fluid and chemicals and has a thickness determined by the relative influence of zymogen transport by radial diffusion and axial convection. As time progresses, platelets enter the region and adhere to the subendothelium or aggregate on already adherent platelets, and this increases the total thickness and volume of the shell. Flow and diffusion combine to bring proteins into the shell from upstream and to carry them downstream away from the damaged wall. These effects are incorporated into the model in a simple way by assuming that each *plasma-phase* protein is delivered to or removed from the shell at a rate proportional to the difference between its concentration within the shell and its concentration far outside of the shell. Unactivated platelets are also transported by flow and diffusion, and, in addition, can undergo an activation process and can adhere to the damaged vessel wall and to other adherent platelets. The chemical transport rate coefficient depends on flow and diffusion parameters, the size of the injury, and the number of platelets deposited on it. The system of equations is obtained by using the Law of Mass Action for binding/unbinding and for the enzymatic reactions and by treating transport to and from the reaction shell as just described. An additional important feature of the model is that it allows for a decrease in the activity of subendothelial reactions (TF:VIIa activation of IX and X) as the subendothelium becomes covered with platelets.

Some results from the model are shown in Figure 4. We see in frame (a) that a small change in TF density can induce a 3 to 4 order of magnitude increase in thrombin production. Thrombin concentrations for TF densities just above threshold are sufficient to cause substantial platelet activation. If the TF density is above threshold, the density of specific binding sites on the platelet surfaces serves as a second 'switch'. Frames (b) and (c) show the very different time course of thrombin concentration for TF above and below threshold. All of our calculations were done with 1% of plasma VII initially activated as suggested by experimental measurements. The switch-like response of the system to TF allows VIIa to circulate at levels too low to trigger significant thrombin production in the face of constant low-level trauma to the vasculature, but at levels that lead to substantial thrombin production upon more serious injury.

Frames (d) and (e) are simulations with deficiencies in Factor VIII or IX that cause the two most common forms of hemophilia. Frame (e) shows the results when only what are believed to be the main chemical inhibitors of coagulation are included in the model. At normal and even significantly higher than normal concentrations of the inhibitors, 100-fold reductions in the concentration of Factor VIII or IX caused virtually no change in thrombin levels at 600 s in contrast to the severe effects these deficiencies have clinically. The reason is that the concentration of the complex TFPI:Xa, which is the inhibitor of the TF:VIIa complex on the subendothelium, is far too low to effectively inhibit the activation of X to Xa by TF:VIIa. Since this production continues, the alternate pathway that requires the tenase complex formed from Factors VIIIa and IXa is not needed. This led us to make the suggestion that as platelets deposit on the subendothelium, they cover the TF:VIIa molecules, physically inhibiting them. Frame (d) shows results when we assume that TF:VIIa activity is reduced in proportion to the fraction of the subendothelium covered by platelets. We see that a 100-fold reduction in Factor VIII or IX causes a substantial reduction in thrombin production, and that the reduction with Factor VIII deficiency

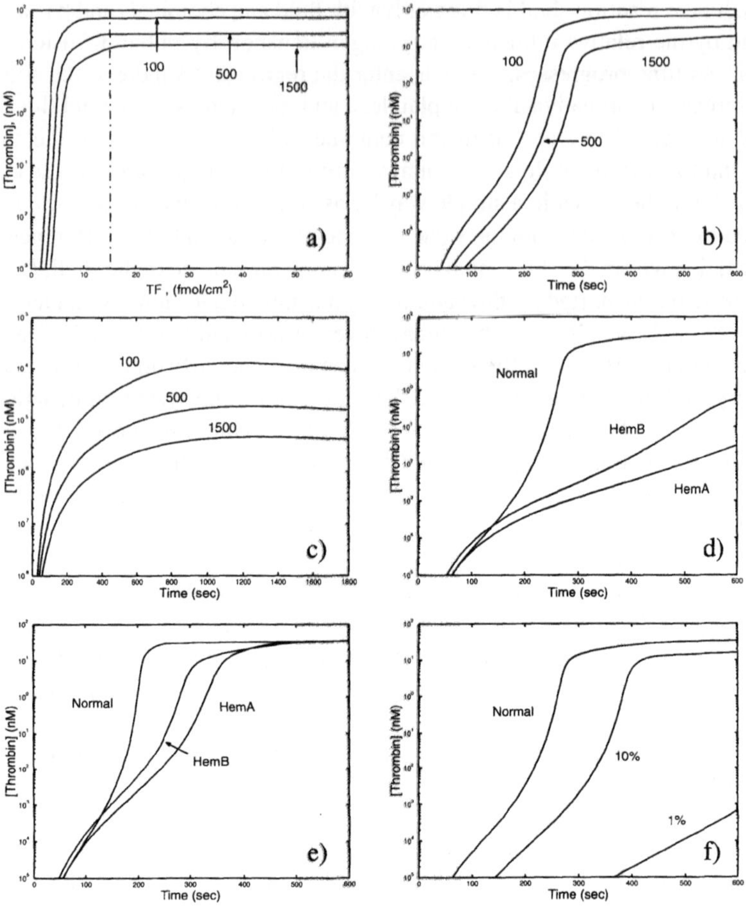

Figure 4. (a) Dependence of thrombin concentration at 600 s on the density of exposed TF molecules at various shear rates. (b) Thrombin concentration vs. time at various shear rates and a TF density of 15 fmol/cm^2. (c) Thrombin concentration vs. time at various shear rates and a TF density of 1 fmol/cm^2. (d) Hemophilia: Time course of thrombin concentration for different levels of Factor VIII or IX at shear rate 500 /sec and TF density 15 fmol/cm^2. Normal denotes normal levels of VIII and IX. HemA denotes 1% of normal VIII and normal IX. HemB denotes 1% of normal IX and normal VIII. (e) Same as (d), but not allowing platelet deposition to block TF:VIIa activity on the subendothelium. (f) Thrombin concentration vs. time for normal platelet count (250,000 plt/mm^3), 10% of normal, and 1% of normal at shear rate 500 /sec and TF density of 15 fmol/cm^2.

is more pronounced, as observed clinically. So it may well be that the role of the indirect pathway to Factor Xa activation is to continue Factor Xa activation on the surfaces of platelets after platelet deposition on the subendothelium prevents further TF:VIIa activity.

Reductions in platelet count (the number density of platelets in the plasma) by a factor of 10 cause little or no clinical symptoms, while more severe reductions lead to bleeding problems. In frame (f) we see that the model predicts only a modest reduction in thrombin levels when the platelet count is 10% of normal, while there is a substantial dropoff when the platelet count is 1% of normal.

4. Conclusions

We have presented two examples of modeling of blood clotting. The first emphasizes fluid and cell interactions, while the second simplifies the mechanics and emphasizes the biochemistry. Both types of models can give important insights about the clotting process. With the first we are calculating the forces on wall-adherent platelets and small aggregates and the fluid-dynamic and chemical environment near the aggregate. With the second we have given possible answers to the coagulation questions posed above. A future goal is to combine the 3D platelet aggregation simulations with a more comprehensive treatment of biochemistry and chemical transport.

Acknowledgments

This work was supported, in part, by NSF grants DMS-9307643 and DMS-9805518 and by a John Simon Guggenheim Foundation Fellowship to A. Fogelson. An allocation of computer time from the Center for High Performance Computing at the University of Utah is gratefully acknowledged. CHPC's SGI Origin 2000 system is funded in part by the SGI Supercomputing Visualization Center Grant.

References

[1] R.W. Colman, J.J. Cook and S. Niewiarowski (1994) Mechanisms of platelet aggregation. In: *Hemostasis and thrombosis: Basic principles and clinical practice* (Colman, Hirsch, Marder, and Salzman, eds.) J.B. Lippincott Company, Philadelphia, 508–523

[2] A.L. Fogelson (1992) Continuum models of platelet aggregation: Formulation and mechanical properties. *SIAM J Appl Math* **52**, 1089

[3] A.L. Fogelson (1993) Continuum models of platelet aggregation: Mechanical properties and chemically induced phase transitions. *Contemp Math* **141**, 279–294

[4] A.L. Fogelson and A.L. Kuharsky (1998) Membrane binding site density can modulate activation thresholds in enzyme systems. *J Theor Biol* **193**, 1–18

[5] H.C. Hemker and H. Kessels (1991) Feedback mechanisms in coagulation. *Haemostasis* **21**, 189–196

[6] J. Jesty and Y. Nemerson (1995) The pathways of blood coagulation. In: *Williams Hematology* (E. Beutler, M. Lichtman and B. Coller, eds.) McGraw-Hill, New York, 5th ed., 1227–1238

[7] A. Kuharsky (1998) Mathematical modeling of blood coagulation. *PhD thesis* University of Utah

[8] A. Kuharsky and A. Fogelson (2001) Surface-mediated control of blood coagulation: The role of binding site densities and platelet deposition. *Biophysical Journal* **80**, 1050–1074

[9] R.J. LeVeque (1996) High-resolution conservative algorithms for advection in incompressible flow. *SIAM J Numer Anal* 33, 627–665. Algorithm available from http://www.amath.washington.edu/~claw

Polymer and Cell Dynamics – Multiscale Modeling
and Numerical Simulations, 155–167
ed. by W. Alt, M. Chaplain, M. Griebel und J. Lenz

Space- and Time-resolved Simulations of Processes in Biofilm Systems on a Microscale

Hans-Joachim Bungartz[1], Martin Kühn[3], Miriam Mehl[2] and Stefan Würtz[4]

1) Institut für parallele und verteilte Systeme, Universität Stuttgart, Universitätsstraße 138, D-70569 Stuttgart, Germany
2) Institut für Informatik, Technische Universität München, Boltzmannstr. 3, D-85748 Garching, Germany
3) Lehrstuhl für Wassergüte und Abfallwirtschaft, Technische Universität München, Boltzmannstr. 3, D-85748 Garching, Germany
4) Department of Civil and Environmental Engeneering, University of California, Davis, One Shields Avenue, Davis CA 95616, USA

Summary. New experimental and analytical techniques such as confocal laser scanning microscopy (CSLM) or the use of RNA-targeted probes have provided insight into the morphology, architecture, and function of biofilm cultures. The different observations made there suggest that more attention has to be paid to a detailed study of microscale processes such as flow and transport phenomena as well as to the development of the biofilm's primary components, i.e., microbial cells and extracellular polymeric substances (EPS). For that, numerical simulations are a promising approach. However, due to the large variety of different effects and influence factors, strong multiscale characteristics with respect to both time and space, and the need for an explicit high spatial resolution in order to capture the occurring changes of the underlying geometry because of biomass growth 3D simulations have hardly been tackled so far. Actually, most existing simulation tools for biofilm systems are based on strongly simplified model assumptions that have turned out to be not valid in general.

In this work, we report on the first steps towards microscale simulations of flow, transport, reactive, and growth processes in 3D biofilm geometries obtained from CLSM images of a small and defined monoculture biofilm setup. The basic framework is the finite volume CFD solver *Nast++*, to which transport equations (convection-diffusion in the fluid phase, diffusion-reaction in the biofilm) and the cellular automaton *CAsim* for capturing biomass growth are coupled. Some numerical results of realized simulations as well as strategies for an increased numerical efficiency are presented.

1. Introduction

Biofilms are aggregations of microorganisms that adhere to some fixed surface, typically in an aquatic or submerged environment (on pebbles in a lake, for example). Nowadays, their metabolism is used for biological wastewater treatment in special biofilm reactors.

Depending on the substances to be decomposed, certain microbial species (typically a larger number) have to be settled and stimulated to start their metabolic activity. Thus, realistic biofilms are complex biological systems, whose efficacy and efficiency depend on several factors that are sometimes subdivided into *physical* or *physicochemical* factors on the one hand and *biological* ones on the other hand. Concerning the first class, the flow conditions are crucial, since they influence the transport of nutrients and other necessary substances *to* the biofilm as well as the removal of the metabolic products *from* the biofilm or the whole reactor, respectively. Furthermore, the flow field seems to be important for the growth and the mechanical stability of the biofilm. However, in addition to these aspects, the reactor's function also may be influenced by some competition of species caused by the occurrence of grazing protozoa and metazoa, i.e., by microorganisms that rapidly multiply without contributing to the function of the biofilm reactor and hence actually disturb its operation seriously.

Whereas traditionally scaling, operation, and optimization strategies were primarily based upon heuristic conjectures and approaches, the last decade has seen the development and use of new methods that allow us to study the biofilm *in situ* on a microscale and thus to obtain a more fundamental understanding of biofilm systems. As an example, the application and combination of confocal laser scanning microscopy (CLSM) and of fluorescently labeled RNA-targeted probes led to a complete change of the predominant common idea of biofilms' morphology, architecture, and function. Channels and a highly porous structure became evident where homogeneity had been supposed, species that had been seen as dominant turned out to be rather unimportant, and convective transport entered the scene where so far only diffusive processes had been taken into account [13].

These results also gave new impetus to the (mathematical) modeling and numerical simulations of biofilm systems. Especially, they show that simulations without a spatial resolution of the underlying scenario can hardly describe or even predict the processes within biofilm systems in detail. Consequently, although the question of how and to what extent an increased microscale understanding of biofilms can result in more efficient wastewater plants has not yet been answered, the microscopic simulation of flow, transport, and reactive phenomena [7] has become a promising and challenging sector of biofilm research.

The main reason that the available advanced analytical or computational methods often cannot be used directly for improved operation of a biofilm reactor is the dilemma of different scales. On the one hand, both a microscope and a supercomputer allow only small regions of concern – the first due to practical reasons, the latter due to computing time limitations. On the other hand, the complete reactor is much bigger than such sections and is far beyond the scope of a high-resolution capture. Concerning time, there is a similar problem: processes on the microscale such as fluid flow or convective transport are fast ones and go off within milliseconds, whereas the relevant processes for the operating engineer such as biomass growth or decreasing metabolic efficacy may last hours or even days (which makes it impossible to use sampling rates that detect even the fastest processes). Nevertheless, with respect to both time and space, there is a strong inter-scale dependence: coarse-scale phenomena visible there can be explained or predicted only with the help of fine-grain phenomena that are not visible on the coarse level. Due to the typical heterogeneity of biofilm systems and due to their changes in time, the classcial remedy – some averaging or homogenization approach – can be of only limited use in the biofilm context.

In this paper, we present a combined experimental and computational approach to microscale simulations of some important aspects of biofilm systems: fluid flow, transport processes within both the fluid phase and the biofilm, reactive processes within the film, and biomass growth. For that, a CLSM-based method for semi-automated image aquisition and processing, a 3D CFD solver and a cellular automaton have been coupled. In Section 2, the target system, a small flow channel with a monoculture laboratory biofilm, is described. Section 3 is dedicated to a short summary of the underlying models and computational tools used, followed by some simulation results. Finally, we give an outlook over ongoing numerical enhancements.

2. The Target System: A Microscopic Biofilm Domain

Because biofilm morphology has turned out to be highly heterogeneous [20] and dependent on flow and nutrient/oxygen supply [11, 19], it is generally assumed that the microscale characteristics have an important effect on the overall behavior of a biofilm reactor (e.g., on the evenness of biofilm growth and, thus, on the efficiency of pollutant reduction). Hence, simulation approaches on the reactor scale that neglect details of the biofilm geometry and the ongoing microscopic changes and processes have to be completed by the inclusion of phenomena at the level of single microbes. Since the whole spectrum of orders of magnitude cannot be captured within a single computation, we start with studying the processes on the microscale (i.e., the level of continuum mechanics), which is the target scale of this paper. Moreover, the composition of realistic reactor biofilms is far too complex and too fluctuant for both a proper analysis by CLSM or related techniques and numerical simulations (at least for the time being).

Therefore, instead of a realistic wastewater plant's reactor, we examine a biofilm cultivated in a laboratory reactor (a small glass flow cell) under defined conditions to gain some new insight in the basic processes and their interaction. That is, the cell is supplied with a defined nutrient solution from a stirred and aerated reservoir. Constant flow conditions are ensured by a pressure-independent displacement pump [10]. Concerning the nutrient, we concentrate on monocultures (e.g., *Sphingomonas* sp. LB126 or *E. coli* DH5α). The part of the flow cell to be captured by CLSM is chosen following heuristic strategies from [9] in order to ensure that we consider a section that is representative under the given flow and nutrient conditions.

CLSM produces a stack of 2D images focusing on one horizontal plane through the biofilm each. After an image-processing step for the enhancement of the data, this image stack can be directly used to create a 3D fluid-obstacle geometry by applying a suitable threshold of brightness in order to map grayscale onto black-and-white images and by stacking the discrete slices according to their vertical distance (see Fig. 1 or [4]). Note that the term obstacle simply denotes the targeted and labeled type of cells captured by microscopy.

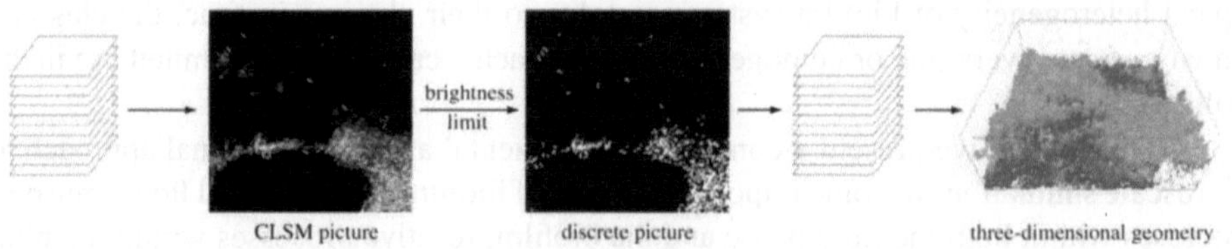

Figure 1. Construction of a 3D biofilm geometry from 2D grayscale CLSM images (see color plate 5)

Besides pure geometric snapshot information, experiments can provide the evolution of the biofilm geometry during a certain period of time as well as selected local velocities *via* particle tracking (see Fig. 2 and [12]).

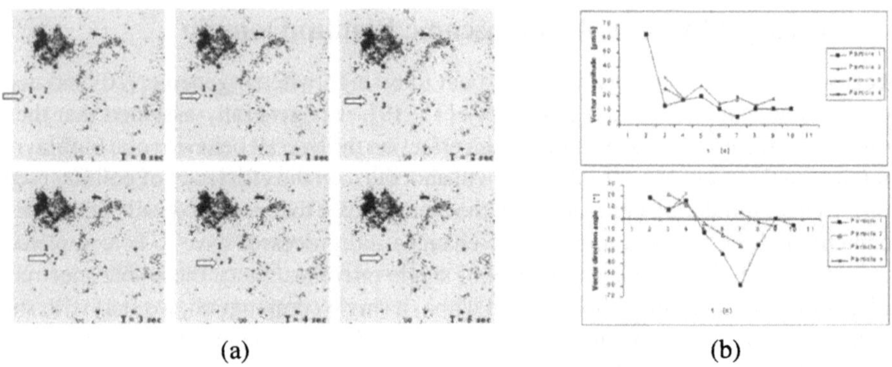

Figure 2. Time-resolved measurements: sequence of snapshots tracking particles in a horizontal slice through a monoculture biofilm (a) and experimentally captured local velocities of four tracked particles (b) (see color plate 5)

In this paper, we restrict our microscopic simulations to fluid flow, solute transport, and biofilm growth, i.e., to some of the crucial physicochemical processes. However, it is obvious that this can reflect only a part of the story. Other physical effects such as electrostatic phenomena or, in particular, processes such as competition among different species or microbes to be regarded from a more biological point of view are of a certain importance, too, and will have to be taken into account for an increased level of accuracy.

3. Simulation of Flow, Transport, and Growth Based on a Hybrid Continuous-Discrete Model

3.1. Continuum Mechanics Equations for Fluid Flow and Solute Transport

For the microscopic simulation of fluid flow and nutrient transport in biofilm systems, we use the standard model directly based on the conservation laws of continuum mechanics, i.e., first, the 3D incompressible Navier-Stokes equations

$$\frac{\partial \vec{u}}{\partial t} + (\vec{u} \cdot \nabla)\, \vec{u} = \frac{1}{Re} \Delta \vec{u} - \nabla p + \vec{g}\,, \tag{1}$$

and

$$0 = \operatorname{div} \vec{u} \tag{2}$$

for the flow velocity $\vec{u}$ and the pressure p. Here, $\vec{g}$ denotes external forces such as gravity, and Re is the Reynolds number. Note that in biofilm systems like those studied here, Re is typically very small (between 0.1 and 5.0).

The concentration C of some nutrient or pollutant in the bulk fluid and in the biomass itself is modeled by a convection-diffusion equation in the fluid domain Ω_f, by a diffusion-reaction equation in the biofilm domain Ω_b, and by Fick's Law describing sorption at the interface Γ of Ω_f and Ω_b (similar to [16]):

$$\frac{\partial C}{\partial t} + (\vec{u} \cdot \nabla)\, C = D_f \Delta C \qquad \text{in } \Omega_f\,, \tag{3}$$

$$\frac{\partial C}{\partial t} = D_b \Delta C + r(C) \qquad \text{in } \Omega_b\,, \tag{4}$$

$$D_f \left.\frac{\partial C}{\partial \vec{n}}\right|_{\Omega_f} = D_b \left.\frac{\partial C}{\partial \vec{n}}\right|_{\Omega_b} \qquad \text{at } \Gamma\,. \tag{5}$$

In Eq. (5), $\left.\frac{\partial C}{\partial \vec{n}}\right|_{\Omega_f}$ is the first normal derivative of the concentration with respect to the interface Γ at the fluid side of Γ, and $\left.\frac{\partial C}{\partial \vec{n}}\right|_{\Omega_b}$ is analogously defined as the first normal derivative at the biofilm side of Γ. D_f and D_b are the diffusion coefficients in the fluid and in the biofilm, respectively, and r denotes the reaction term within the biofilm. As a first and commonly used approximation of the ongoing reactions within the biofilm, we use a monod kinetic reaction term [8, 14]

$$r(c) = \frac{\alpha \cdot c}{\beta + c}, \tag{6}$$

with some – up to present heuristic – reaction coefficients α and β depending on the biofilm species and the particular solute considered. Monod kinetic reaction terms can easily be generalized to the situation of several solutes [14]. Note that Eq. (5) causes the normal derivative of C to be discontinuous at the interface Γ if $D_f \neq D_b$.

For the numerical solution of these model equations, we use the object-oriented simulation software *Nast++* [3, 17]. *Nast++* is based upon a finite volume discretization on a staggered grid. For the coupling of Eqs. (1) and (2), we apply Chorin's projection method, which guarantees mass conservation in each time step. In terms of time discretization, we use an explicit finite difference scheme for both the Navier-Stokes equations (1) and (2) and the transport equations (3)–(5). Thus, the Poisson equation derived from Eq. (2) according to Chorin's projection method (see [6]), is the only equation that has to be solved in each time step of the numerical simulation and for which, in consequence, we have to provide efficient numerical solvers.

Taking into account that biofilm geometries can be very complex and heterogeneous, we have to perform our simulations on highly resolved grids implying a large number of unknowns. For this purpose, *Nast++* has been parallelized for clusters of workstations and high performance computers, such as the vector parallel computer VPP700, by decomposing the underlying domain (Fig. 4). In addition to parallelization, we need an efficient multigrid (MG) solver and adaptive grid refinement to obtain acceptable computing times. For the time being, we use a simple geometric multigrid approach based on Galerkin operators for cell-centered grids, constant interpolation, and linear restriction (see Fig. 3). In pure fluid cells, operators are the direct analogon of the fine-grid operators. Thus, additional coarse-grid operators have to be stored in coarse-grid cells consisting of obstacles as well as fluid fine-grid cells, exclusively. Figure 4 shows some first results for this multigrid method, adaptive refinement, and parallelization.

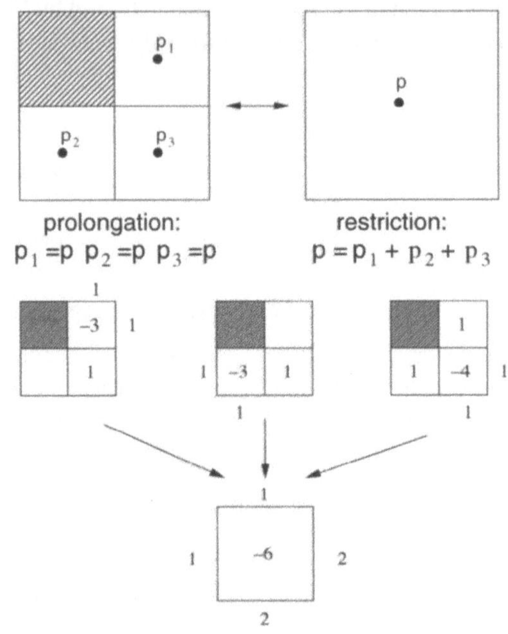

Figure 3. Generation of the Galerkin operator for mixed fluid-obstacle coarse-grid cells

Figure 5 shows the result of flow and transport simulations in a monoculture biofilm geometry obtained by the CLSM-based process described above.

3.2. Discrete Model for Biofilm Growth

Compared with fluid flow and transport phenomena, biofilm growth – with respect to the size of the grid cells – is a much slower process. In our restricted model world, the evolution of the biofilm's morphology is mainly driven by the concentration of the substances taken into account, which means by the increase of biomass and by the expansion behavior of the latter. Since this behavior can be described by a relatively small set of rules, it seems to be sufficient to model biofilm growth on a discrete basis, for example, with

nr. of proc.	computing time	parallel efficiency
1	7.50 h	100 %
2	3.75 h	100 %
4	2.25 h	80 %

(a)

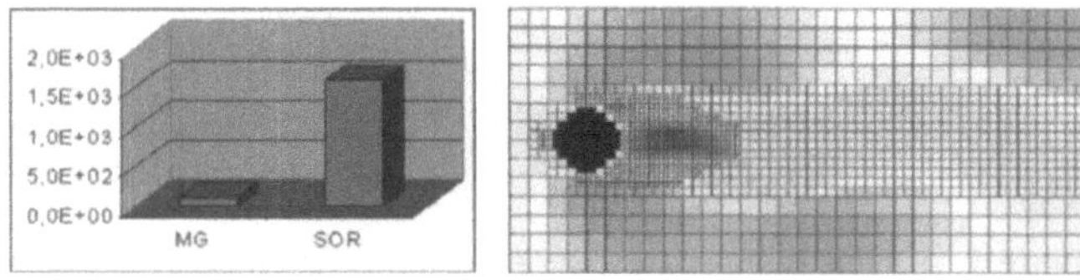

(b) (c)

Figure 4. Increasing numerical efficiency by (a) parallelization of the simulation software (biofilm simulation, computing times for one, two, and four processors of an SGI Origin 200); (b) multigrid techniques for the fast solution of the pressure Poisson equation (biofilm simulation, $64 \times 64 \times 32$ grid cells); (c) adaptive grid refinement (flow round a circular obstacle) (For colored picture of (c) see color plate 6.)

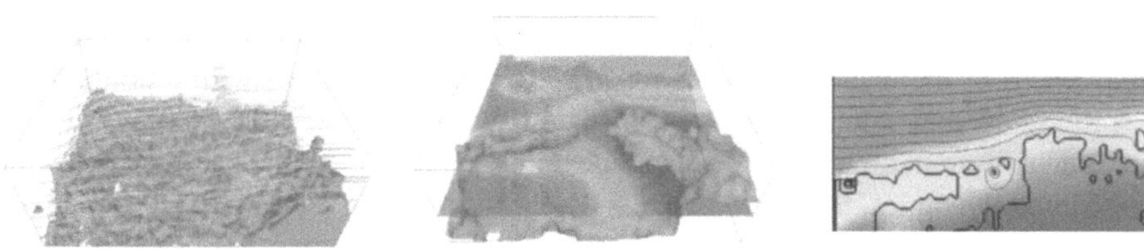

(a) (b) (c)

Figure 5. (a) Streamlines, (b) nutrient concentration in the fluid phase, and (c) nutrient concentration in a vertical cut through an experimentally defined section ($512 \times 512 \times 50\,\mu$m) of a four-day old *E. coli* DH5α monoculture biofilm (see color plate 6).

cellular automata. A cellular automaton consists of a finite system of possible states of the computational grid cells on the one hand and of (deterministic or stochastic) local rules for state transitions on the other.

Since the mechanisms of biofilm growth are still not completely known and are too complex to be captured by numerical simulation, any useful software must be designed as a flexible tool that allows the simple modification or addition of both states and rules, in order to be able to adapt both models and codes to any progress in the understanding of biofilm growth. The framework *CAsim* developed for the rapid implementation of arbitrary cellular automata fulfils these requirements. Within *CAsim*, states and rules can be easily defined using a script language [13]. This flexibility enables us to test and compare different sets of rules (i.e., different growth models) with varying sets of parameters.

For our present simulations, we use a rather simple automaton with only three states (fluid cells, stable biofilm cells, and biofilm cells that are ready to divide) to show the general applicability of the automaton framework to simulate biofilm growth. The ability to divide is assumed to be determined by the time integral over the nutrient concentration C taken as an indicator for biomass growth. Finally, the direction of cell division (which fluid neighbor is to be occupied, or which biofilm neighbor is to be invaded?) is the result of some random process based on a probability distribution depending on the result of a search for the closest free fluid cell. For a simple example, Figure 6 shows how such a

Figure 6. *CAsim:* cellular automaton iterations for one growth step in the 2D biofilm model (dark: cells ready for dividing; see color plate 7)

cellular automaton works. Due to the high flexibility of *CAsim*, those preliminary rules and states can easily be enhanced and adapted to more realistic models.

3.3. Coupling of Flow, Transport, and Growth Simulations

If we want to bring together the two parts of our model – continuous flow and transport and discrete growth – into one simulation context, we are once more confronted with a problem of scales [15]. While biofilm growth, as already mentioned, is a very slow process in terms of the resolution of the underlying grid and happens within hours or even days, fluid flow and solute concentration change within milliseconds. In many situations where such different time scales occur, some homogenization in time has to be used (think of molecular dynamics, for example). In our case, however, it seems that we can do without that and use a simpler approach as suggested in [15]: Since fluid flow and solute concentration tend to equilibrate very fast (compared to the time step of biofilm growth) after each biofilm growth step, we neglect this adaptation phase of flow and concentration and compute only the steady-state solution for the Navier-Stokes and transport equations. This quasi-stationary approach results in an equilibrium solute concentration for the current biofilm geometry. Now, the cellular automaton is started with cells above a certain predefined concentration threshold labeled as 'dividing cells'. The automaton generates a new geometry, for which the whole cycle is started again (see Fig. 7). Thus, only biofilm growth is simulated time dependently. Figure 7 shows the result of a 2D cut through a single species biofilm.

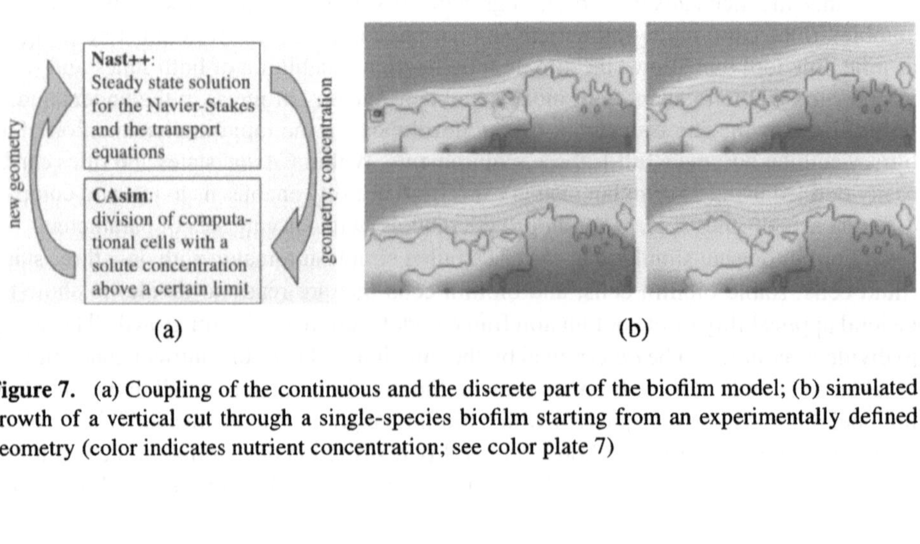

Figure 7. (a) Coupling of the continuous and the discrete part of the biofilm model; (b) simulated growth of a vertical cut through a single-species biofilm starting from an experimentally defined geometry (color indicates nutrient concentration; see color plate 7)

4. Outlook

4.1. Enhancement of the Underlying Model

In the previous section, we described the simulation of flow, solute transport, and growth in a microscopic monoculture biofilm domain. The underlying model is based on the conservation laws from continuum mechanics for the simulation of fluid flow and solute transport on the one hand and on some more-or-less heuristic discrete rules of biomass expansion for capturing biofilm growth on the other. This simple hybrid model should be seen as the starting point for a more sophisticated biofilm modeling, including further phenomena such as reactions in multiculture biofilms, mechanical stress and detachment due to fluid-structure interaction, and cell-to-cell interactions between different species of microbes. The latter can be integrated into the discrete part of the model by including suitable additional rules – which have to be determined in an outstanding modeling process – into the cellular automaton. The necessary additions on the continuous side that come along with more species or mechanical considerations are more challenging from the numerical point of view, of course, since the coupling between the transport equations for different species has to be taken into account or even a new software package for mechanical equations has to be integrated in the overall framework consisting of the Navier-Stokes and transport solver and the cellular automaton for the time being.

Another important aspect is the presence of extracellular polymeric substances (EPS). Depending on the biofilm's age and other influence factors, the percentage of the dry biofilm mass accounting for the EPS strongly varies and may be up to 90% (see Fig. 8 and [12]). Apart from the evident influence of the EPS on the system's morphology and on fluid flow and transport in it, phenomena within the EPS have to be understood, modeled, and taken into account in the simulation tools. Experimentally, the EPS can be detected by fluorescently labeled lectines, at least for certain species of microbes (e.g., *Sphingomonas*, [12]). However, the knowledge concerning composition and function of EPS is not yet sufficient to allow for realistic simulations. At present, we consider EPS simply as a second kind of obstacle (besides the microbial cells), but it might be better to treat them as a second fluid phase or as a viscoelastic porous medium themselves. In the latter case, one would have to supplement the existing programs with a solver for flow and transport equations in porous media (possibly within the framework of *Nast++*) and a solver for structural mechanic equations to simulate the fluid-structure interactions resulting from the visoelasticity of the EPS matrix.

To make microscale simulations useful for the examination and/or optimization of whole biofilm reactors (and, thus, of wastewater plants), we additionally have to close the gap between micro- and macroscale simulations of flow and transport. This means that, first, we have to couple the convection-diffusion processes (i.e., flow and transport). For this purpose, we have to derive appropriate nonlinear boundary conditions (at the surface of the substratum) for fluid flow and solute transport on the macroscale from several microscale simulations under varying flow and solute supply conditions (see Fig. 9). Second, the biofilm growth simulated on the microscale has to be integrated into the macroscale simulations. This will be one focus of our future work.

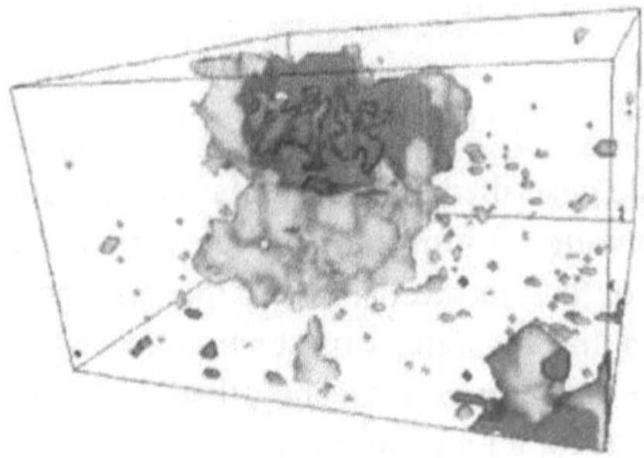

Figure 8. Experimentally defined cluster (30 × 30 μm) of cells of *Sphingomonas* sp. LB126 and its embedded *Sphingomonas* exopolysaccharides (sphingans) (For explanation see color plate 8).

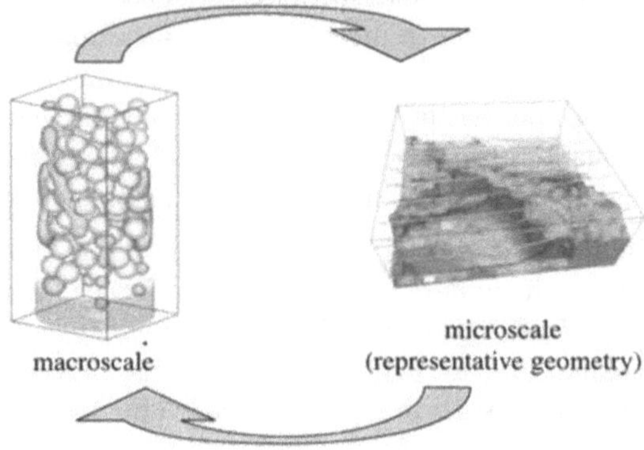

Figure 9. Schematic illustration of the coupling between micro- and macroscale computations for the convection-diffusion problem (see color plate 8).

4.2. Enhancement of the Numerical Efficiency

As mentioned above, the complexity of biofilm geometry requires computations on highly resolved grids and, thus, efficient numerical methods. To fulfil the special demands of biofilm simulation – areas with high geometrical heterogeneity and very homogeneous areas (e.g., in the bulk fluid or deep layers of the biofilm) – adaptivity is an important issue, in particular. In the case of adaptive Cartesian grids, data reasonably have to be organized into trees (i.e., quadtrees in 2D, octrees in 3D). In contrast to arrays, the access to directly neighboring grid points is more complicated and expensive in tree structures. Thus, the discretization of *Nast++* will be reorganized to obtain a cell-based discretization (comparable to FE methods) instead of the current point-based discretization (see Fig. 10).

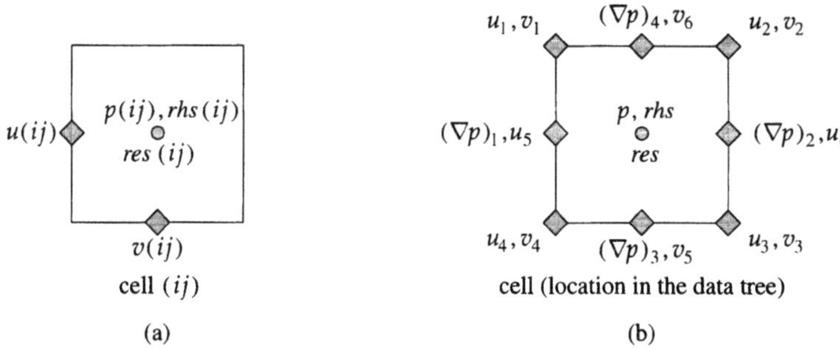

Figure 10. Discretization of the velocity, pressure, and residual data: point-based discretization (a) and cell-based discretization (b)

If we use the old, point-based discretization, the Poisson equation's residual $res = \delta p - rhs$ is computed according to the following algorithm:

```
for all (i,j)
```
$res(i,j) := (p(i+1,j) + p(i-1,j) + p(i,j+1) + p(i,j-1) - 4p(i,j))/h^2$
$-rhs(i,j)$

Using a cell-based discretization, the same residual is computed according to

```
for all cells
```
$(\nabla p)_1 := 0, (\nabla p)_2 := 0, (\nabla p)_3 := 0, (\nabla p)_4 := 0$
```
for all cells
```
$(\nabla p)_1 + = p, (\nabla p)_2 - = p, (\nabla p)_3 + = p, (\nabla p)_4 - = p$
```
for all cells
```
$res := ((\nabla p)_2 - (\nabla p)_1 + (\nabla p)_4 - (\nabla p)_3)/h^2 - rhs$

and, thus, without accessing data outside of the current cell.

The tree-based data storage implies an inherent hierarchy of cells. This can be exploited in a straightforward way for data hierarchization and, in a further step, for multigrid methods encompassing not only the Poisson equation for pressure but also the whole time-dependent Navier-Stokes equations. For this purpose, waveform relaxation [18] can be used, which is in some sense similar to full multigrid (FMG, [2]), but can be used for time-dependent problems instead of stationary equations only. Due to the nonlinearity of the time-dependent Navier-Stokes equations with regard to the flow velocities, we have to use a full approximation scheme for the flow velocities. If our data hold an inherent hierarchy, as in the case of octree-based storage, this can be performed easily and cheaply regarding storage requirements and computing time with the help of hierarchization and dehierarchization of data, as we do not need any further data structures for coarse-level data besides the fine-level tree [5].

Besides those algorithmic aspects, the applied multigrid method has to be enhanced in order to obtain a truely robust and efficient solver. First, the "mergence" of separate domains on coarse levels (see Fig. 11) has to be prohibited – e.g., by methods related to algebraic multigrid, for which the selection of coarse-grid points and interpolation is

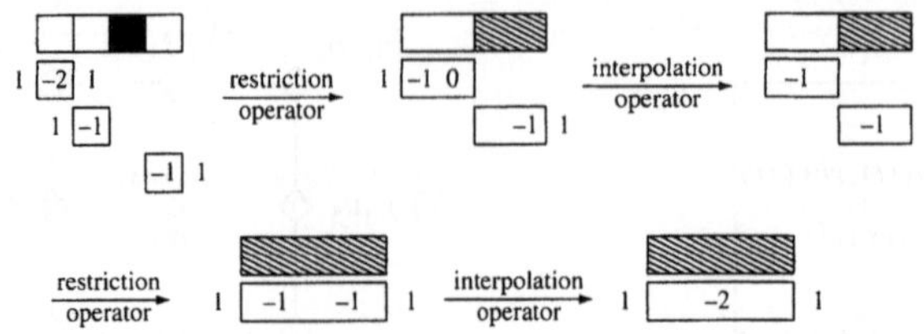

Figure 11. "Mergence" of separate domains for the one-dimensional Poisson equation discretized with the three-point stencil in the course of level coarsening using cell-centered grids, Galerkin operators, first-order (constant) interpolation, and second-order (linear) restriction

highly related to the strength of coupling between fine-grid points. Second, we have to use higher-order interpolation and restriction, in particular in the case of waveform relaxation [2, 18], to embrace the second-order discretization of the differential equations on the fine level.

Acknowledgments

This work was supported by the Research Centre for Fundamental Studies of Aerobic Biological Wastewater Treatment (SFB 411), München, Germany.

References

[1] Bishop, P. L. (1997) Biofilm structure and kinetics. *Water Sci. Tech.* **36**(1), 287–194

[2] Brandt, A. (1977) Multi-level adaptive solutions to boundary value problems. *Math. Comput.* **31**, 333–390

[3] Brück, B. (1998) Nast++: An objectoriented framework for modular flow simulations (german). *Diploma thesis* Institut für Informatik, TU München: `http://www5.in.tum.de/publikat/dipl/brueck.ps.gz`

[4] Bungartz, H.-J., Kühn, M., Mehl, M., Hausner, M., Würtz, S. (2000) Experiments and numerical simulations on a microscale. *Wat. Science and Tech.* **41**(4-5), 331–338

[5] Griebel, M. (1989) Zur Lösung von Finite-Differenzen- und Finite-Element-Gleichungen mittels der Hierarchischen-Transformations-Mehrgitter-Methode. *Doctoral thesis* TU München

[6] Griebel, M., Dornseifer, T., Neunhoeffer, T. (1997) *Numerical Simulation in Fluid Dynamics – a Practical Introduction.* SIAM, Philadelphia

[7] Hermanowicz, S. W. (1999) Two-dimensional simulations of biofilm development: Effects of external environmental conditions. *Water Sci. Tech.* **39**(7), 107–114

[8] Horn, H., Hempel, D. C. (1997) Substrate utilization and mass transfer in an autotrophic biofilm system: Experimental results and simulation. *Biotech. Bioeng.* **53**(4), 363–371

[9] Korber, D. R., Lawrence, J. R., Hendry, M. J., Caldwell, D. E. (1992) Programs for determining statistically representative areas of microbial biofilms. *Binary* **4**, 204–210

[10] Kühn, M., Hausner, M., Bungartz, H.-J., Wagner, M., Wilderer, P. A., Würtz, S. (1998) Automated confocal laser scanning microscopy and semiautomated image processing for analysis of biofilms. *Appl. Env. Microbiol.* **64**(11), 4115–4127

[11] van Loosdrecht, M. C. M., Picioreanu, C., Heijnen, J. J. (1997) A more unifying hypothesis for biofilm structures. *FEMS Microbiol. Ecol.* **24**, 181–183

[12] Kühn, M., Mehl, M., Hausner, M., Bungartz, H.-J., Würtz, S. (2000) Time-resolved study of biofilm architecture and transport processes using experimental and simulation techniques: The role of EPS. *Water Sci. Tech.* **43**(8), 143–151

[13] Mehl, M. (2001) An interdisciplinary approach to three-dimensional numerical simulation of flow, transport, and growth in biofilm systems on a microscale (german). *Doctoral thesis* TU München

[14] Picioreanu, C. (1999) Multidimensional modeling of biofilm structure. *Doctoral Theses* Delft University

[15] Picioreanu, C., van Loosdrecht. M. C. M., Heijnen, J. J. (1998) A new combined differential-discrete cellular automaton approach for biofilm modeling: Application for growth in gel beads. *Biotech. Bioeng.* **57**(6)

[16] Picioreanu, C., van Loosdrecht, M. C. M., Heijnen, J. J. (1999) Discrete-differential modeling of biofilm structure. *Water Sci. Tech.* **39**(7), 115–122

[17] Poegl, M. (1999) Enhancement of Nast++: Three spatial dimensions, representation of the geometry and parallelization (german). *Seminar paper* Institut für Informatik, TU München: `http://www5.in.tum.de/publikat/fopra/poegl.ps.gz`

[18] Vandewalle, S. (1993) *Parallel Multigrid Waveform Relaxation for Parabolic Problems.* B. G. Teubner Verlag, Stuttgart

[19] Wimpenny, J. W. T., Colasanti, R. (1997) A unifying hypothesis for the structure of microbial biofilms based on cellular automaton models, *FEMS Microbiol. Ecol.* **22**(1), 1–16

[20] Zhang, T. C., Bishop, P. L. (1994) Density, porosity and pore structure of biofilms. *Water Res.* 28, 2267–2277

Polymer and Cell Dynamics – Multiscale Modeling
and Numerical Simulations, 169–203
ed. by W. Alt, M. Chaplain, M. Griebel und J. Lenz

On Selected Individual-based Approaches to the Dynamics in Multicellular Systems

Dirk Drasdo

Max-Planck-Institute for Mathematics in the Science, Inselstr. 22–26, D-04103 Leipzig, Germany

Summary. In recent years a number of different individual-based models for spatiotemporal dynamics in multicellular organisms or parts of them have been established. Individual-based models (here: individuum = cell) become necessary if (i) one is interested in understanding the organization principles in tissues down to length scales of the order of a cell diameter in order to link the microscopic dynamics with a collective phenomenon, and (ii) the phenomenon under study includes variations of material or kinetic properties on length scales of the order of the cell diameter. In this article we give a brief overview over a number of individual-based model approaches.

1. Introduction

The understanding of principles underlying the complicated spatialtemporal processes during morphogenesis and tissue maintenance requires a model or a class of models capable of adequately considering the many degrees of freedom specific to multicellular systems. An essential difference from physical or chemical multiparticle systems is the particular ability of each individual biological cell to change its properties by differentiations. The rules that control differentiation on the time scale of ontogenesis involve the much longer time scale of evolution on which species with these rules were selected on the phenotypic level of multicellular colonies or organisms. Growth and differentiation leads to different cell types in multicellular assemblies on length scales of several cell diameters. These requirements favor single-cell-based models as a general framework.
The use of single-cell-based models has a number of further advantages: (i) They in principle allow one to explicitly incorporate sub-cellular structures such as the cytoskeleton. (ii) Biological growth processes always have stochastic components. These can naturally be included into individual cell descriptions. (iii) Since single-cell-based models represent an *in silico* experimental situation, they naturally include many different aspects of a biological problem that would have to be considered by separated approaches in a continuum description. For example, for tumor growth, single-cell-based models provide a machinery to adequately mimic bulk and surface properties, while continuum descriptions usually require separate descriptions for the bulk and the surface.
Clearly, single-cell-based models also have a number of disadvantages. (i) For most multicellular systems, not enough is known to establish a detailed model, or part of the model behavior may depend on (unknown) details. However, part of the system behavior often

is generic, i.e., does not depend on model details. This in principle allows a model classification according to generic behavior. The number of classes extracted in this way is typically much smaller than the degrees of freedom of the system and is determined by, e.g., certain symmetries, the interaction range, e.g., between cells and between cells and substrate, or the space dimension. (ii) It is often not possible to identify generic behavior solely based on computer simulations since (a) generic regimes are often too short to be clearly separated from transients or, if they show up in the asymptotic limit of infinitely long times or large lengths, they are not accessible to but at best only indicated by a computer simulation, (b) it often cannot be ruled out that for parameter combinations different from those used in the computer simulations the observed behavior may change. (iii) For multicellular systems that consist of many cells and reveal only very small changes on the length scale of a single cell, continuum descriptions often have the advantage that model mechanisms can be identified more easily. (iv) For systems with many cells, single-cell approaches often have a high CPU time requirement.

In this paper we briefly report on recent single-cell models and how they may fit into a model hierarchy according to their spatial resolution. We distinguish between two types of approaches, lattice and off-lattice models. Among the term lattice model, we subsume all models that imply an underlying regular or random grid.

2. Lattice Models

We shall distinguish between three types of lattice models: (i) those for which an individual cell is represented by many connected lattice sites, (ii) those that represent an individual cell by a single site of a lattice, (iii) those where the lattice spacing is large compared to the typical size of an individual cell so that one lattice site (or, more precisely, lattice square in the case of a square, lattice hexagon in the case of a hexagonal lattice, or Voronoi polygon in the case of a Voronoi tessellation) can be occupied by many (approximately homogeneously distributed) biological cells.
The first approach (i) may be viewed as a discretization of cell shape, while in approach (ii) cell shape and tissue arrangement are uniquely determined by the lattice symmetry. Accordingly, the cell volume may be identified with the lattice spacing (which fixes the cell size as equal for all cells). The last approach (iii) does not represent the cell shape at all but is often useful in extracting large-scale properties of migrating or growing cell populations, since it often provides a good starting point when looking for a continuum description of the problem.

2.1. One Cell, One Lattice Site

Due to the vast literature in this field, only a short overview of selected models is presented. Nevertheless, the presented choice should give an impression of important advantages and shortcomings of this type of model. Models of multicellular systems defined on regular lattices have been extensively studied and are still very popular, mainly because they are easier to program and have much smaller CPU time requirements than do off-lattice models. Eden [30] suggested a model of avascular tumor growth on a square lattice where each lattice point can at most be occupied by one individual cell. This assumption takes into account the excluded volume effect: each cell occupies a particular volume in space. Thus, two cells cannot be at the same place at the same time. Different versions of the Eden model have been suggested, distinguished by the way cell divisions are performed (see, e.g., [57] and refs. therein). In each of these versions, a cell division can only take place on an unoccupied lattice site adjacent to an occupied lattice site. Usually, only lattice sites in horizontal or vertical direction are considered as adjacent, not in diagonal direction. In version A a free site in the neighborhood of all occupied lattice sites is randomly chosen with equal probability of being occupied. In Eden's original model (version B), all possible paths from occupied to adjacent free lattice sites are determined, i.e., each occupied site is linked with each of its adjacent free sites. Then one of the distinct paths is chosen at random with equal probability. In version C first an occupied lattice site with at least one free neighbor site is randomly chosen and then one of its free neighbor sites is randomly occupied, both random variables being selected according to a uniform distribution. A number of modifications and generalizations of the Eden model have been proposed. In the William-Bjerkness [103] model for tumor growth, a pair of neighbor sites, one being occupied and one being free, has a probability of α to become occupied and of β to become unoccupied in the next instant of time. For $\alpha/\beta > 1$ an existing cluster grows. Williams and Bjerkness originally believed that their findings based on computer simulations showed a fractal surface structure, but this interpretation has been disproved analytically by Molison [70]. The surface structure was self-affine, hence not fractal, which means smooth, on large spatial scales. In the Richardson model [84] each free site adjacent to an occupied site becomes occupied with probability p (independent for each free site). For $p = 1$ the growth process is completely deterministic, and for $p \to 0$ combined with an appropriate rescaling of time, the Eden model is recovered. Large Eden clusters contain growth sites in only a thin surface layer that occupies a negligible fraction of the cluster volume in the limit $N \to \infty$. This implies that the clusters attain a compact shape as they become larger (e.g., [57, 84]).

Of particular interest are generic properties of these models, i.e., those properties that are independent of the detailed choice of the microscopic growth rules. One motivation to search for quantities insensitive to model details is the limited availability of quantitative data in biological systems and the wide spread in those data, which makes it in principle impossible to include all potential influences on a certain biological process and, hence, to provide a model quantitative in detail. For the growth models mentioned above, insensitive quantities are, e.g., generic regimes of a growth law of the population size, $N(t)$ (t denotes time), the (global) fractal bulk dimension d_f, defined, e.g., by $N \propto R^{d_f}$, and some robust surface properties used to characterize growing surfaces in physical systems.

If the initial state was one single cell, a compact cluster emerges. For the growth models above, computer simulations and simple arguments suggest $N(t) \propto e^{t/\tau}$ (τ being the mean cell cycle time) for $t \to 0$ and $N(t) \propto t^d$ in d dimensions as $t \to \infty$ (Fig. 1). A line of argument is the following. For small times each cell has free adjacent sites; hence, all cells are able to divide and contribute to the cluster growth, resulting in exponential growth. For large times only cells at the surface are able to divide, i.e., $dN \propto N_s dt$

with N_s being the number of boundary cells. On large scales, the global fractal bulk dimension equals the space dimension, $d_f = d$, the fractal surface dimension is $d - 1$, hence $N_s \propto N^{d-1}$ resulting in $dN \propto N^{d-1}dt$ and $\Rightarrow N(t) \propto t^d$ after integration (for the William-Bjerkness model a rigorous proof of a trivial fractal surface dimension ($= d-1$) has been given by Molision, [70]).

Eden-like models are able to describe growth curves of avascular tumors also quantitatively. To show this, the cell division rule is generalized (Drasdo, unpubl.) by assuming that each dividing cell is able to shift at most k cells (the above described Eden-like models correspond to the special case $k = 0$; see Fig. 1). The direction of cell division is chosen at random among those directions into which not more than k cells would have to be shifted if a cell division would take place. Hence, for $k > 0$ also cells inside the cluster are able to divide if their distance from the surface is at most k lattice spacings.

Further model classification is facilitated by studying the surface properties of growing clusters (e.g., [10, 57, 59, 66]). Such studies are motivated by the observation that the surface of a growing cluster often is rough, revealing spatial undulations of typical longitudinal and vertical size. The structure of the surface does not change if longitudinal and vertical length scales are changed by an appropriate self-affine transformation. Self-affine objects can be characterized by growth exponents. For example, the growth exponent β measures how the width w (Fig. 2) of the surface scales with time, $w \propto t^\beta$ for $t \ll t_x$. t_x is the saturation time at which the typical horizontal size of a typical undulation becomes of the order of the total surface length. How such a limitation occurs can be seen for a surface of fixed longitudinal size L in x-direction growing into y-direction (growth in $1+1$ dimensions). In this case, the width of the surface saturates to $w = w_{sat}$ (at $t = t_x$). The dynamic exponent measures how the crossover time scales with L, i.e., $t_x \propto L^z$. The roughness exponent α measures how the saturation width scales with system size L, i.e., $w_{sat} \propto L^\alpha$ (at $t \gg t_x$). The different growth exponents (including the local fractal dimension) are not independent but are related by scaling relations.

Especially for the Eden model, the critical surface properties have been extensively studied in different euclidian dimensions (e.g., [42, 57, 66]). What makes these concepts so appealing is that a growing system can be classified by the exponents introduced above into "universality classes" which depend only on certain characteristic features of the dynamics. For example, if cell diffusion at the surface driven by a chemical gradient determines the surface structure, the growth exponents are different, as they would be in the absence of cell surface diffusion. It is often possible to conclude from the (macroscopic) exponents certain processes of the (microscopic) dynamics.

Different attempts have been made to model tumor growth of avascular and vascular tumors quantitatively. Qi et al. [81] established a cellular automaton (CA) model to explain how the Gompertz-growth law (i.e., exponential population growth with exponentially decaying growth rate, compare [34] with, e.g., [33]) may emerge. Düchting and Vogelsänger [27] and Düchting et al. [28] modeled the growth of vascularized tumors for different chemotherapy schemes, while Anderson [2] modeled tumor invasion and metastasis formation. All these models use square lattices.

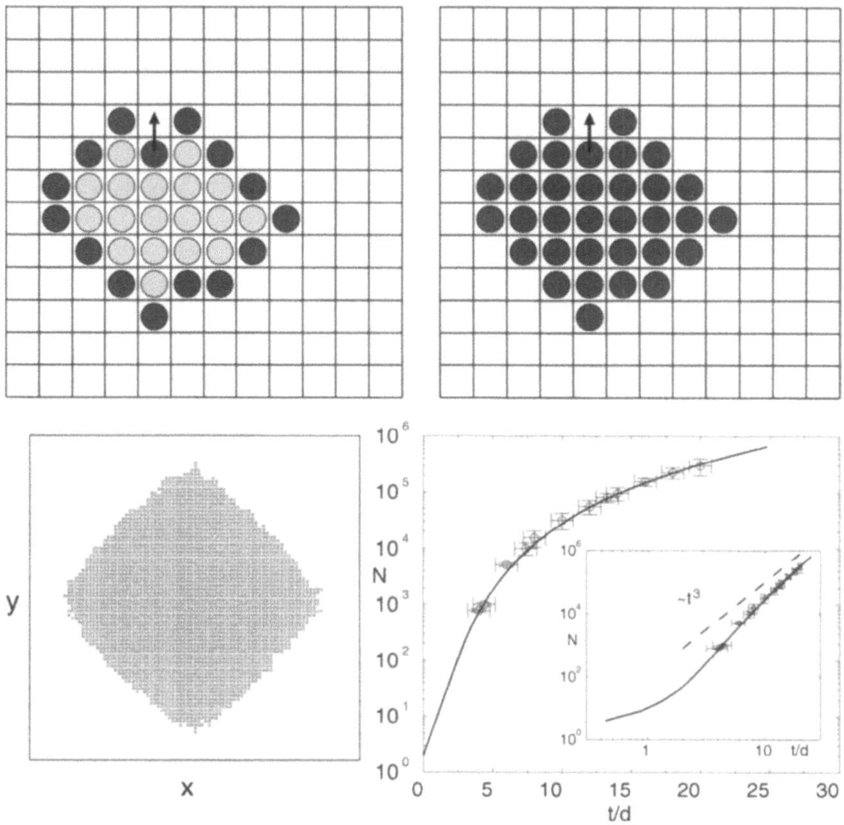

Figure 1. Schematic representation of a "tumor" consisting of $N = 33$ cells on a square lattice. Each cell is represented by a filled circle. The black circles denote cells that are able to divide. *Upper left*: Only cells with free adjacent lattice sites are able to divide (e.g., the arrow indicates a cell that can only perform a division in a vertical direction). This rule describes the principle of cell division in standard Eden-like models. *Upper right*: A cell is able to push at most $k = 2$ cells towards the horizontal or vertical direction during cell division. This division rule may be considered a generalization of the cell division rule in traditional Eden-like models. The larger k is, the longer is the initial growth regime where all cells are able to divide. *Lower left*: Typical snapshot of a configuration in a computer simulation at $N = 10^5$ cells in $d = 3$ dimensions (here, $k = 1$). Shown is a crosssection of the $3d$-aggregate at $z = 0$. The quadratic tumor shape results from anisotropies due to the lattice symmetry. *Lower right*: comparison between experimental data (points taken from [33]) and computer simulations (full line). As in the experiment, the initial growth is exponentially eventually crossing over into a power law at $t \approx 5d$.

Models on regular lattices have also been set up for different biological situations. Steinberg [89, 90] used a single-cell approach on a hexagonal lattice to explain a sorting experiment by Townes and Holfreter [96].

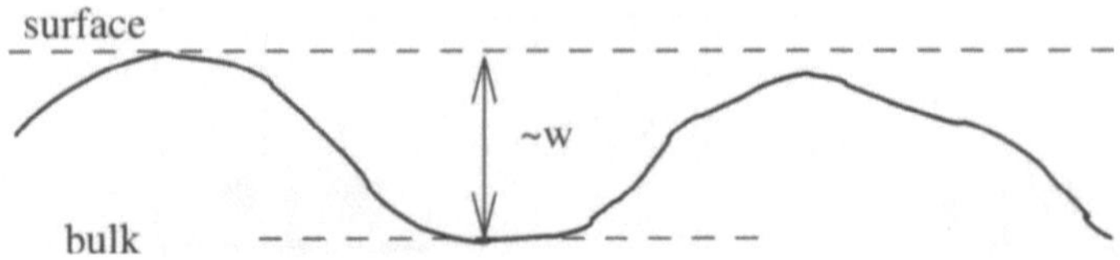

Figure 2. Sketch of a piece of surface separating a free space (above) from the bulk (below). w is the surface width.

These authors had observed that dissociated cells from an amphibian embryo at the neurula stage re-aggregate in suspension such that presumptive epithelial cells are surrounded by neural plate cells (see Fig. 3 for the result of cell sorting in a computer simulation on a three-dimensional square lattice). Starting with a random mixture of two cell types,

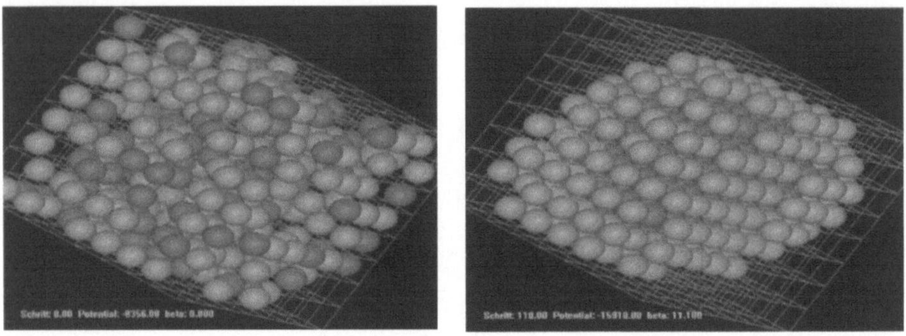

Figure 3. Cell sorting on a square lattice with two cell types A (red), B (green) in $d = 3$ dimensions in a computer simulation assuming periodic boundary conditions in x,y-and z-direction. Be $V_{AA} < 0$, $V_{BB} < 0$, $V_{AB} < 0$ the adhesion energies between two A cells, two B cells, and one A and one B cell, respectively. Starting with a random mixture of A and B-cells (*left*), a configuration where one cell type is surrounded by the other corresponds to the configuration of minimum energy if $V_{AA} < V_{AB} < V_{BB}$ (*right*). (See color plate 9.)

Steinberg demonstrated that cells sort out as in the experiment if the attraction between epithelial cells is the strongest, between neural plate cells is the weakest, and between epithelial and neural plate cells is intermediate. In his model a cell was allowed (i) to jump on free adjacent lattice sites and (ii) to exchange the place with its neighbor cells. Cell sorting due to cell-specific (*differential*) adhesion has very recently been found also *in vivo* [37].

Furthermore, cellular automata approaches on regular lattices have been used to model the cell kinetics in intestinal crypts (e.g., [62, 78]); similar model approaches to intestinal crypts have been studied by Meinzer et al. [68] and Isele and Meinzer [51]. Crypts are pear-shaped, one-cell-thick pockets in the intestinal wall responsible for maintenance of the high cell turnover in the intestine [79]. Most cell divisions take place in the lower 2/3 of the crypts; cells then migrate towards the crypts' orifice from where they are released into the intestinal lumen. In the model an individual crypt is represented by a cylindrical array of cells, with periodic boundary conditions in horizontal direction (Fig. 4). The objective of these models was to identify cell pedigrees in order to quantitatively explain labeling and mitotic index curves, experimentally observed in mice. The approach again uses the excluded volume assumption. Since crypt cells are densely packed, an arbitrary rule that allows cells only to migrate on free lattice sites cannot be appropriate. Instead, cell migration is controlled by cell division, i.e., cells move only as a consequence of a cell division according to two alternative rules: (i) After the direction of cell division has been chosen randomly with equal probability either upward or towards the right or left side, a dividing cell pushes the above, right, or left column of cells by one lattice unit towards the crypt orifice (Fig. 4). (ii) The direction of cell division is chosen in such a way that always the oldest neighbor cell (and its above column) of a dividing cell is shifted by one lattice spacing towards the crypt orifice and its space is occupied by one of both daughter cells (one daughter cell always occupies the space of the mother cell). The latter assumption has been shown to yield a better fit to experimental data [62]. Cells that reach the crypt orifice are assumed to be released into the lumen and removed from the system.

Cellular automaton approaches where each individual cell is identified with one lattice site has also been used by many other authors (for a review, see, e.g., [15]). For example, Kaandorp et al. [53] used a square lattice approach to model the growth of stony corals in a flowing fluid. The growth results from aggregation of nutrients at the surface of the coral, and the fluid is modeled by a lattice Boltzmann approach. Bussemaker et al. [11] studied a 2D cellular automaton model on the formation of cooperative swarming movement, e.g., in myxobacteria or birds. Each node on a square lattice can be occupied by at most four particles (cells or birds) with unit velocities into either the $-x$, $+x$, $-y$, or $+y$-direction. The number of particles at each node is conserved during interaction; hence, the spatially averaged cell density is constant in time. The particles on each node favor alignment in the direction of the local average flux with a certain probability given by a Boltzmann factor where the inverse temperature plays the role of a sensitivity parameter. The model shows a continuous phase transition at a critical value of the sensitivity parameter. Below the critical value, the spatially averaged velocity is zero, while above it is positive, revealing a cooperative movement in a certain direction. Ben-Jacob and others have studied the pattern formation in bacterial colonies in a number of papers using continuum approaches and individual based models, in particular cellular automata (for review, see [5]). Bretschneider et al. [9] used CAs to model cell movement during Dictyostelium mound formation, Stevens and Schweitzer [92] and Stevens [93] used a stochastic CA to model the gliding and aggregation of microorganisms based on chemotaxis. Börner et al. [7] use a CA model to describe the rippling in myxobacteria aggregates. Agur et al. [1] represented the spatial structures in the bone marrow in hematopoiesis by a CA.

There are a number of pitfalls and limitations that have to be kept in mind when working with automaton models on regular lattices.

The resulting spatial structures of many models on regular lattices often show effects from the underlying lattice symmetry. An example is the shape anisotropy of growing clusters in the type of growth models reported on above at sufficiently large length scales (e.g., Wolf [104]). Batchelor and Henry [4] have shown for cluster growth models of

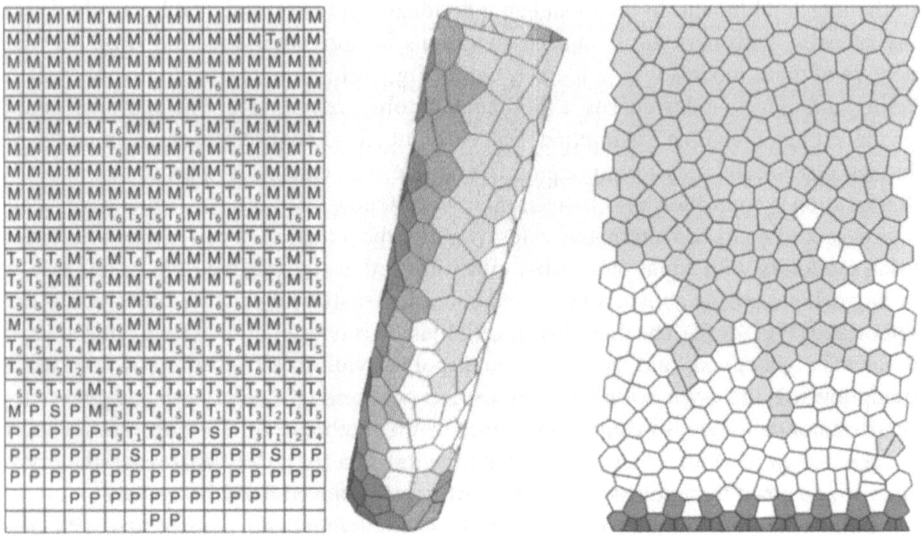

Figure 4. Model design for studying the cell dynamics in an intestinal crypt. *Left*: cellular automaton approach on a regular lattice (after [62, 78] and (*middle/right:*) using a Voronoi diagram (with courtesy of Meineke and Loeffler, see [67]). *Left*: P, S, T_i, and M denote paneth, stem, transient, and mature cells, respectively. Only S and T_i cells are able to proliferate. *Middle/right*: The paneth cells are dark, S and T_i cells are light gray, and mature cells are dark gray. The geometry is fixed, represented by periodic boundary conditions in horizontal direction as indicated by the cylinder. S and T_i cells divide and push their neighbors towards the upper boundary at the top of the crypt (which represents the crypt orifice). In both approaches the paneth cells are fixed; hence, the other cells are pushed towards the crypt orifice. As they reach the crypt orifice, they are removed from the system (see also color plate 9).

the Eden type that noise reduction in these models displays anisotropies earlier (see also Fig. 1). (*Vice versa*, increasing the noise typically shifts anisotropies towards larger length scales. Symmetry-induced anisotropies can also be reduced or sometimes be eliminated by increasing the range of interaction on a lattice, e.g., [65].

Cellular automata models are rule-based. Rivet and Boon [85] classify the paradigm established by a lattice-gas cellular automata (LGCA) approach (LGCA are a special type of cellular automata) as "a fictitious oversimplified micro-world constructed as an automaton universe based not on a realistic description of interacting particles (as in molecular dynamics), but merely on the laws of symmetry and of invariance of macroscopic physics", where nevertheless their dynamics (of the LGCA) can show a remarkable correspondence to the behavior of real systems. This view may be adopted also for many CA models or related models of multicellular systems. For example, during the compression of multicellular aggregates, cell distortions and displacements occur on length scales small compared to the cell diameter [31]. Clearly, this experiment cannot be modeled in a direct physical way by a lattice model where each lattice site is identified by an individual cell. However, it may be possible to construct additional rules to account for this and other experimentally observed behavior on scales sufficiently large compared to the cell diameter so that the resulting dynamics become very similar to that of the real system. Still, since the model approach is not based on a direct realistic physical description of the interacting particles, other physical or physiological effects may be overlooked or described in an inappropriate way.

For some CA models equations for the density of particles have been rigorously derived in some limits (see, e.g., [85, 88]).

Nevertheless, also in cases where such a proof does not exist, lattice models in which each lattice site can be occupied by only one single cell can be very useful because such models are easier to implement than more direct physical individual-based models and work at moderate computer power, so that they are worthwhile to be considered at least for a first guess. Furthermore, they often provide a good starting point to deduce a continuum description from microscopic dynamics and hence may contribute to the understanding of a dynamical process at least qualitatively.

Several attempts have been made to circumvent some of the problems described in the last paragraphs. One was to use a Voronoi tessellation instead of a regular lattice in order to avoid the emergence of anisotropies (see Fig. 1). Biologically, this may be motivated by the findings of Honda [45, 46], who has shown that, e.g., cultured mono-layer cells and epithelial cells in tissue can well be approximated by a Voronoi tessellation. In order to generate a Voronoi diagram for a given (e.g., random) distribution of N cells, to each of the N center points (*generators*) in space the region is assigned which is closer to this point than to any other point. The construction rule is the same as for the first Brillouin-zone of crystals (e.g., [3]). The difference is in the location of the generators which have a discrete translational and rotational symmetry for crystals, which is usually not the case for cells. An individual cell within a Voronoi tessellation is represented by a single point. Its cell shape is completely determined by the location of its nearest neighbor cells and is convex by construction (see Fig. 4). Cells are nearest neighbors if and only if they share a common Voronoi edge. By construction, a Voronoi diagram is space filling. For example, Kansal et al. [55] used CA on a Voronoi tessellation to model the growth of brain tumors (unfortunately in this work the translation invariance in the average cell size is violated; thus, their results should be considered with caution). If the position of the Voronoi cell generators is chosen randomly in space and is fixed prior to the modeling of any dynamics, a Voronoi tessellation yields a random lattice (as opposed to a regular lattice; in such a lattice the artifacts due to discrete lattice symmetries disappear; Drasdo and Trebels [unpubl.]). Still, the Voronoi approach keeps the lower length scale in the order of a cell diameter.

2.2. One Cell, Many Lattice Squares

Steinberg [89, 90], Graner and Glazier [40], and Glazier and Graner [36] suggested consideration of differential surface energies between cells and cells and their surrounding extracellular medium as the driving forces for the cell-sorting phenomena observed by Townes and Holfreter [96]. In contrast to Steinberg, Graner and Glazier represented each cell at a higher spatial resolution. To do this, they extended the Q-states Potts model (e.g.,

[58]) to include different cell types and (approximately) take into account the conserved cell volume. They suggested the following model energy:

$$H = \underbrace{\sum_{\langle \underline{i},\underline{j}\rangle} J_{\tau(\sigma)\tau(\sigma')}\left[1-\delta_{\sigma_{\underline{i}},\sigma_{\underline{j}}}\right]}_{surface\ energy} + \underbrace{\lambda \sum_{\sigma} \left[v(\sigma,t) - v_0(\tau(\sigma),t)\right]^2 \Theta(v_0(\sigma,t))}_{volume\ constraint}, \tag{1}$$

defined on a square lattice. Each biological cell is labeled by an unique index σ. Many lattice sites are assigned to each individual cell; hence, many lattice sites carry the same index σ (see Fig. 5). For N cells under consideration, $\sigma \in \{1,\dots,N\}$. $\underline{i},\underline{j}$ labels the lattice squares, $\underline{i} = (i_1, i_2, \dots, i_d)$ with d being the space dimension. $\sigma_{\underline{i}}$ denotes the (biological) cell to which lattice square $\underline{i}$ belongs. Each cell has an associated cell type $\tau(\sigma)$. $J_{\tau(\sigma)\tau(\sigma')} > 0$ is the coupling constant that specifies the surface energy between neighboring sites of type $\tau(\sigma)$ and $\tau(\sigma')$. λ is a Lagrangian multiplier and specifies the strength of the volume constraint, $v(\sigma,t)$ the volume of cell σ at time t, while $v_0(\sigma,t)$ denotes the target volume. v_0 is the volume of an isolated cell. The sum runs over all neighbor sites $\langle \underline{i},\underline{j}\rangle$. $\delta_{\sigma_{\underline{i}},\sigma_{\underline{j}}}$ is the Kronecker-symbol. If adjacent lattice sites belong to the same cell σ, they do not contribute to the surface energy; otherwise, the corresponding surface energy contribution is positive. The extracellular medium can be treated as a different cell type $\tau = M$ with unconstrained volume achieved mathematically by setting $v_0(M,t) < 0$. $\Theta(x)$ is the Heavyside function ($\Theta(x) = 1$ for $x > 0$ and 0 otherwise).

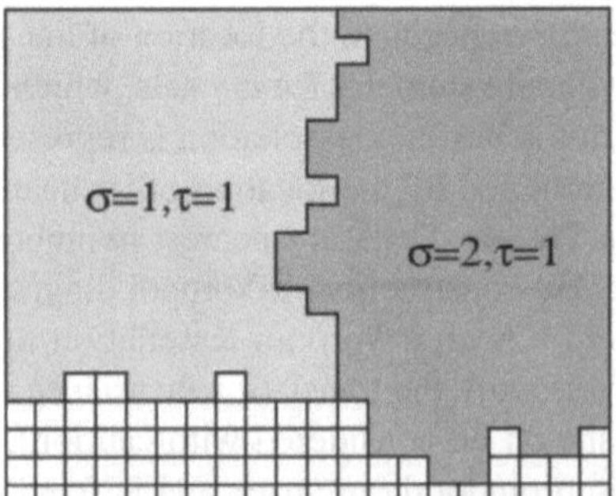

Figure 5. Sketch of two neighboring cells in the model of Graner and Glazier [40]. Each cell occupies many sites on a lattice. Both cells have the same cell type $\tau = 1$. The index σ marks the individual cells. $\tau = M = 2$ denotes the extracellular medium (small squares).

For $\lambda = 0$, $N = Q$, and $J_{\tau\tau'} \equiv J = const$, we obtain the original Q-states Potts Hamiltonian used in the theory of critical phenomena to describe order-disorder transitions in percolation problems and soap froths. Potts models on periodic lattices are known to have first-order phase transitions for $Q > 4$ in two and for $Q > 2$ in three space dimensions. The case $Q = 2$ corresponds to the celebrated Ising model (see, e.g., [58]).

Assuming an over-damped dynamics, i.e., friction terms to be large compared to inertia terms, cell movement can be modeled by the Metropolis algorithm ([69]): a cell i is chosen at random with equal probability and its state σ_i is changed into another state $\sigma'_{\underline{i}}$. The variable change $\sigma_{\underline{i}} \to \sigma'_{\underline{i}}$ at lattice site $\underline{i}$ is accepted with probability

$$p(\sigma_{\underline{i}} \to \sigma'_{\underline{i}}) = \begin{cases} 1 & if \quad \Delta H < 0, \\ e^{-\beta \Delta H} & if \quad \Delta H \geq 0. \end{cases} \tag{2}$$

β is a proportionality factor, which for fluids is $\beta = 1/(k_B T)$ (k_B: Boltzmann constant, T: temperature). This approach to the dynamics corresponds to the numerical integration of a corresponding master equation with transition rates $\propto p(\sigma_{\underline{i}} \to \sigma'_{\underline{i}})$ (see, e.g., [58]). Cell shape and dynamics rely on surface energies only. Apart from lattice effects, an isolated cell adopts a circular minimum energy shape. Graner and Glazier [40] considered two cell types, a low (l) and high (h) adhesive cell type in a medium M in $d = 2$. Furthermore $J_{\tau,\tau'} = J_{\tau',\tau}$, $J_{l,M} = J_{h,M}$, and $0 < J_{h,h} < J_{l,h} < J_{l,l} < J_{l,M} = J_{h,M}$. If $J_{h,M}$, $J_{l,M}$ are not larger than the cell-cell coupling constants, the cells dissociate. $1/\beta$ must not exceed $J_{\tau,\tau'}$ in order to avoid a too-strong dissociation of individual cells into several unconnected parts (which is unbiological). If $(J_{h,h} + J_{l,l})/2 < J_{l,h}$, cell sorting or engulfment are observed; otherwise, a checkerboard is found [36]. To see this, imagine a 1D chain of N cells, all having a fixed target volume, and assume periodic boundary conditions. If h and l cells alternate, forming a checkerboard, the total energy is $N J_{l,h} A$, A being the interface that neighbor cells have in common. If h and l cells are separated, the total energy is

$$A\left(\frac{N-1}{2}(J_{h,h} + J_{l,l}) + 2J_{h,l}\right). \tag{3}$$

Hence, for large N, separated phases are energetically favored if $\frac{J_{hh}+J_{l,l}}{2} < J_{h,l}$. In $d = 2$, since $J_{hh} < J_{l,h}$, the energy is minimized if the number of $h-h$ contacts are maximized, which favors a circular cluster of h cells being enclosed by l cells. For an initially random mixture of h and l cells, this leads to cell sorting. For initially separated phases with a minimum interface between h and l cell clusters, this leads to engulfment. It is essential to choose the temperature T appropriately [36]. Above the critical temperature T_{c2} for spinodal decomposition (e.g., [58]) defined by $N_{squares}\delta E = N T_{c2}$, entropic effects induce a total mixing of h- and l-cells. Here,

$$\delta E \equiv \left(J_{hl} - \frac{J_{hh} + J_{ll}}{2}\right) \tag{4}$$

($= \gamma_{h,l}$), $N_{squares}$ is the total number of squares (sites) of the lattice. $\gamma_{h,l}$ is the surface tension between a cell of type l and a cell of type h. Clearly $T < T_{c2}$ is essential. At $T \approx n J_{ll}$, membranes separating neighboring l-cells reveal a (unbiological) crumpling; at $T \approx n J_{hh}$ this happens with h-cells (n is the number of neighbors on the lattice, for a next-nearest-neighbor lattice in $d = 2$ dimensions, $n = 8$). If $T = T_{c1} \approx\sim n\gamma_{h,l}$ cells often dissociate into single squares surrounded by neighbors not belonging to the same cell. On the other hand, at too-low temperatures, the configurations may be trapped in local minima of the energy landscape. By this reasoning, T should be smaller than T_{hh} and larger than zero during the simulation. Within the model, T is just a parameter. In order to quantitatively link the theory with the experiments, its precise value must be known. Very recently, Beysens et al. [6] collected experimental evidence suggesting that cells can behave as viscous fluids and deduced an effective temperature F_T in multicellular systems as $k_B T$-analogue in fluid systems. This finding provides a natural basis for a quantitative modeling.

The original model approach by Graner and Glazier [40] has been extended in several directions and by different authors. Jiang et al. [52], Savill and Hogeweg [86], and Maree et al. [63, 64] have modeled different aspects of the Dictostelium development, Stott et al. [95] and Turner and Sherrat [97] extended the approach to model the growth of avascular tumors, and Zajac et al. [106] modeled convergent extension by mediolateral intercalation during *Xenopus* (a frog species) gastrulation. During mediolateral intercalation, a tissue narrows along one axis and elongates along another at right angles. At the same time the cells become elongated at right angles to the elongation of the tissue. In order to provide a potential explanation for this effect, the authors introduced polar cells by assuming an anisotropic surface adhesion from different types of surface molecules on the "short" (s) and "long" (l) side of the cells, yielding a different interaction strength between ll, ss, and ls sides of adjacent interacting cells. If the interaction strengths are tuned appropriately, cell movements driven by energy minimization result in a global tissue shape change. Hogeweg [44] studied morphogenesis in model multicellular organisms linking the strength of cell-cell and cell-substrate adhesion to the intracellular genetic machinery. The latter is represented by a boolean network. Cell differentiation changes the rules and topology of the boolean network in an evolutionary process according to a predefined fitness criterion, namely, the amount of difference in gene expression among all cells.

Since defined on a lattice, this model again shows the usual limitations of many lattice models, namely, that the lattice symmetry shows up in the resulting patterns. During cell sorting simulations, this results, e.g., in fluctuations in the local contact angle which causes boundary pinning. The pinning can partly be avoided by choosing a sufficiently high temperature T. As discussed above, a too-high T leads to unbiological effects such as cell fragments detached from the main body. Thus, T must be tuned carefully in order to avoid problems by too small or too large temperatures.

3. Off-lattice Models

Since cell migration, cell growth, and cell deformations occur continuously, a direct, realistic approach to the dynamics of cell growth and motion should be free of any predetermined lattice structure that imposes an unnatural lower (cutoff) length scale as in the lattice models presented in Section 2.1 above. We subsume models without a predetermined underlying lattice under the term off-lattice models. Still this does not exclude the existence of an artificial lower length scale, e.g., if single moves or cell division are not performed continuously, as illustrated below.

Odell et al. [74] suggested a mechanical model in $d = 2$ dimensions for gastrulation in echinoderms, neural tube formation in urodele amphibians, and ventral furrow formation in *Drosophila*. The initial situation is a circular cell monolayer, each cell being joined at its side membranes to its right and left neighbor (Fig. 6).

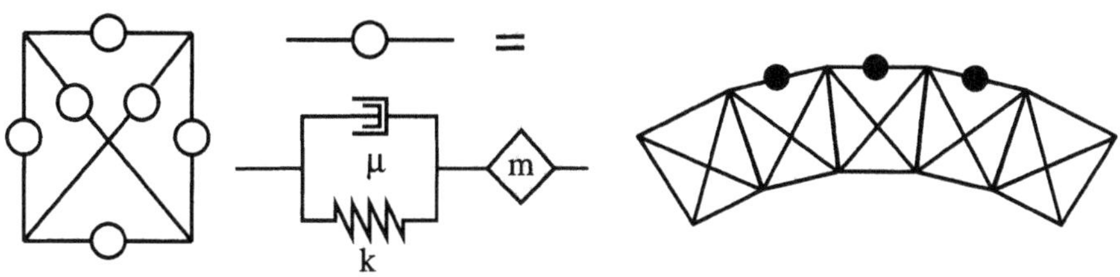

Figure 6. *Left/middle*: Model design of cells in the model by Odell et al. [74]. *Left:* Each individual cell is represented by a square. Along the membranes and the diagonals are viscoelastic elements (denoted by circles). *Middle*: Each viscoelastic element consists of a spring with spring constant k, a dash-pot with viscosity μ, and a mass m. *Right*: A subset of five cells from a (closed) circular one-cell-thick configuration. Neighboring cells are tight to share a common edge. The black filled circles denote nonlinear springs that contract after being over-stretched.

Cells are assumed to be cuboidal and are represented by a rectangle. Each of the four corners of the rectangle is linked to every other corner by a viscoelastic (Voigt-) element (Fig. 6) to mimic the effect of filament bundles underneath the cell membrane and of the cytoskeleton. Hence, there is one element along the basal, the apical, the two-side membranes, and along the diagonals of a cell. The length of each viscoelastic element obeys the equation:

$$m\frac{d^2L}{dt^2} = \underbrace{-k(L - L_0)}_{F_{\rm elast}} \underbrace{-\mu\frac{dL}{dt}}_{F_{\rm visc}} + F_{\rm load}\,, \tag{5}$$

where m represents the net mass moved by a length change of the filament bundle, L_0 is the equilibrium length of the bundle, and $F_{\rm elast}$ and $F_{\rm visc}$ denote the elastic and viscous forces, respectively. $F_{\rm load}$ is the force exerted on the bundle, parallel to the bundle by neighboring systems. In a limited connected region (contractile region), the apical membranes contains a nonlinear spring which after over-stretching contracts, i.e., in a limited region

$$\frac{dL_0}{dt} = G(L, L_0)\,. \tag{6}$$

Cell volume conservation was (approximately) insured by an additional equation for the internal cell pressure. $G(L, L_0)$ models the nonlinear spring. If by any signal the apical membrane of a cell within the contractile region is over-stretched, it immediately contracts and by this causes the apical membranes of its neighbor cells to be stretched as well (which then themselves are contracted). This perturbation propagates from the first stretched cell, from cell to cell, and towards the margins of the contractile region and induces a macroscopic invagination. More recently, Kam et al. [54] disproved this mechanism by 3D fluorescence microscopy, which showed that all cells within the contractile region change shape synchronously instead of a propagating wave-like cell contraction. Later, a number of competing models of sea urchin invagination based on finite element simulations were proposed by partly the same group [14].

A simple approach that describes cells similar to particles in a strongly viscous fluid and also takes into account cell division has been proposed by Drasdo et al. [16] and Drasdo [18, 19] for growing tissue cultures and non-vascular tumors. Guided by the observation that isolated cells in cultures or suspensions often have a spherical shape, the model assumes that each cell is spherical directly after cell division and deforms into a dumbbell shape during mitosis (Fig. 7), i.e., actively changes its "equilibrium shape".

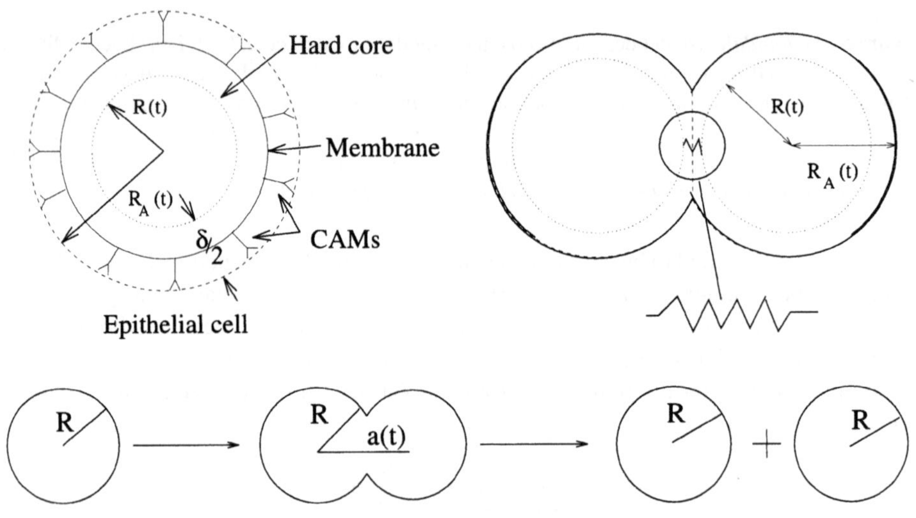

Figure 7. Model design in the off-lattice approach by Drasdo et al. [16] and Drasdo [18]. *Upper left*: Sketch of a model cell. *Upper right*: Construction instruction of the interface between adjacent cells. *Lower*: Cell division algorithm. In the approach an isolated cell is approximated by a perfect sphere. The approach approximates cells with limited deformability. The limited deformability of a cell is considered, assuming the existence of a hard core repulsion indicated by the innermost circle (dotted) with radius $R(t)$ in the *upper left picture*. The interaction range of a cell is illustrated by the radius $R_A(t)$. The position of the cell membrane is not explicitly specified by the approach but may be constructed as indicated in *upper right picture*. *Lower panel*: During cell division a cell deforms into a dumbbell by increasing its axis by small steps from $a(t) = 0$ to $a(t) = 2R$ (shown is the hard core part of the cell).

To each assembly of cells a ("total") energy is assigned according to $V^{\text{tot}} = \sum_{i<j} V_{ij}^{NN}$, where V_{ij}^{NN} summarizes nearest-neighbor (NN) interactions that result from the competition between attractive interactions due to adhesion molecules anchored in the cell membranes and repulsive contributions first from the limited cell deformability and compressibility and second from the loss of membrane steric entropy. The detailed shape

of the potential V_{ij}^{NN} has only minor influence on the model results and leaves generic growth regimes invariant [18]. A reasonable ansatz for V_{ij}^{NN} is

$$V_{ij}^{NN} = \begin{cases} \infty & \text{if} \quad \tilde{d}_{ij}(t) < 0 \\ \epsilon\left(\left[\frac{2\tilde{d}_{ij}(t)}{\delta} - 1\right]^2 - 1\right) & \text{if} \quad 0 \le \tilde{d}_{ij}(t) \le \delta \\ 0 & \text{otherwise.} \end{cases} \tag{7}$$

Here, $\tilde{d}_{ij}(t) = d_{ij}(t) - 2R$. For the definition of R see Figure 7. d_{ij} denotes the distance between the nearest circles of the neighboring dumbbells i and j (a circular cell is a dumbbell with axis length $a = 0$). V_{ij}^{NN} has a minimum of $\tilde{d}_{ij} = \delta/2$. The lower cutoff at $\tilde{d}_{ij} = 0$ mimics the excluded volume effect. δ determines the range over which a cell can be stretched or compressed in a certain direction and is estimated to $\delta \approx 0.1R$. $\epsilon > 0$ measures the resistance against deformations. (Note also that $V_{ij}^{NN} < 0$ for $0 \le \tilde{d}_{ij}(t) \le \delta$, which models short-range attraction.) The hardcore repulsion takes into account that cells cannot be arbitrarily compressed or deformed. Active cell deformations during mitosis cause pressure on the neighbor cells in the direction of the deformation. This leads to an increase of the total energy V^{tot}. It is assumed that the neighbor cells then either move their center of mass or change their orientation in order to minimize V^{tot}. A further assumption is that (i) inertial terms are negligibly small compared to dissipative terms and (ii) processes not explicitly considered, such as the cell metabolism, intra-cellular movements of the cytoplasm, and reorganization of the cytoskeleton, give rise to a stochastic component in the displacement of the cells. Accordingly, for isolated cells a purely random movement is assumed. Although growth is intrinsically a non-equilibrium problem, the dynamics have been modeled by the Metropolis method, which corresponds to the numerical integration of a master equation [16, 69]. This may be justified, noting that after each growth step all cells move to relax the configuration at least into a local equilibrium [16]. In the simulations, cells are randomly chosen to perform either a small translation ($\ll R$), orientation change ($\ll \pi$), or growth ($\ll R$) trial. Each translation or rotation is accepted with probability $P_a = 1$ if the energy change $\Delta V^{tot} = V_{t+\Delta t}^{tot} - V_t^{tot} < 0$ and with probability $P_a = \exp(-\Delta V^{tot}/F_T)$ if $\Delta V^{tot} \ge 0$ (hence isolated cells move diffusively in correspondence with the findings of Schienbein et al. [87] and Mombach et al. [71]). t is time, Δt the time span between two migration or rotation trials. F_T [6] is an effective energy and the $k_B T$ analogue in cellular systems (T: temperature, k_B:Boltzmann const.). It can be absorbed into a redefinition of ϵ, by $\epsilon \equiv \epsilon/F_T$. A growth trial is accepted only if it does not result in too-strong cell deformations, i.e., if $d_{ij} > 2R$. Between two growth trials a cell performs $n_g \gg 1$ translation and rotation trials. A cell not subject to any excluded volume interactions has an average cycle time $\tau \propto n_g$. If excluded volume interactions occur, some growth trials are rejected and the real average cycle time becomes $\tau_R \ge \tau$.

Figure 8 shows a computer simulation of cell sorting with the model of Drasdo [17] and Drasdo et al. [16]. (A physically very similar model approach to cell sorting has been developed independently by Graner and Swada [41]. These authors integrated

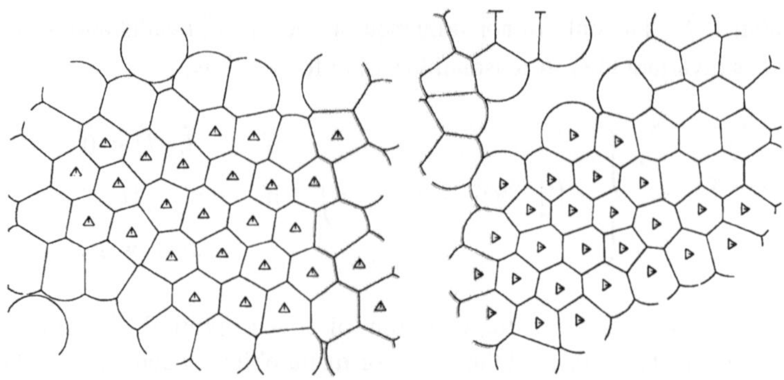

Figure 8. Cell sorting in $d = 2$ dimensions in the off-lattice model of Drasdo [17] and Drasdo et al. [16] with two cell types A, B in a computer simulation assuming periodic boundary conditions in vertical and horizontal direction. The cell boundaries are constructed by Voronoi tessellation with a cutoff radius of $R_c = R + \delta/2$. The triangle denotes cells of type A. Be $V_{AA} < 0$, $V_{BB} < 0$, $V_{AB} < 0$ the adhesion energies between two cells both of type A, both of type B, and one of type A, the other of type B, respectively. Starting with a random mixture of A and B cells, a configuration where one cell type is surrounded by the other is obtained if $V_{AA} < V_{AB} < V_{BB}$ (*left*) (compare also Fig. 3). The final configuration results from an energy minimization. For $V_{AA} < V_{AB} > V_{BB}$, de-mixing occurs (*right*).

the equation of motion for each cell in the limit of zero noise instead of using energy minimization.)

Figure 9 (left picture) shows a typical picture of a growing cell aggregate in $d = 2$ dimensions, Figure 9 (right picture) the corresponding growth law $N(t)$. The growth regimes are the same as in Figure 1, but the geometry of the cluster does not show artifacts due to a lattice symmetry. The model results suggest that a power-law-like growth regime can also be a consequence of biomechanical effects, namely, if cells are impeded from further cell divisions by a too-large local pressure. This mechanism is able to describe the constant growth velocity of the tumor diameter by Bru et al. [10], where the experimental setting suggests that it might not be the shortage of nutrients that limits cell proliferation to a surface layer (Bru, priv. communication). The approach is also capable of a quantitative description of the data of Freyer and Sutherland [33] (see Drasdo and Höhme [23, 24]).
The model has been extended to describe generic and specific dynamic situations in one-layered tissues. Examples are blastula formation [20, 21], the cell kinetics in intestinal crypts [22] and wound healing [16, 17].

Figure 10 shows a time series of different scenarios for a caricature of a wound healing process. The model includes epithelial and connective tissue cells. Epithelia cells are characterized by (i) a strong attractive interaction and (ii) anisotropic growth preferring linear topologies. This has been achieved by introducing internal orientations of cells corresponding to polar exprimation of CAMs. Consequently, the intercellular interaction depends on relative cell orientations. In this way, linear or planar topologies (important for 3D assemblies) are stabilized. Furthermore, the growth rate of epithelia is chosen

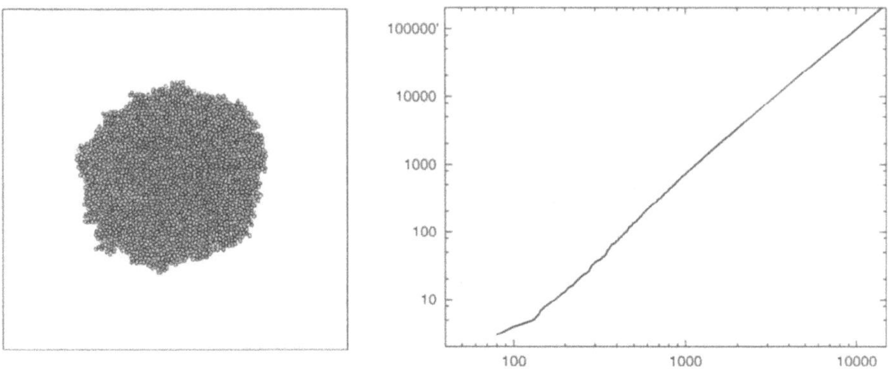

Figure 9. *Left*: Geometry of a growing cell aggregate. *Right*: Corresponding growth law N vs. t in $d = 2$ dimensions [18, 19]. The asymptotic growth regime is determined by growth only within a stripe of finite size at the surface, hence $N(t) \propto t^2$. The initial growth regime is an exponential.

large enough to ensure that the cut can be closed before the fibroblasts can grow out of the wound. In the initial configuration, cells at the surface of the lesion are switched into the proliferating (p)-state, as is known from observation [29]. Without any other regulation mechanism, the wound is closed; however, the macroscopic shape is distorted by a budding of the epithelial layer. Furthermore, proliferating cells remain in (p)-state (see Fig. 10c). Thus, a regulating (p)→(n) switch ("n": non-proliferating) is necessary for correct reconstruction. Within the first scenario (see Fig. 10a), epithelia switch from (p) to (n) if attractive polar regions come into contact. A fibroblast i stops proliferating if it has more than four direct interaction partners j ($V_{ij} \neq 0$). In the second scenario (Fig. 10b) the (p)→(n) switch takes place if the intrinsic cell growth velocity falls below a threshold due to hard core interactions between nearest neighbors.
Figure 11 shows typical time series of simulation results during blastulation and gastrulation and crypt fission, For blastulation, gastrulation, and crypt fission (i) cell division is assumed to be directional, and (ii) cell polarity has been assumed to give rise to a bending force, both to ensure that a single-layered tissue structure is maintained [20, 21, 22]. Due to the very fast cell turnover in intestinal crypts, directed influences on cell migration are large compared to stochastic influences. Hence, the stochastic component of cell migration may be neglected, leading to the following equation of motion for each individual cell i (again, neglecting inertia forces):

$$\gamma \frac{d\underline{r}_i}{dt} = \underline{F}_i \,, \tag{8}$$

where $\underline{r}_i$ is the position of cell i, γ is the friction constant, and $\underline{F}_i$ is the total force exerted on cell i containing nearest-neighbor cell-cell interaction forces, bending forces, and forces exerted by the basal membrane and the surrounded connective and muscle tissue [22].

An essential advantage of this approach is that the model parameters can be linked to experimentally measurable kinetic and biomechanical quantities [20, 22, 23, 24]. This

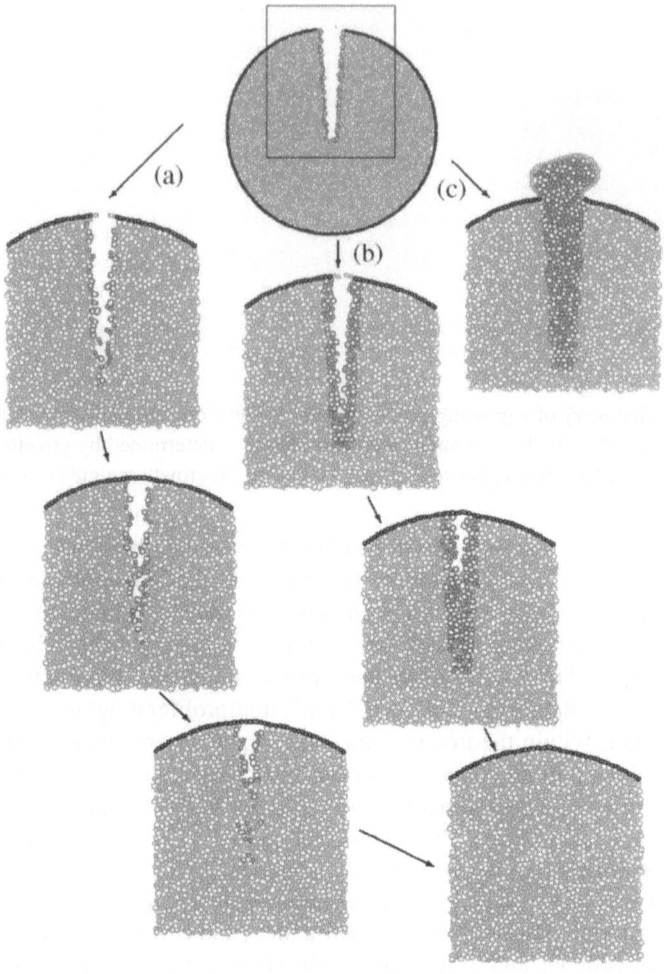

Figure 10. Regeneration after a cut-like lesion of a 2D model tissue, composed of epithelial cells surrounding fibroblasts. Cells may switch from proliferating (p) to differentiated, non-proliferating (n) state as described below. (For color codes description see color plate 10.) The scenarios (a)–(c) show different mechanisms to close the lesion. In order to prevent budding (shown in (c)) and obtain a correct closure of the lesion as in (a), (b), an active switch from (p) → (n) is necessary. For a detailed explanation of the different scenarios, see text. Scenario (a) leads to a gradual differentiation of fibroblasts, and (b) implies a (p)→(n) switch of most of the regenerated cells only after the cut has been filled.

opens the door to a quantitative model of the behavior of isolated cells and cells in a tissue within the same approach. The link between model and experimental parameters can be established by first defining a length, time, and energy scale (e.g., the cell diameter,

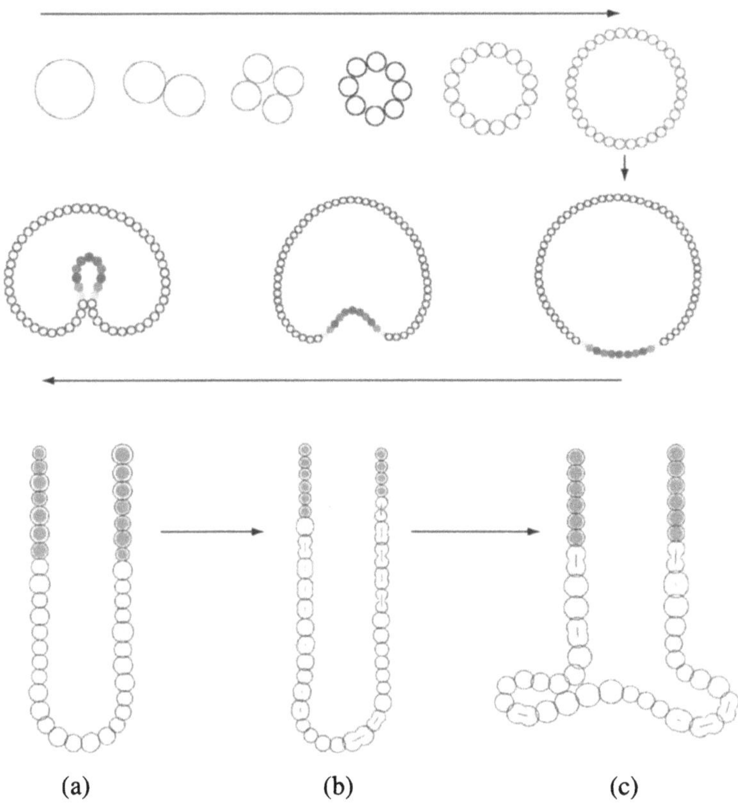

Figure 11. *Upper panel*: Blastulation and gastrulation in a 2D model of *Synapta digita* or sea urchin [20]. Colored cells have a negative spontaneous curvature and are responsible for the invagination during gastrulation. *Lower panel*: Typical fission scenario in a crypt [22]. Dark cells denote mature cells, hollow circles denote cells capable to divide. Between (a) and (b) is a time period of five cell cycle times. The crypt is stable. At (b) the elastic Young modulus of the one-layered tissue is reduced by a factor of 5. This destabilizes the crypt and results in fission (c). The fission is a consequence of a buckling instability [21]. (See color plate 11.)

the cycle time, and the effective energy F_T [6]). The final step is to express the diffusion constant, the potential shape parameters, and the step sizes for cell growth and cell movements as functions of these scales and the elastic moduli. Computer simulations performed with this model have shown a surprisingly good quantitative correspondence to experimental findings in growing avascular tumor spheroids.

However, usually only cells with moderate deformability can be described appropriately within this approach (in one-layered tissues, strongly deformable cells may nevertheless often be modeled since there a cell is able to maintain its volume by elongating perpendicular to the layer). The position of cell membranes is not explicitly specified but may be approximated by Voronoi polygons using the centers of the cells as generators (Figs. 7, 8, see also [16].

Meineke et al. [67] studied a lattice-free individual-based model to study cell kinetics in intestinal crypts (Fig. 4). A model crypt was approached by a cylinder of constant size and geometry (Fig. 4). The model assumes a deterministic, friction-dominated cell migration dynamics (i.e., contributions to the cell migration due to noise are neglected as well as effects due to inertia) and elastic springs between neighboring cells (as in [22]). The interaction partners of an individual cell were determined by a Voronoi tessellation, which is updated after each move (either a cell migration, growth, or division step). In this case, only cells that share a common edge are subject to physical interactions. The interaction partners that are found in a Voronoi tessellation are nearly identical to those found by cell-cell distance in the limit $\delta \ll R$ (see Eq. (7)). A Voronoi tessellation in principle allows arbitrary shaped convex cells; hence, if cell-cell (or other) interactions result in significant changes in the cell volume, cell volume conservation must be insured explicitly. In the Voronoi approach this could in principle be achieved by adding a force term $\sim \frac{\partial(\sum_j (v_j - v_0)^2)}{\partial \underline{r_i}}$ to the equations of motion (such a term is also included in the model equations of Graner and Glazier [40], see above). Here, v_j is the momentary volume of cell j in the Voronoi tessellation, and v_0 is the volume of an isolated cell. But, since the shape of an individual cell is exclusively determined by its own position and the position of its neighbors, the description of cells at surfaces of cell assemblies require further extensions (see Fig. 8 and [17]). A further complication for the Voronoi-tessellations is that they are difficult to implement in $d > 2$ space dimensions; see also Alt et al. (this volume, Part III). There, the Voronoi-tessellation method is mentioned as a particular case of a general model framework defining cell-cell interactions *via* certain interaction fields.

An approach that does obey volume conservation (and therefore allows for large cell deformations) and takes into account the viscoelastic properties of cells has been proposed by Palsson and Othmer [76], and Palsson [77]. See Figure 12 for a sketch of their model cell (in $d = 2$ dimensions).

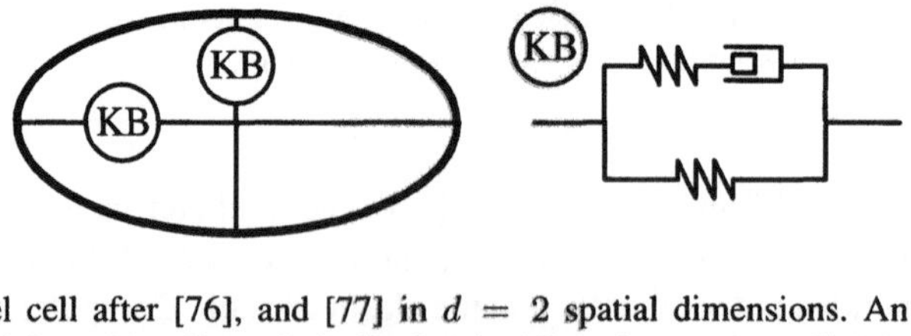

Figure 12. Model cell after [76], and [77] in $d = 2$ spatial dimensions. An individual cell is represented by a deformable ellipse (*left*). Each principle axis contains a Kelvin body (*right*). The size of the principle axis is a result of the cell-cell interaction.

Each cell is modeled as an ellipse (or by an ellipsoid in $d = 3$ dimensions). Along each axis of the ellipse is a Kelvin element that models the viscoelastic behavior of an individual cell. The basic equation of motion for the center of mass of cell i is

$$\underbrace{\mu_s \frac{A_{is}}{A} \underline{v}_i}_{(1)} + \underbrace{\mu_c \sum_j \frac{A_{ij}}{A} (\underline{v}_i - \underline{v}_j)}_{(2)} = \underline{F}^{act}_{i(j/s)} + \sum_j \left(\underline{F}^{pass}_{ij} - \underline{F}^{act}_{ji} \right) \tag{9}$$

Term (1) is a damping term modeling friction with the surrounding medium, while term (2) models friction between cells. Term (2) is of particular importance, e.g., in 3D cell sorting as in Figure 13, where cells are in direct contact and move on each other instead of migrating on a substrate. A_{is} is the common surface between cell i and the substrate, A_{ij} between cells i and j, A is the total cell surface, μ_c is the cell viscosity, μ_s is the substrate viscosity, and $\underline{v}_i$ is the velocity of cell i. $\underline{F}^{act}_{i(j/s)}$ is the reactive force caused by the active force cell i exerts on cell j or on the surfaces, and $\underline{F}^{pass}_{ij}$ and $\underline{F}^{act}_{ji}$ summarize adhesive, elastic, and active forces from neighbor cells on cell i. One possibility to solve this equation is to define an iterative procedure sweeping through the array of cells and performing at least one iterative step for each Δt. (The full model equations include an equation for the torque.) Figure 13 shows a figure of a sorting process in $d = 3$ dimensions simulated by this model [76, 77].

A limitation of the model is that the orientation of the principle ellipsoid axes leads to a non-biological anisotropy of the model demonstrated in Figure 14.

One possibility to overcome this problem is to completely triangulate the cell surface, as has been done for vesicles (e.g., [38, 39], see Fig. 15). In two space dimensions, such an approach has been used by Bottino et al. [8] for the crawling of nematode sperms. The authors of this article placed on every edge that connects two neighboring grid points of the triangular mesh a viscoelastic (Voigt-) element parallel to a tensile stress element. However, due to the high computational demands, only populations of a very small number of cells could be treated in that way.

Weliky and Oster [101] proposed a model for the morphogenesis of epiboly in the teleost fish *Fundulus* based on 2D polygonal cells. A very similar approach was used by Weliky et al. [102] for modeling the cell behavior underlying tissue convergence and extension during Notocord morphogenesis in *Xenopus laevis*. Epiboly describes the movement of epithelial sheets (usually of ectodermal cells), which spread as a unit to enclose the deeper layer of the embryo [35, 105]. (In our model description we follow closely the original article by Weliky and Oster.) In each model cell, Weliky and Oster account for circumferential elastic forces and the force generated by hydrostatic and osmotic pressure. The mechanical forces are applied to the polygonal nodes of the cell boundary. A cell node protrudes or contracts when the nodal forces are not in mechanical equilibrium. In an epithelial sheet, adjacent cells share common boundary nodes; in this way, mechanical force is transmitted from cell to cell, mimicking junctional coupling. These junctional nodes can slide, and nodes may appear or disappear, so that the number of polygonal sides is variable. A schematic picture of an individual model cell is shown in Figure 16.

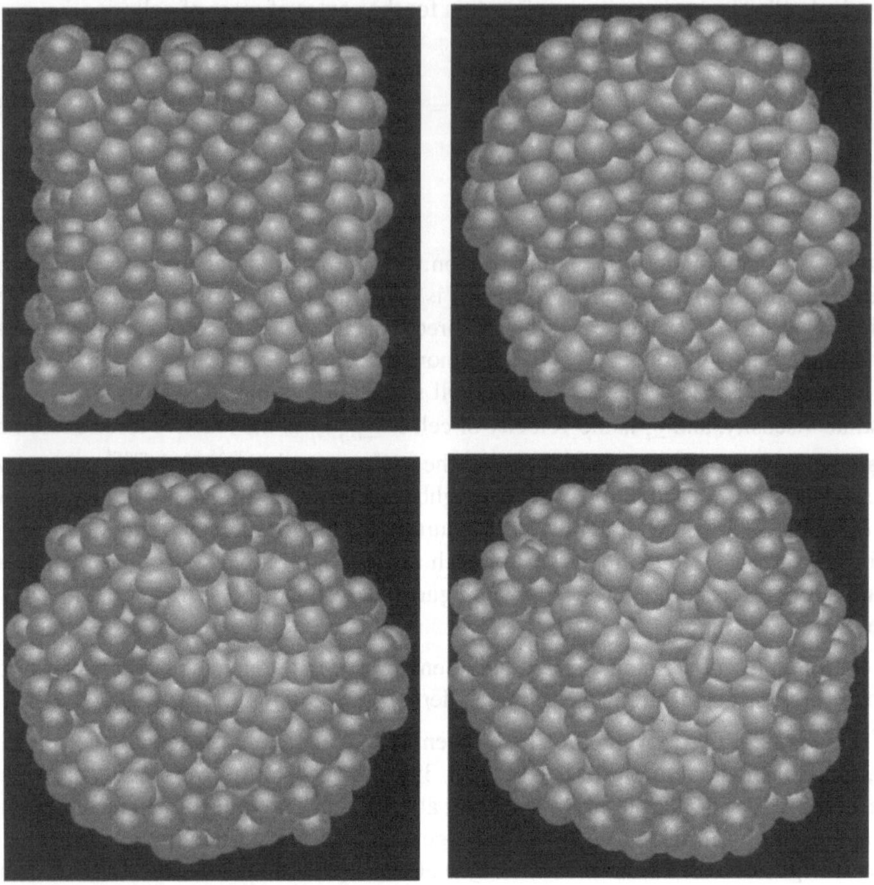

Figure 13. Typical time series of a cell-sorting process with the model of Palsson and Othmer [76] (pictures with courtesy by E. Palsson). (For description see color plate 12.)

The total force acting on an epithelial node is the sum of all net forces from all cells sharing this node plus the external forces:

$$\underline{F} = \sum_{i=1}^{N} \underline{F}_i + \underline{F}_{ext}. \tag{10}$$

Here,

$$\underline{F}_i = \underline{P} + \underline{T}_{i,i-1} + \underline{T}_{i,i+1} \tag{11}$$

is the net force acting at a cell node i. $\underline{T}_{i,i-1}$ and $\underline{T}_{i,i+1}$ denote the elastic forces from nodes $i-1$ and $i+1$ on node i, respectively. The elastic forces summarize the effects of both the contractile machinery of circumferential actomyosin microfilament bundles and

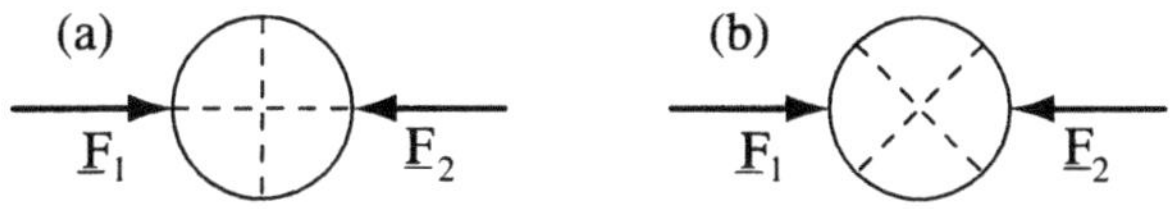

Figure 14. Illustration of an anisotropy of cell deformation in the model by Palsson and Othmer [76], and Palsson [77] (shown for the analogue in $d = 2$ dimensions where each cell is represented by an ellipse instead of an ellipsoid) for different orientations of the principle axes of the ellipse (broken lines). In both pictures the cells are subject to the same two forces $\underline{F}_1$ and $\underline{F}_2$. Although both situations are physically equivalent, the cell deformation would not be the same in the situation presented in the two pictures, since it depends on the orientation of the principle axes. In the situation presented in the *left* picture (a) $\underline{F}_1$ and $\underline{F}_2$ would result in a compression of the cell along the horizontal axis of the ellipse because the horizontal axis is parallel to the forces. At the same time, the cell would elongate into the vertical axis in order to maintain its volume. In the *right* picture (b) the forces act equally along both axes thus would compress both axes at the same time and reduce the cells' volume.

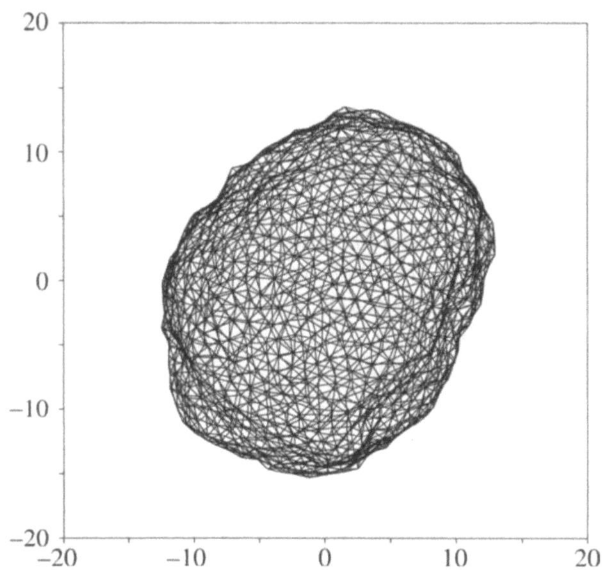

Figure 15. Triangulated surface of a vesicle (courtesy of G. Gompper, see also [38]).

the elastic forces of the cortical cytogel. The elastic forces on a node i are defined as

$$\underline{T}_{i,i-1} = \underline{\hat{T}}_{i,i-1} k_{elast} U \tag{12}$$

$$\underline{T}_{i,i+1} = \underline{\hat{T}}_{i,i+1} k_{elast} U. \tag{13}$$

Here, the index i enumerates the nodes of a cell in clockwise direction, $\underline{T}_{i,i-1}$ is a vector pointing from node $i-1$ to node i, $\underline{T}_{i,i+1}$ is a vector pointing from node $i+1$ to node i, U

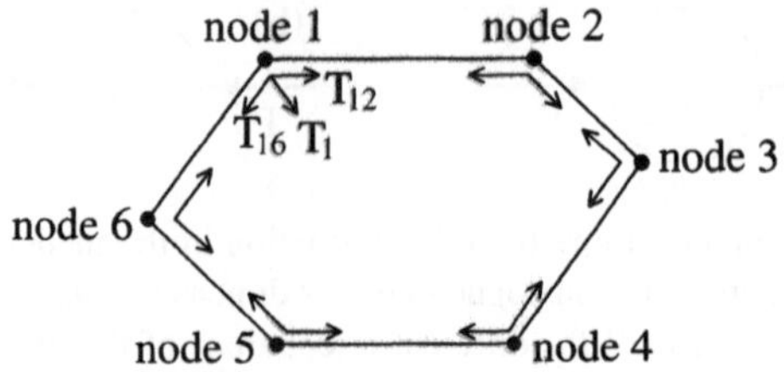

Figure 16. Model cell in Weliky and Oster [101]. An individual cell has a number of variable nodes (six in this picture) connected by springs. T_{ij} denotes the elastic force from node i on node j ($j \in \{i-1, i+1\}$). T_i is the resulting elastic force on node i from both of its neighbor nodes, $j = i-1$ and $j = i+1$. The elastic forces try to minimize the boundary length of the cell but are balanced by an internal pressure which tries to maximize the cell area (see text).

is the cell perimeter, and k_{elast} is the elastic modulus of the cell cortex. $\hat{\underline{T}}_{i,i-1}$ and $\hat{\underline{T}}_{i,i+1}$ are the corresponding unit vectors. The elastic forces tend to contract the cell boundary. The pressure comprises hydrostatic and osmotic contributions and is represented by a single 2D vector

$$\underline{P} = k\frac{\hat{\underline{P}}}{A} \tag{14}$$

directed outward from the cell boundary and bisecting the angle between the two line segments joined at a node, i.e., the pressure acts "normal" to the polygonal surface at each node. A is the cell area, and $\hat{\underline{P}}$ is a unit vector that describes the direction of the nodal pressure. The movement of a cell results from the movement of the nodes of the polygonal cell circumference. In the equation of motion, for each node a friction-dominated dynamics is assumed. Accordingly,

$$\gamma\frac{d\underline{x}}{dt} = \underline{F}(\underline{x}) \tag{15}$$

An isolated cell with many nodes would adopt a nearly circular shape, where the force due to the internal pressure balances the forces due to the elastic boundaries. This approach describes a cell as a polygon, as in a Voronoi tessellation. Different from the Voronoi tessellation, the cell shape is not solely determined by the positions of its neighbors. In the limit of infinitely many edges, an isolated cell becomes a perfect sphere in accordance with experimental findings in culture or suspension. Nevertheless, it is questionable whether the approach that assumes that the cell shape and size are a consequence of the balance of forces from elastic bundles in the cell periphery and the internal cell pressure is correct, or if the cytoskeleton does not determine the size and shape of an isolated cell (which would favor a model as in [74]).

Also other approaches have been studied. Honda [45, 46], Honda et al. [47], and Honda and Yamanaka [48] have modeled the effect of cell loss, cell division, and cell migration in a number of epithelial tissues and cell assemblies from early developmental stages by a Voronoi tessellation. In this approach, cell division is modeled by the center-gravity center method. A generator splits into two generators along the long axis of the polygon, resulting in one additional Voronoi cell. This (primary) Voronoi diagram provides the starting point for a secondary Voronoi diagram deduced from it by placing the new generators for the secondary Voronoi into the centers of gravity of the Voronoi cells of the primary diagram. The procedure is repeated until the Voronoi diagram eventually converges towards a stationary diagram. A similar geometric approach has been proposed by Ransom and Matela [82] in three space dimensions.

An off-lattice computer toolbox for simulating basic biological and pathological processes (in $d = 2$ dimensions) has been proposed by Stekel et al. [91] and Vawer and Rashbass [99]. In the former paper, they consider static squamous epithelia. In their model an individual cell is represented by a circle performing an over-damped movement under the influence of forces from neighbor cells and chemicals. They included the possibility of different cell types, physical forces, cell division and differentiation (e.g., the existence of pedigrees), chemical factors, aging, and cell death. Unfortunately, their papers lack a precise model description. Czirok et al. [13] (see also the contribution in this volume) studied the formation of complex bacterial colonies modeling bacterias as self-propelled, hard-core particles capable of moving in continuum space. They take into account nutrient diffusion, reproduction, and sporulation of bacteria. A lattice-free growth model (applicable, e.g., to avascular tumors) has been studied by Wang et al. [100], who used an off-lattice version of the Eden model (version C) where each cell is represented by a circle in continuous space. Cell division is implemented by adding a circle (daughter cell) to the neighborhood of an existing circle (mother cell) in such a way that the daughter and the mother cell touch each other. Thereby, only those cell divisions are allowed that do not result in overlaps with other circles. Although this model does not operate on a lattice, it still keeps the lower length scale of a cell diameter. A similar off-lattice approach has been used by Kreft et al. [56] to model bacterial colony growth where an emphasis has been put on the metabolism of the bacteria.

4. Alternative Approaches

Clem et al. [12] proposed a dynamic simulation model of epithelial tissue renewal in nasal epithelia, also approximating a cell by a hard core sphere. Different from the other approaches, during cell division one daughter cell takes the location of the mother cell, and the second daughter is placed into the nearest empty space large enough so that no overlaps to other cells occur. The nearest neighbors of the mother cell are defined using a *Gabriel graph*. Unfortunately, from their model description it is unclear whether the algorithms obey physical laws and principles.

A purely topological approach has been proposed by Dubertret and Rivier [25], and Dubertret et al. [26] for the renewal of the epidermis based on 2D foams. A monolayer of cells is described as a froth where each cell is characterized by the number of its edges. Elementary moves are the exchange of neighbors, in which two cells gain a side while two other cells lose a side, cell division, and detachment of cells. The time evolution of the probability density of the number of sides $p_s(t) = N_s(t)/N(t)$ (N_s: no. of s-sided cells, N: total no. of cells) is presented in an equation of a mean field type (for the one-point function).

$$\frac{dp_s}{dt} = \sum_{k\geq 3} p_k P_m(k) D_m(k) \left\{ \underbrace{-\delta_{ks}}_{(i)} + \underbrace{\Gamma(k \to s)}_{(ii)} + \underbrace{\frac{2}{k} M_{s-1}(k)}_{(iii)} - \underbrace{\frac{2}{k} M_s(k)}_{(iv)} - \underbrace{p_s}_{(v)} \right\}$$
$$+ \sum_{k\geq 3} p_k P_d(k) D_d(k) \left(\underbrace{-\delta_{ks}}_{(vi)} - \underbrace{M_s(k)}_{(vii)} + \underbrace{\sum_{i=-1}^{s-3} M_{s-i}(k) a_i(k)}_{(viii)} + \underbrace{p_s}_{(ix)} \right) \tag{16}$$

(i) describes division of an s-sided cell, (ii) a k-sided dividing cell that has an s-sided daughter ($\Gamma(k \to s)$ is the conditional probability that a k-sided cell divides and has an s-sided daughter), (iii) the effect of a dividing cell on a neighboring (s-1)-sided cell, (iv) the effect of a dividing cell on a neighboring s-sided cell, (v) the production of an extra cell during mitosis (even if no s-sided cell is affected by a mitosis, the total number of cells increases hence the fraction of s-sided cells $p_s = N_s/N$ decreases). The subscript "m" describes mitosis, the index "d" detachment. $P_m(k)$ is the conditional probability that an existing k-sided cell divides, $D_m(k)$ the number of mitosis of k-sided cells per unit of time. (viii) describes affected neighbors that have (s-i) sides ($-1 \leq i \leq s-3$) before disappearance, where $a_i(k)$ is the conditional probability that a k-sided detaching cell gives i sides ($-1 \leq s \leq k-4$) to one of its neighbors. For the origin of the terms (vi), (vii), and (ix), compare to the explanation of the terms (i), (ii), (iv), and (v), respectively. $M_s(k)$ is the mean number of k-sided cells neighboring an s-sided cell and depends on a structural parameter σ. p_6 (the probability, that a cell has six edges) and the second and third cumulants (variance μ_2 of the number of sides, skewness μ_3 of the distribution) can be shown to characterize the stationary solutions of the equation for different choices of parameters. Because the $\mu_2(p_6)$ is supposed to be universal for $p_6 > 1/3$, p_6 and μ_3 become the essential quantities to characterize one-layered tissues. The approach is applied to epithelia of cucumber, amnion cells of the early embryo, and cultured MDCK cells. Due to a unique relationship between μ_3 and p_6 for a given division kernel and transition rate found in their simulations, the authors claim that from knowing μ_3 and p_6, which are accessible to experiments, one can conclude the dynamics.

A completely different approach has been proposed by Lindenmeyer [60, 61], and Herman and Rozenberg [43] and extended by Fraccia et al. [32] and Prusinkiewicz [80] based on *generative grammars*. The application of generative grammars to the description of growing spatial structures was originally formulated as a growing cellular array of finite automata. Figure 17 shows an illustrative example applied to the development of *Callithamnion roseum*, a particular red algae (Lindenmeyer, [60, 61]). Lindenmeyer defined a production system by (i) a function $\delta(p, q) = r$, which yields the next state r to a given internal state p and an input sequence q; (ii) a function $\lambda(p, q) = u$, which maps a given internal state p and an input sequence q on an output sequence u; (iii) a set of internal states $p \in \{a, b, c, \dots\}$; (iv) a number of input states $q \in \{1, 2, 3, \dots\}$ and (v) a set of output states $u \in \{\alpha, \beta, \gamma, \dots\}$. The sets are usually finite but may be infinite as well. An example for bud formation is shown in Figure 17.

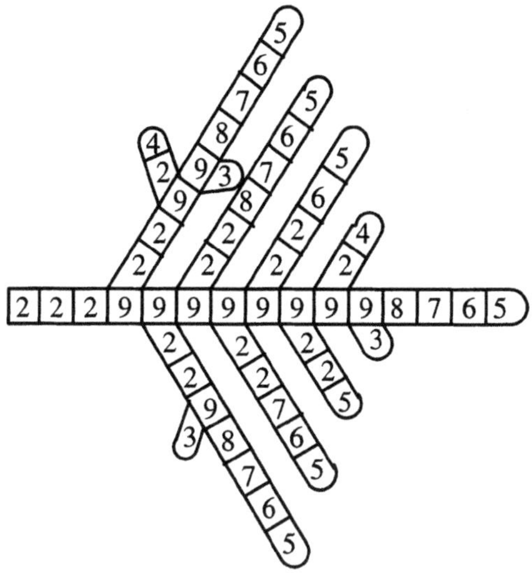

Figure 17. Bud formation by a Lindenmeyer system. Given is a number of states $p \in \{1, 2, 3, \dots, 9\}$ and the production rules $\delta(p = 1) = 2 * 3$, $\delta(p = 2) = 2$, $\delta(p = 3) = 2 * 4$, $\delta(p = 4) = 2 * 5$, $\delta(p = 5) = 6 * 5$,$\delta(p = 6) = 7$, $\delta(p = 7) = 8$, $\delta(p = 8) = 9 * [3]$, $\delta(p = 9) = 9$. The symbols "[", "]" describe the opening and closing of a new branch. The initial state was $p = 1$.

5. Discussion

In this article we briefly reviewed a number of different individual-based models of different granularity to multicellular systems. Individual-based models describe the local dynamics of individual cells, as opposed to continuum models that describe a locally averaged quantity over many individuals (a density). A continuum approach is reasonable if the density varies only smoothly on the length scale of a single cell. For example, the quantity "cell density" is in general not appropriate to model the noisy but sharp surface structure of a solid tumor, while the distance $\underline{r}$ from the center of mass is often an appropriate quantity. Hence, the study of surface properties and bulk properties of growing cell aggregates usually requires the consideration of different quantities in separate continuum approaches for the bulk and the surface. Cellular automaton models often provide a reasonable starting point if one wants to construct a continuum description from microscopic dynamics. They are able to model surface as well as bulk properties. In any case, individual-based models represent a direct approach that allows us to take into account system changes on a length scale of an individual cell or below. The appropriate selection

of a particular model approach depends on the given biological situation and the precise question that should be answered. Those cellular automata, where one cell is identified with one lattice site, often are appropriate at least as a first approach, despite their shortcomings described in the article, since they are often not too difficult to implement and have only moderate CPU time requirements. A more physical approach favors lattice-free models without a fixed lower length scale of a cell diameter. Such models often allow us to include elastic or viscoelastic properties of individual cells at least to some extent. This in indispensable for modeling all of those biological situations in which the shape or shape changes of individual cells can influence the dynamics or geometry of a large population of cells. Clearly, a finer resolution results in a larger number of model parameters, and requires more programming effort and significantly more CPU time.
It is appealing if an individuum-based approach is able to model isolated cells as well as cells interacting in a tissue. This is the case, e.g., for the approaches of Weliky and Oster [101], Drasdo et al. [16], and Stekel et al. [91]. The polygonal cell shape in epithelial tissues is often well presented, for instance, by Voronoi tessellation as used, e.g., by Honda et al. [45], Meineke et al. [67], and in modified versions by Ransom and Matela [82], Graner and Swada [41], and Drasdo et al. [17]. A "pure" Voronoi approach does not allow the description of isolated cells by construction since the shape of a single Voronoi cell is solely defined by the position of its neighbors. Odell et al. [74] and Weliky and Oster [101] used biomechanical motivated cell descriptions. Graner and Glazier [40] used an approach from statistical physics (the Potts model) modified and extended in order to model multicellular systems. Within their approach, a cell's shape is determined by the cell surface tension and its attractive interactions to its nearest neighbors and the substrate. A very fine spatial resolution of cell shape would require a triangulation of the cell surface as done for vesicles, e.g., by Gompper and Kroll [38] combined with a representation of the cytoskeleton network anchored in the cell surface.
One might be confused about the various different approaches to cell dynamics. Some authors use a Monte-Carlo approach (e.g., [16, 40]), others solve equations of motion either for each individual cell (e.g., [22, 50, 76, 77]) or for each node of the polygon that characterizes the cell membrane (e.g., [74, 101, 102]). However, all approaches assume a migration dynamics that assumes that the effect of friction dominates that of inertia. Friction results from interactions of a cell with surrounding molecules and neighbor cells. Isolated, cultured cells [87] as well as embryonic cells [72] in a sea of other cells perform a random movement. The origin of the random movement of an individual cell may be due to an active random movement of a cell, which by this active type of movement explores its environment, or due to collisions with surrounding cells or molecules. On time and length scales large compared to the microscopic time and length scales of molecules surrounding the cells, a natural mathematical description is to consider each cell as a separate object, model the interaction with other cells explicitly, and subsume effects due to collisions with molecules surrounding each cell by a noise term (for a similar line of argument in complex fluids, see, e.g., Rapaport, [83]). The noise term can be neglected if, e.g., cells move along chemotactic gradients or if the influence of directed forces is sufficiently large compared to the random effects by collisons of a cell with surrounding molecules.

The dynamics of quantities that average over many degrees of freedom (such as the number of cells) can often be modeled by continuum equations. In principle they can be obtained from solving a hierarchy of moment equations obtained from the dynamical equations for the joint probability for a many-particle system (if the latter is known) (e.g., [73, 98]). In practice the equations for different moments are often coupled, which requires termination of the hierarchy at some point. Even then, the equations are mostly very complicated. However, in the case of independent particles, a system of equations for particle densities can be derived rigorously from an individual particle description in the limit where the number of particles tends to infinity. The same is true if the particle/particle interactions decay sufficiently fast in space and the interaction range (given, e.g., by the width of an interaction weight function) scales in an appropriate way with the particle number (e.g., [75, 88, 94]).

What can be achieved by a computer model? Clearly, it is not possible to give any (mathematical) proof by means of a computer model; the possibility of a proof is reserved to analytical investigations alone. A computer simulation may be regarded as an *experiment in silico*, i.e., a caricature of an experimental situation focused on particular system properties. This allows for testing of biological hypotheses and study of experimental situations undisturbed by unknown or uncontrolled influences. Despite, or better, because of, the many possibilities opened by computer simulations, it is very important to verify simulation results whenever possible by mathematical analysis or at least by simple mathematical arguments. (Otherwise one cannot completely exclude studying of an error in the program code instead of a biological effect.)

Often biologists complain that a computer model cannot be realistic at all since it does not include all known details about a given biological system or because it does not deal with living material. Both are clearly true to some extend. A mathematical or computer model can only give hints saying which mechanisms agree or disagree with a specific biological situation. The hallmarks of a model are correct predictions that allow us to identify those experiments that are particularly suited to distinguish between competing hypotheses. A model cannot prove which biological mechanism out of a number of possible mechanisms is correct in a specific biological situation. The identification of the correct mechanism is reserved to an appropriate experiment, usually in the target organism, alone. This, however, is also true for animal models to humans. For example, the mouse model as an approach to humans also very often can only give hints. Nevertheless, nobody generally questions that a mouse model is useful. In this sense it is to some extent also a matter of habituation what is considered to be useful or not.

An essential question is which quantities in models and experiments should be compared. An important class of candidates are generic properties, such as generic regimes of growth curves, surface exponents of growing tumors, etc. The generic properties are very robust and do not depend on model details. Instead, they belong to a whole class of systems with common properties, e.g., range and type of interaction, spatial dimension, and symmetries. The study of this class of candidates is particularly suitable if biological details are not known (which is usually the case) or if essential processes should be identified and separated from processes that are irrelevant for a studied biological effect. The study of generic properties also implies to apply the concept of universality. This is to classify biological system behavior due to certain characteristic features, often on a microscopic length scale, that show up in the macroscopic behavior, e.g., in the value of an exponent.

Another class of candidates for the comparison of experiments and models is quantities that allow us to distinguish between very specific situations. This requires much more detailed knowledge about the system.

Acknowledgments
I thank W. Alt, T. Bretschneider, A. Deutsch, J. Galle, and A. Stevens for a critical reading of the manuscript or parts of it, and M. Kruspe for performing the computer simulation that leads to Figure 3. Support by the Deutsche Forschungsgemeinschaft under Grant No. LO 342/4-3 is gratefully acknowledged.

References

[1] Agur, Z., Daniel, Y. and Ginosar, Y (2002) The universal properties of stem cells as pinpointed by a simple discrete model. *J Math Biol* **44**(1), 79–86

[2] Anderson, A (2000) Mathematical modeling of tumor invasion and metastasis. *J Theor Med* **2**, 129–154

[3] Ashcroft, N.W. and Mermin, N.D (1976) *Solid state physics.* (Hold Sounders International)

[4] Batchelor, M.T., Henry, B.I (1991) Limits to Eden growth in two and three dimensions. *Phys Lett A* **157 (4.5)**, 229–236

[5] Ben-Jacob, E., Cohen, I. and Levine, H (2000) Cooperative self-organization of microorganisms. *Adv in Phys* **49 (4)**, 395–554

[6] Beysens, D., Forgacs, G. and Glazier, J.A (2000) Cell sorting is analogous to phase ordering in fluids. *PNAS* **97 (17)**, 9467–9471

[7] Börner, U., Deutsch, A., Reichenbach, H. and Bär, M (2001) Rippling in myxobacterial aggregates – Patterns arising from cell-cell collisions. *Preprint Max-Planck-Inst. f. Physics of Complex Systems* (Dresden)

[8] Bottino, D., Mogilner, A., Roberts, T., Steward, M., and Oster, G. (2002) How nematode sperm crawl. *J. Cell Sci.* **115**, 367–384

[9] Bretschneider T, Vasiev B, Weijer CJ (1997) A model for cell movement during *Dictyostelium* mound formation. *J Theor Biol* **189**(1), 41–51

[10] Bru, A., Pastor, J.M., Fernaud, I., Melle, S., and Berenguer, C. (1998) Super-rough dynamics on tumor growth. *Phys Rev Lett* **81**(18), 4008–4011

[11] Bussemaker, H.J., Deutsch, A. and Geigant, E (1997) Mean-field analysis of a dynamical phase transition in a cellular automaton model for collective motion. *Phys Rev Lett* **78**(26), 5018–5021

[12] Clem, C.J., König, D., and Rigaut, J.P (1997) A Three-dimensional dynamic simulation model of epithelial tissue renewal. *Anal Quant Cytol Histol* **19**(2)

[13] Czirok, A., Ben-Jacob, E., Cohen, I., and Vicsek, T (1996) Formation of complex bacterial colonies via self-generated vortices. *Phys Rev E* **540**(2), 1791–1801

[14] Davidson, L.A., Koehl, M.A.R., Keller, R. and Oster, G.F (1995) How do sea urchins invaginate? Using bio-mechanics to distinguish between mechanisms of primary invagination. *Development* **121**, 2005–2018

[15] Deutsch A (1999) Cellular automata and biological pattern formation *Habilitation Theses* University of Bonn, Germany

[16] Drasdo, D., Kree, R. and McCaskill, J.S (1995) A Monte Carlo model to tissue cell populations. *Phys Rev E* **52 (6)**, 6635–6657

[17] Drasdo, D (1994) Monte Carlo Simulationen in zwei Dimensionen zur Beschreibung von Wachstumskinetik und Strukturbildungsphänomenen in Zellpopulationen. *PhD-thesis* (Göttingen, Nov. 1993 and Verlag Shaker) Aachen (ISBN 3-86111-785-1)

[18] Drasdo, D (1996) Different growth regimes found in a Monte Carlo model of growing tissue cell populations. In: *Self organization of complex structures: From individual to collective dynamics.* (F. Schweitzer editor) Gordon and Breach, London, 281–292

[19] Drasdo, D (1998) A Monte-Carlo approach to growing solid non-vascular tumor. In: *Networks in Biology and Physics.* (Beysens, D., Forgacs, G. eds.) Springer, 171–185

[20] Drasdo, D. and Forgacs, G (2000) Modeling generic and genetic interactions in Cleavage, Blastulation and Gastrulation. *Dev Dyn* **219 (2)**, 182–191

[21] Drasdo, D (2000) Buckling instabilities in one-layered growing tissues. *Phys Rev Lett* **84**, 4424–4427

[22] Drasdo, D. and Loeffler, M (2001) Individual-based models to growth and folding in one-layered tissues: Intestinal crypts and blastulation. *Nonl Analysis* **47**, 245–256

[23] Drasdo, D. and Höhme, S (2001) Towards a quantitative single-cell based model approach to growing multicellular spheroids. *Int. Workshop on Deformable Modeling and Soft Tissue Simulation.* (accepted)

[24] Drasdo, D. and Höhme, S (2002) Individual Based Approaches to Birth and Death in Avascular Tumors. *Math Comp Mod* (in press)

[25] Dubertret, B. and Rivier, N (1997) The renewal of the epidermis: A topological mechanism. *Biophys J* **73**, 38–44

[26] Dubertret, B., Aste, T, Ohlenbusch, H.M., and Rivier, N (1998) Two-dimensional froths and the dynamics of biological tissues. *Phys Rev E* **58**(5), 6368–6378

[27] Düchting, W. and Vogelsänger, Th (1985) Recent progress in modelling and simulation of three-dimensional tumor growth and treatment. *BioSystems* **18**, 79–91

[28] Düchting, W., Ulmer, W. and Ginsberg, T (1996) Cancer: A challenge for control theory and computer modelling. *Europ J Cancer* **32A**(8), 1283–1292

[29] Dunphy, J.E (1978) *Wound healing.* MedCom-Press, New York

[30] Eden, M (1961) In: *Proc. of the 4th. Berkeley Symposium on Mathematics and Probability, Vol. IV.* ed. by. J. Neyman (University of California Press)

[31] Forgacs, G., Foty, R.A., Shafrir, Y., and Steinberg, M.S. (1998) Viscoelastic properties of living embryonic tissues: a quantitative study. *Biophys J* **74**, 2227–2234

[32] Fracchia, F.D., Prusinkiewicz, P., and De Boer, M.J.M (1990) Animation of the development of multicelluar structures. In: *Computer Animation '90* (Magnenat-Thalmann, N. and Thalmann, D., eds.) Springer-Verlag Tokyo

[33] Freyer, J.P. and Sutherland, R.M (1985) A reduction in the in situ rates of oxygen and glucose consumption of cells in EMT6/Ro spheroids during growth. *J Cell Physiol* **124**, 516

[34] Gompertz, B (1825) On the nature of the function expressive of the law of mortality. *Phil Trans Roy Soc (Lond.)* **27**, 513–585

[35] Gilbert, S.F (1997) *Developmental Biology.* Sinauer Associates, Sunderland

[36] Glazier, J.A., Graner, F (1993) Simulation of the differential adhesion driven rearrangement of biological cells. *Phys Rev E* **47**(3), 2128–2154

[37] Godt, D. and Tepass, U (1998) Drosophila oocyte localization is mediated by differential cadherin-based adhesion. *Nature* **395**, 387–391

[38] Gompper, G., Kroll, D.M (1995) Driven transport of fluid vesicles though narrow pores. *Phys Rev E* **52**(4), 4198–4208

[39] Gompper, G., Kroll, D.M (1997) Fluctuations of polymerized, fluid and hexatic membranes: Continuum models and simulations. *Cur. Op. in Colloid & Interfacial Sci.* **2**, 373–381

[40] Graner, F., Glazier, J.A (1992) Simulation of biological cell sorting using a two- dimensional extended Potts model. *Phys Rev Lett* **9**(13), 2013–2016

[41] Graner, F. and Sawada, Y (1993) Can surface adhesion drive cell rearrangement? Part II: A geometrical model. *J Theor Biol* **164**, 477–506

[42] Halpin-Healy, T., Zhang Y.C (1995) Kinetic roughening phenomena, stochastic growth, directed polymers and all that. *Phys Rep* **254**, 215

[43] Herman, G.T. and Rozenberg, G (1975) *Developmental systems and languages.* (North Holland/American Elsevier)

[44] Hogeweg, P (2000) Evolving mechanisms of morphogenesis: On the interplay between differential adhesion and cell differentiation. *J Theor Biol* **203**, 317–333

[45] Honda, H (1978) Description of cellular patterns by dirichlet domains: The two-dimensional case. *J Theor Biol* **72**, 523–543

[46] Honda, H (1983) Geometrical models for cells in tissues. *International Review of Cytology* **81**, 191–248

[47] Honda, H., Kodama, R., Takeuchi, T., Yamanaka, H., Watanabe, K., Eguchi, G (1984) Cell behaviour in a polygonal cell sheet. *J Embryol exp Morph Supplement* **83**, 313–327

[48] Honda, H., Yamanaka, H. Dansohkawa, M. (1984) A computer simulation of geometrical configurations during cell division. *J Theor Biol* **106**(3), 423–435

[49] Honerkamp (1998) *J Statistical Physics* (Springer, Berlin, Heidelberg)

[50] Igoshin, O.A., Mogilner, A., Welch, R.D., Kaiser, D., and Oster, G (2001) Pattern formation and traveling waves in myxobacteria: theory and modeling. *Proc Natl Acad Sci* **18**; **98** (26), 14913–14918

[51] Isele, W.P. and Meinzer, H.P (1998) Applying computer modeling to examine complex dynamics and pattern formation of tissue growth. *Comput Biomed Res* **31**, 476

[52] Jiang, Y., Levine, H., Glazier, J (1998) Possible cooperation of differential adhesion and chemotaxis in mound formation of *Dictyostelium. Biophys J* **75**(6), 2615–2625

[53] Kaandorp, J.A., Lowe, C.P., Frenkel, D., Sloot, P.M.A. (1996) Effect of nutrient diffusion and flow on coral morphology. *Phys Rev Lett* **77**(11), 2328–2331

[54] Kam, Z., Minden, J., Agard, D., Sedat, J.W. and Leptin, M. (1991) Drosophila gastrulation: analysis of cell shape changes in living embryos by three-dimensional fluorescence mircroscopy. *Development* **112**, 365–370

[55] Kansal, A.R., Torquato, S., Harsh IV, G.R., Chiocca, E.A., Deisboeck, T.S. (2000) Simulated brain tumor growth dynamics using a three-dimensional cellular automaton. *J Theor Biol* **203**, 367–382

[56] Kreft, J.U., Booth, G. and Wimpenny, J.W.T (1998) BacSim, a simulator for individual-based modelling of bacterial colony growth. *Microbiology* **144**, 3275–3287

[57] Krug, J. and Spohn, D (1991) Kinetic roughening of growing surfaces. In: *Solids Far From Equilibrium* (ed. by C. Godreche) Cambridge Univ., NY

[58] Landau, D.P. and Binder, K. A (2000) *Guide to Monte Carlo simulations.* Statistical Physics, Cambridge Univ. Press

[59] G. Landini and J.W. Rippin (1993) Fractal fragmentation in replicative systems. *Fractals* **1**(2), 239–246

[60] Lindenmeyer, A (1968) Mathematical models for cellular interactions in development. I. Filaments with one-sided inputs. *J Theor Biol* **18**, 280–299

[61] Lindenmeyer, A (1968) Mathematical models for cellular interactions in development. II. Simple and branching filaments with two-sided inputs. *J Theor Biol* **18**, 300–315

[62] Loeffler, M., Stein, R., Wichmann, H.E., Potten, C.S., Kraur., P., Chwalinski, S (1986) Intestinal cell proliferation I. A comprehensive model of steady-state proliferation in the crypt. *Cell Tissue Kinet* **19**, 627–645

[63] Maree, A.F.M., Panfilov, A.V., Hogeweg, P (1999a) Migration and thermotaxis of Dictyostelium discoideum slugs, a model study. *J Theor Biol* **199**, 297–309

[64] Maree, A.F.M., Panfilov, A.V., Hogeweg, P (1999b) Phototaxis during the slug stage of Dictyostelium discoideum slugs, a model study. *Proc R Soc Lond Ser* **266**, 1351–1360

[65] Markus, M. and Hess, B (1990) Isotropic cellular automaton for modelling excitable media. *Nature* **347**(6288), 56–58

[66] Meakin, P (1993) The growth of rough surfaces and interfaces. *Phys Rep* **235**(4 & 5), 189–289

[67] Meineke, F.A., Potten, S.C. and Loeffler, M (2001) Cell migration and organization in the intestinal crypt using a lattice-free model. *Cell Prolif* **34 (4)**, 253–266

[68] Meinzer, H.P., Sandblad, B., and Baur, H.J (1992) Generation-dependent control mechanisms in cell proliferation and differentiation – the power of two. *Cell Prolif* **25**, 125

[69] Metropolis, N., Rosenbluth, A.W., Rosenbluth, M.N., Teller, A.H., and Teller, E (1953) Equation of state calculations by fast computing machines. *J Chem Phys* **21**, 1087–1092

[70] Molison, D (1972) Conjecture on the spread of infection in two dimensions disproved. *Nature* **240**(22), 467–468

[71] Mombach, J.C.M., Glazier, J.A., Raphael, R.C. and Zajak. M (1995) Quantitative comparison between differential adhesion models and cell sorting in the presence and absence of fluctuations. *Phys Rev Lett* **75**(11), 2244–2247

[72] Mombach, J.C.M., Glazier, J (1996) Single cell motion in aggregates of embryonic cells. *Phys Rev Lett* **76**(16), 3032–3035

[73] Nicolis, G. and Prigogine, I., (1967) *Self-organization in nonequilibrium systems.* (Wiley, New York)

[74] Odell, G.M., Oster, G., Alberch, P. and Burnside, B (1981) The mechanical basis of morphogenesis. *Dev Biol* **85**, 446–462

[75] Öhlschläger, K (1989) On the derivation of reaction-diffusion equations as limit dynamics of systems of moderately interacting stochastic processes. *Probab Th Rel Fields* **82**, 565–586

[76] Palsson, E. and Othmer, H.G (2000) A model for individual and collective cell movement in *Dictyostelium discoideum Proc Nat Acad Sc* **12**(19), 10448–10453

[77] Palsson, E (2001) A three-dimensional model of cell movement in multicellular systems. *Future Generation Computer Systems* **17**, 835–852

[78] Paulus, U., Loeffler, M., Zeidler, J., Owen, G. and Potten, C.S (1993) The differentiation and lineage development of goblet cells in the murine and small intestinal crypt: experimental and modeling studies. *J Cell Sci* **104**, 473–484

[79] Potten, C.S., Loeffler, M (1990) Stem cells: attributes, cycles, spirals, pitfalls and uncertainties. Lessons for and from the crypt. *Development* **110**, 1001–1020

[80] Prusinkiewicz, P (1994) Visual models of morphogenesis. *Artificial Life* **1**, 64–74

[81] Qi, A.-S., Zheng, X., Du, C.-Y., and An, B-S (1993) A cellular automaton model of cancerous growth. *J Theor Biol* **161**, 1–12

[82] Ransom, R. and Matela, R.J. (1984) Computer modelling of cell division during development using a topological approach. *J Embryol exp Morph Supplement* **83**, 233–259

[83] Rapaport, D.C (1995) *The art of molecular dynamics simulation.* (Cambridge Univ. Press)

[84] Richardson, D (1973) Random growth in a tessellation. *Proc Camb Phil Soc* **74**, 515–528

[85] Rivet, J.P., and Boon, J.P (2001) *Lattice Gas Hydrodynamics.* Cambridge University Press

[86] Savill, N. J., Hogeweg, P (1997) Modelling morphogenesis: From single cells to crawling slugs. *J Theor Biol* **184**, 229–235

[87] Schienbein, M., Franke, K., and Gruler, H (1994) Random walk and directed movement: comparison between inert particles and sel-organized molecular machines. *Phys Rev E* **49**(6), 5462–5471

[88] Spohn, D (1991) *Large scale dynamics of interacting particles.* Springer, Berlin, Heidelberg

[89] Steinberg, M.S (1964) The problem of adhesive selectivity in cellular interactions. In: *Cellular membranes in Development.* (M. Locke, editor) Academic Press, New York, 321–366

[90] Steinberg, M.S (1970) Does differential adhesion govern self-assembly processes in histogenesis? Equilibrium configurations and the emergence of a hierachy among populations of embryonic cells. *J Exp Zool* **173**, 395–434

[91] Stekel, D., Rashbass, J., Williams, E.D (1995) A computer graphic Simulation of squamous epithelium. *J Theor Biol* **175**, 283–293

[92] Stevens, A. and Schweitzer, F (1997) Aggregation induced by diffusing and non-diffusing media. In: *Dynamics of Cell and Tissue Motion.* (W.Alt, A.Deutsch, and G. Dunn, eds.) Birkhäuser, Basel, Switzerland, 183–192

[93] Stevens, A (2000a) A stochastic cellular automaton modeling gliding and aggregation of myxobacteria. *SIAM J APPL MATH* **61 (1)**, 172–182

[94] Stevens, A (2000b) The derivation of chemotaxis equations as limit dynamics of moderately interacting stochastic many-particle systems. *SIAM J Appl Math* **61**(1), 183–212

[95] Stott, E. L., Britton, N. F., Glazier, J. A., Zajac, M (1999) Stochastic simulation of benign avascular tumor growth using the Potts model. *Mathematocal and computer Modelling* **30**, 183–198

[96] Townes, P.L. and Holfreter, J (1955) Directed movements and selective adhesion of embryonic amphibian cells. *J Exp Zool* **128**, 3–120

[97] Turner, S. and Sherrat, J.A (2002) Intercellular adhesion and cancer invasion: A discrete simulation using the extended Potts model. *J Theor Biol* **216**(1), 85–100

[98] van Kampen, N.G (1992) *Stochastic Processes in Physics and Chemistry.* Elsevier, North Holland

[99] Vawer, A. and Rashbass, J (1997) The biological Toolbox: A computer program for simulating basic biological and pathological processes. *Computer Methods and Programs in Biomedicine* **22**, 203–211

[100] Wang, C.Y., Liu, P.L., and Bassingthwighte, J.B (1995) Off-lattice Eden-C cluster growth model. *J Phys: Math Gen* **28**, 2141–2147

[101] Weliky, M. and Oster, G (1990) The mechanical basis of cell rearrangement. I. Epithelial morphogenesis during Fundulus epiboly. *Development* **109**, 373–386

[102] Weliky, M., Minsuk, S., Keller, R., Oster, G (1991) Notochord morphogenesis in *Xenopus laevis*: simulation of cell behavior underlying tissue convergence and extension. *Development* **113**, 1231–1244

[103] Williams, T., Bjerknes, R (1972) Stochastic model for abnormal clone spread through epithelial basal layer. *Nature* **236**, 19–21

[104] Wolf, D (1989) Fractal Growth. In: *Computersimulations in Physics* Research Centre Juelich [in German]

[105] Wolpert, L (1998) *Principles of Development.* Oxford Univ. Press, Oxford

[106] Zajac, M., Jones, G.L. and Glazier, J.A (2000) Model of convergent extension in animal morphogenesis. *Phys Rev Lett* **85**(9), 2022–2025

Polymer and Cell Dynamics – Multiscale Modeling
and Numerical Simulations, 205–219
ed. by W. Alt, M. Chaplain, M. Griebel und J. Lenz

Individual-based Models of Cohort Migration in Cell Cultures

András Czirók[1,2], Andreas Deutsch[3] and Michael Wurzel[3]

1) Dept. of Anatomy and Cell Biology, Kansas University Medical Center, 3901 Rainbow Blvd WHW 1008, Kansas City, KS 66160, USA
2) Dept. of Biological Physics, Eötvös University, Budapest, Hungary
3) Center for High Performance Computing, Dresden University of Technology, D-01062 Dresden, Germany

Summary. With the aim of understanding the emergence of collective migration from local interactions of organisms, we study biologically inspired, inherently non-equilibrium models consisting of self-propelled particles. We focus on the implications of both an intrinsic motility and a local alignment interaction on pattern formation.

We study an off-lattice and a cellular automaton (CA) model and two basic interactions between neighboring oriented cells. Density-dependent interaction (CA only) provides a model of differential adhesion, while orientation-dependent interaction (both models) yields a model of collective motion or swarming. Analysis by correlation function (off-lattice) and linear stability of the Boltzmann equation (CA) are shown.

Both analyses permit us to deduce important orientational and spatial aspects of simulation outcomes.

1. Introduction

Flocking is a typical example of cooperative animal behavior. Organisms in flocks are characterized by active migration – thereby adjusting their movement according to the distribution and velocities of surrounding animals. From an abstract perspective, flocks may therefore be viewed and modeled as "active self-propelled particles". A similar phenomenon, cohort formation and alignment, can also be observed at the level of cells, e.g., *in vitro*, during the aggregation of bacteria or when groups of eukaryotic cells migrate in a correlated fashion over long times and distances (Fig. 1). Cohort migration of cells is also particularly prevalent *in vivo* during embryogenesis, when groups of cells have to migrate to their final destination within organs or tissues.

Short- and long-range interactions can be considered as triggers of cell alignment and pattern formation. Typical long-range interactions can involve mechanical stress fields or diffusing chemicals [16, 19]. Short-range interactions usually involve adhesive forces and cell-cell signaling. In many cases of interest (when the lack of extracellular matrix and the contact with a rigid substrate do not allow the buildup of mechanical tension

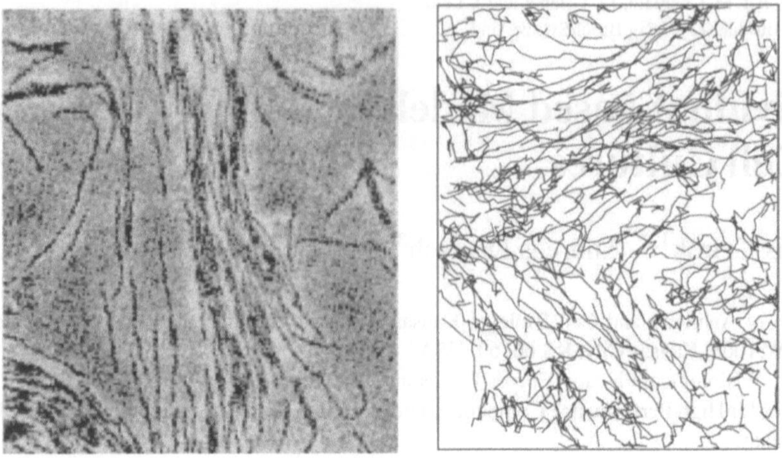

Figure 1. Left: Cell swarms of Myxobacteria (*Myxococcus xanthus*); individual bacteria (length $\approx$ 3–5 μm) and streets of aligned bacterial cells can be distinguished (after [17]). Right: *In vitro* trajectories of cells obtained from the surgical specimen of a highly malignant *Glioblastoma multiforme* brain tumor (from [17]). At high cell densities, a partially ordered migration is apparent. The horizontal extension of the field is 560 μm, trajectories for 24 hours are shown (after [11]).

fields), cell alignment can be a result of short-range interactions: it is merely based on the influence of the cell's nearest neighbors [9]. Here we focus on the implications of *both* an intrinsic motility *and* a local alignment interaction on pattern formation. A corresponding continuum theory was also introduced in the form of generalized Navier-Stokes equations [21, 22]. While hydrodynamic equations are adequate for extended systems with huge cell numbers, if one is interested in the behavior of small cell populations (see, e.g., Fig. 1), discrete (individual-based) models have to be chosen. Here we review individual-based models of locally interacting self-propelled and oriented entities.

The self-propelled particle models can be viewed as transport related, non-equilibrium analogues of ferromagnetic models. In the statistical physical description of magnetic materials, an energy value (E) is associated with each configuration of the magnetic moments (spins). This energy value determines the probability of a certain spin configuration through the relation $p \sim \exp -E/kT$, where T is the temperature and k is a conversion factor. The mapping between energy values and spin configurations, the *Hamiltonian* function, is usually a sum of terms. Each of these terms is an expression composed of the orientation of neighbor spins, interpreted therefore as a local interaction. In models describing ferromagnetic materials, E is decreased (hence the probability of such states is increased) if adjacent spins are oriented parallel to each other, as

$$E \sim -\sum_{\langle i,j \rangle} \vec{S}_i \vec{S}_j, \tag{1}$$

where the summation goes over adjacent spin pairs, and the direction of the spins is denoted by the vector $\vec{S}$. Whether such local interactions can result in a global order, i.e., non-vanishing probability of spin configurations with the majority of the moments pointing in the same direction, depends on the details of the model. In particular, one can show that two-dimensional spin systems, in which each moment can point into any direction without restrictions, do *not* have an ordered state [12]. Instead, the most probable configuration consists of vortices, with most spins aligned locally.

In the case of self-propelled particle models, the Hamiltonian (1) tendency to align the spins is replaced by a rule of dynamics, and the temperature is replaced by the amplitude of the random perturbations or inverse "sensitivity". Nevertheless, in the case of zero-velocity, i.e., if the cells do not move but only reorient, the model can be reduced to the two-dimensional system of rotating spins described by Eq. (1).

The essential variables in our analysis are the hydrodynamic variables density and velocity, yet the flow fields emerging in these models considerably differ from those usually observed in fluid behavior. This difference is due to the inherent lack of Galilean invariance in the flock models. While the equations of hydrodynamics are invariant to velocity addition (i.e., to the $\vec{v} \longrightarrow \vec{v}+\vec{b}$ transformation whith $\vec{b} = const$), the self-propelled particles have preferred velocities; therefore, the rules of their dynamics cannot have this symmetry.

Groups moving together can be observed in both the off-lattice and cellular automaton models for a certain range of control parameters. While the off-lattice version gives a rather detailed picture of the alignment dynamics, the essential elements are also contained in the coarse-grained cellular automaton description, which, in addition, allows a detailed analysis of the phase diagram on the level of the corresponding Boltzmann equation.

2. The Off-lattice Self-propelled Particle Model

The simplest model of collective motion of self-propelled particles, introduced in [23], consists of particles moving on a plane. The particles are characterized by their (off-lattice) location $\vec{x}_i$ and velocity $\vec{v}_i$ pointing in direction ϑ_i. The self-propelled nature of the particles is manifested by keeping the magnitude of the velocity fixed to v_0. Particles interact through the following local rule: at each time step a given particle assumes the *average direction of motion* of the particles in its local neighborhood $S(i)$ (e.g., in a circle of some given radius centered at the position of the ith particle) with some uncertainty, as described by

$$\vartheta_i(t + \Delta t) = \langle \vartheta(t) \rangle_{S(i)} + \xi, \tag{2}$$

where the noise ξ is a random variable with a uniform distribution in the interval $[-\eta/2, \eta/2]$. The locations of the particles are updated as

$$\vec{x}_i(t + \Delta t) = \vec{x}_i(t) + \vec{v}_i(t)\Delta t \tag{3}$$

with $|\vec{v}_i| = v_0 = const.$

2.1. Numerical Simulations

The model defined through Eqs. (2) and (3) with periodic boundary conditions and circular interaction range was studied by performing large-scale simulations. Due to the simplicity of the model, only two control parameters should be distinguished: the (average) density of particles $\varrho = N/L^2$ (where N is the number of particles and L is the system size in units of the interaction range) and the amplitude of the noise η. Depending on the value of these parameters the model can exhibit various types of behaviors, as Figure 2 demonstrates.

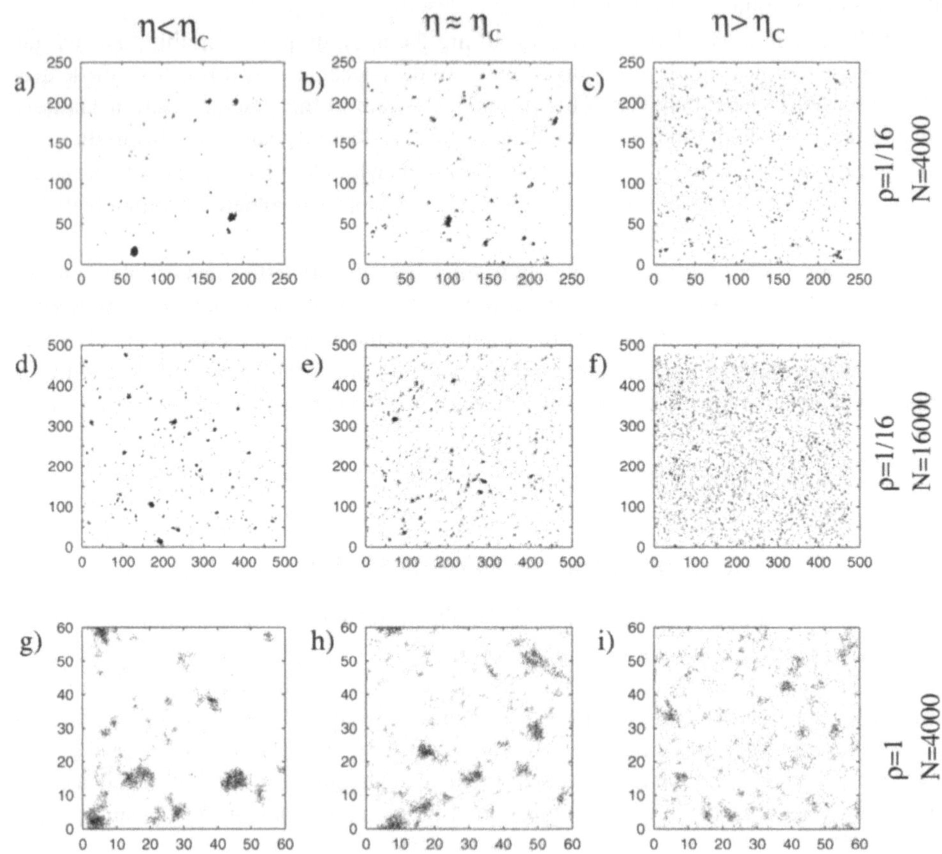

Figure 2. Typical configurations of the positions of self-propelled particles displayed for various values of density and noise. For small enough noise, coherently moving clusters can be identified in the system, which tend to move in different directions and gradually disappear as the noise amplitude is increased. The noise amplitudes applied in the simulations were the following: $\eta = 0.2$ (a,d); $\eta = 0.4$ (b,e); $\eta = 0.8$ (c,f); $\eta = 0.6$ (g); $\eta = 1.2$ (h) and $\eta = 1.8$ (i). The corresponding order parameter values are: $\phi = 0.75$ (a,d); $\phi = 0.19$ (b); $\phi = 0.025$ (c); $\phi = 0.3$ (e,h); $\phi = 0.05$ (f,i) and $\phi = 0.85$ (g). The scales are given in units of the interaction length.

For the statistical characterization of the system, a well-suited order parameter is the magnitude of the *average momentum* of the particles:

$$\phi \equiv \frac{1}{N}\left|\sum_j \vec{v}_j\right|. \tag{4}$$

This measure of the net flow is non-zero in the ordered phase and vanishes (for an infinite system) in the disordered phase, since the simulations were started from a random, disordered configuration, $\phi(t=0) \approx 0$. After some relaxation time, a steady state emerges, indicated by the convergence of the cumulative average $(1/\tau)\int_0^\tau \phi(t)dt$. The stationary values of ϕ are plotted in Figure 3 vs. η for various parameters. For weak noise, the model displays an ordered motion, i.e., $\phi \approx 1$, which disappears in a continuous manner by increasing η. As $L \to \infty$, the numerical results show the presence of a phase transition described by

$$\phi(\eta) \sim \begin{cases} \left(\frac{\eta_c(\varrho)-\eta}{\eta_c(\varrho)}\right)^b & \text{for } \eta < \eta_c(\varrho) \\ 0 & \text{for } \eta > \eta_c(\varrho) \end{cases}, \tag{5}$$

where $\eta_c(\varrho)$ is the critical noise amplitude that separates the ordered and disordered phases. The exponent b (analogous to the critical exponent in equilibrium phase transitions) was found [5] to be $b = 0.42 \pm 0.03$. This large-scale behavior does not depend on the specific choice of the microscopic details, as Figure 3 demonstrates.

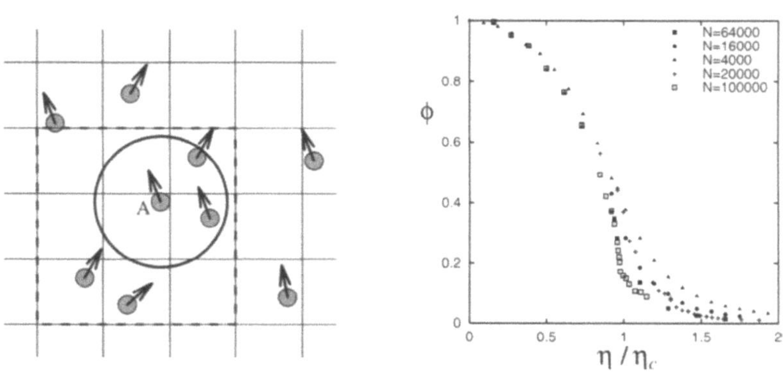

Figure 3. Left: Schematic illustration of the various realizations of the self-propelled particle model. Particles move off-lattice on a plane and interact with other particles located in the surroundings, which can be either a circle (S_1) or 9 neighboring cells in an underlying lattice (S_2). We plot these interaction areas for particle A with a solid and dashed line, respectively. Right: The average momentum of the system in the steady state vs the rescaled noise amplitude η/η_c for various interaction ranges, densities and system sizes. Open symbols refer to interaction type S_1 and $\varrho = 1$, while filled symbols refer to interaction type S_2 and $\varrho = 2$. With increasing η, the order disappears in a continuous manner reminiscent of second-order phase transitions as $N, L \to \infty$.

As a further analogy with equilibrium phase transitions, the fluctuations of the order parameter also increase on approaching the critical line [5]. The tails of the curves are symmetrical and decay as power-laws with an exponent γ close to 2.

Next we discuss the role of density. The long-range ordered phase is present for any ϱ, but for a fixed value of η, ϕ vanishes with decreasing ϱ. The critical line $\eta_c(\varrho)$ in the $\eta - \varrho$ parameter space was found to follow

$$\eta_c(\varrho) \sim \varrho^\kappa, \tag{6}$$

with $\kappa = 0.45 \pm 0.05$ for $\varrho \ll 1$. This critical line is qualitatively different from that of the diluted ferromagnets, since here the critical density at $\eta \to 0$ (corresponding to the percolation threshold for diluted ferromagnets, see, e.g., [20]) is vanishing.

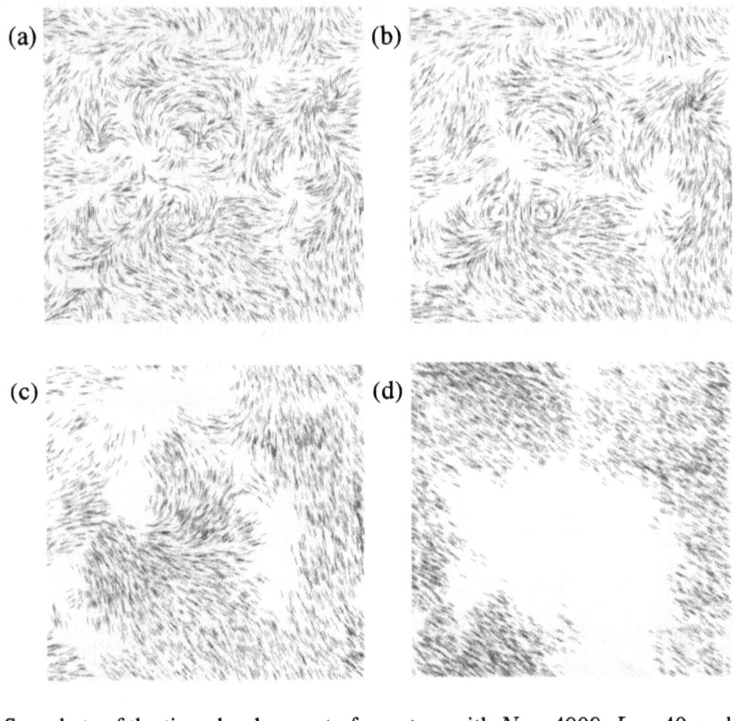

Figure 4. Snapshots of the time development of a system with $N = 4000$, $L = 40$, and $v_0 = 0.01$ at 50 (a), 100 (b), 400 (c), and 3000 (d) steps. First the behavior is reminiscent of the equilibrium XY model, where the long-range order is missing since vortices are present in the system. Here the vortices are unstable, and finally a self-organized long-range order develops. (After [5].)

These findings indicate that self-propelled particle systems can be quite well characterized using the framework of classical critical phenomena, but also show surprising features when compared to the analogous equilibrium systems. The velocity v_0 provides a control parameter that switches between the self-propelled particle behavior ($v_0 > 0$) and an XY ferromagnet ($v_0 = 0$). Indeed, for $v_0 = 0$ Kosterlitz-Thouless vortices [12] can be observed in the system, which are *unstable* (Fig. 4) for any non-zero v_0 investigated in [5].

2.2. Correlation Functions

Besides calculation of the order parameter, further insight can be obtained from characterizing the configurations with correlation functions, such as the velocity-velocity correlation function

$$C(\vec{r}) = \langle \vec{v}(\vec{r} + \vec{r}', t)\vec{v}(\vec{r}', t)\rangle_{\vec{r}',t} \tag{7}$$

where $\vec{v}(\vec{r}, t)$ is the coarse-grained velocity field and the average is taken over all possible values of $\vec{r}'$ and t (in the stationary regime). The system is ordered on the macroscopic scale if $C_\infty = \lim_{|\vec{r}|\to\infty} C(\vec{r}) > 0$. The decay of the fluctuations is given by the connected piece of the correlation function $C_C(\vec{r})$, defined as $C_C(\vec{r}) = C(\vec{r}) - C_\infty$.

One of the major results of the analysis of Toner and Tu [22] was that the attenuation of fluctuations is *anisotropic* in a self-propelled particle cohort. In particular, they predicted that within a certain length scale $|\vec{r}| < \ell$, $C_C(\vec{r})$ decays as

$$C_C(\vec{r}) \sim r_\perp^{-2/5} \tag{8}$$

with $r_\perp$ and $r_{||}$ being the orthogonal and parallel projection of $\vec{r}$ relative to the average direction of motion $\langle \vec{v}(\vec{r}', t)\rangle_{\vec{r}'}$.

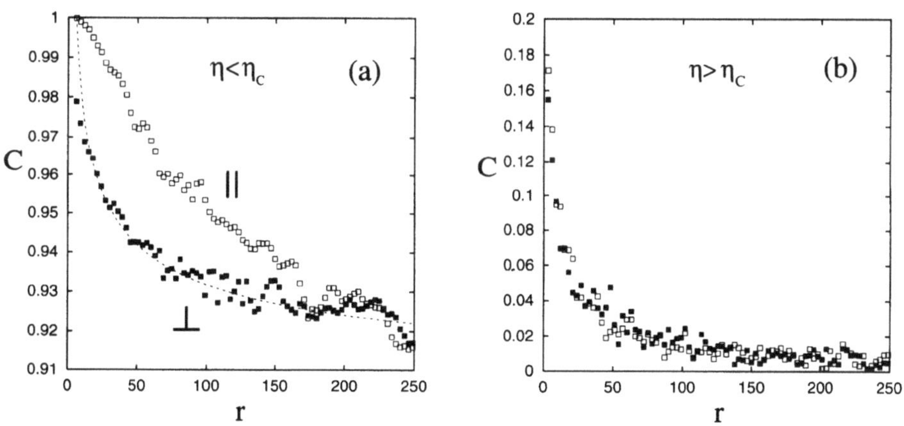

Figure 5. Equal time velocity-velocity correlation functions characterizing the decay of correlations parallel and perpendicular to the average direction of motion of the cohort. $C_{||}$ and $C_\perp$ were calculated for systems with circular interaction range, $N = 4000$, $L = 480$, and $\eta = 0.02$ (a), $\eta = 0.32$ (b). At these parameter values, the critical noise amplitude is at $\eta_c \approx 0.2$. The line in (a) represents a power-law fit on C_C with the predicted exponent $-2/5$.

We calculated equal-time velocity-velocity correlation functions in $(r_\perp, r_{||})$ base for various time moments and averaged them to obtain $C(\vec{r})$. To demonstrate the predicted anisotropy, we averaged $C(r)$ over points for which either the $r_\perp > r_{||}$ or the $r_\perp < r_{||}$ relation holds to obtain

$$C_\perp(r) = \langle C(\vec{r}) \rangle_{|\vec{r}|=r, r_\perp > r_{||}}, \quad C_{||}(r) = \langle C(\vec{r}) \rangle_{|\vec{r}|=r, r_\perp < r_{||}}. \tag{9}$$

For $\eta < \eta_c$ the anisotropy of the correlation functions is clearly seen in Figure 5, while the behavior of $C_\perp$ is consistent with prediction (8). In contrast, for $\eta > \eta_c$ the curves vanish for large r and the system is isotropic as expected.

3. Cellular Automaton Model

Cellular automata are dynamical systems – discrete in space, time, and state. The automaton concept was originally introduced by John v. Neumann and S. Ulam as a model of self-reproducing systems [15] but has been applied extensively to many other biological and also physical, chemical, and even sociological problems [4, 10, 24]. The essential question is always how macroscopic (global) behavior can arise from individual (local) rules. Local configurations are typically updated *simultaneously* and only depend on the local neighborhood configuration, which allows effective parallel processing.

Here, we introduce a cellular automaton mimicking active cellular migration and alignment interaction. From a coarse-grained point of view, the cellular automaton model and the off-lattice model introduced in Section 2 are equivalent. In particular, the noise parameter η defined in Eq. (2) plays a role analogous to the inverse sensitivity $(1/\beta)$ in the automaton model (see below). Other automaton models of swarming have been proposed previously [9, 18]. However, since our cellular automaton is formulated in terms of a lattice gas automaton [8, 13], the nature of the observed "swarming" phase transition can be directly analyzed by examining the approximating nonlinear discrete lattice Boltzmann equation.

The lattice gas cellular automaton model reviewed here is defined on a two-dimensional $L \times L$ square lattice $\mathcal{L}$ with periodic boundary conditions; details can be found in [3]. Each node $\mathbf{r}$ can contain up to four cells in different velocity channels corresponding to nearest neighbor vectors $\mathbf{c}_i = (\cos\phi_i, \sin\phi_i)$ with $\phi_i = \pi(i-1)/2$ and $1 \le i \le 4$ (Fig. 6). The state of the entire lattice at time t is specified by the occupation numbers $s_i(\mathbf{r}, t) \in \{0, 1\}$ denoting the absence resp. presence of a particle in the channel $(\mathbf{r}, \mathbf{c}_i)$. The state of node $\mathbf{r}$ is denoted by $s(\mathbf{r}, t) = \{s_i(\mathbf{r}, t)\}_{1 \le i \le 4}$.

The evolution from time t to time $t+1$ proceeds in two steps: first an interaction step is performed during which the pre-interaction state $\{s_i(\mathbf{r}, t)\}$ is replaced by a post-interaction state $\{\sigma_i(\mathbf{r}, t)\}$ according to stochastic rules that are applied to each node $\mathbf{r}$ independently (simultaneous update); the interaction step is followed by a deterministic migration step during which particles move to nearest neighbor sites in the direction of their velocity, i.e., $s_i(\mathbf{r} + \mathbf{c}_i, t+1) = \sigma_i(\mathbf{r}, t)$. This implies a fixed speed of movement mimicking active cell migration.

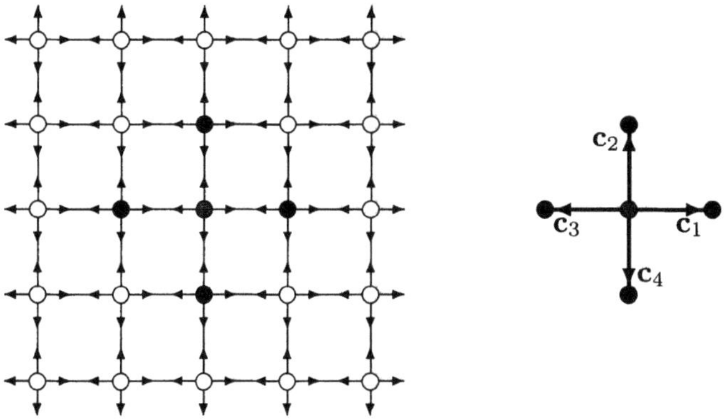

Figure 6. State space and neighborhood in the cellular automaton. Each node (circles) can host up to 4 cells in the channels $c_i (i = 1, \ldots, 4)$. Von Neumann neighborhood consists of 4 nodes (black).

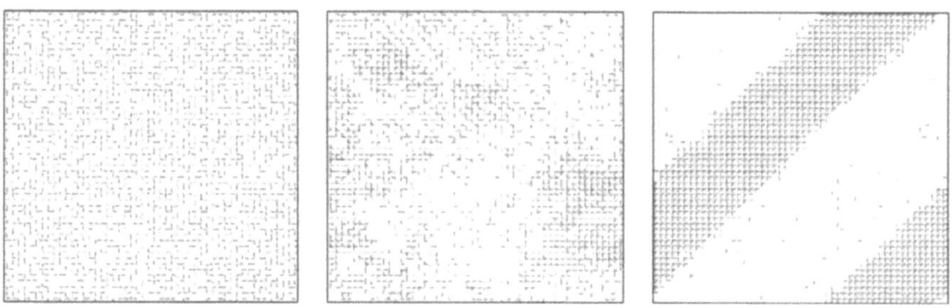

Figure 7. Cohort formation in a cellular automaton model. Shown are snapshots of the systems after 0, 200 and 1000 time steps. Parameters are: sensitivity $\beta = 1.5$, system size $L = 50$, and average density $\bar{\rho} = 0.8$ (after [3]).

3.1. Modeling cell-cell interaction in cellular automata

Due to interaction, cell configurations at individual lattice nodes $\mathbf{r}$ are assumed to instantaneously change from $s(\mathbf{r})$ to $\sigma(\mathbf{r}, t)$. The essential modeling idea is the appropriate formulation of a local field characterizing a given cell interaction. It is assumed that the specificity of a particular cell interaction can be transformed into a corresponding local field $\mathbf{D} = \mathbf{D}(\mathbf{r})$, which is a functional determined from the neighborhood configuration in the automaton. The negative inner product $E = -\mathbf{D} \cdot \mathbf{J}$ of a local flux $\mathbf{J} = \mathbf{J}(\sigma) = \sum_{i=1}^{4} \mathbf{c}_i \sigma_i$ and the field $\mathbf{D}$ can then be interpreted as a corresponding interaction energy. The interaction rules are not governed by a rigorous minimization of E; instead, the post-interaction state σ is chosen probabilistically according to "Boltzmann weights" $e^{-E} = e^{\mathbf{D} \cdot \mathbf{J}}$. The biological interpretation is that a cell tries to minimize the work against the surrounding field by a biased random change of polarization and not by a deterministic approach to an energy minimum.

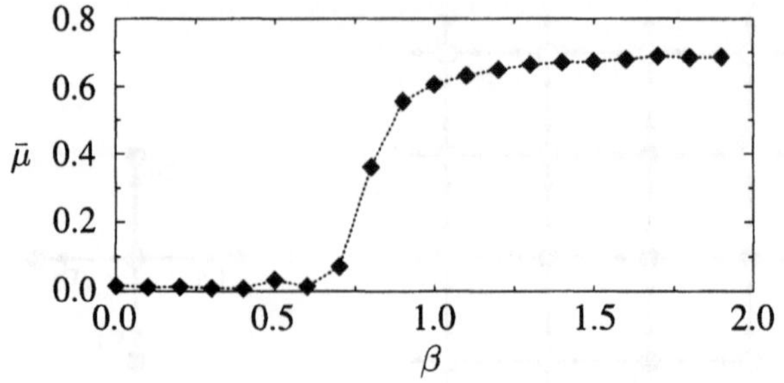

Figure 8. Mean velocity $\bar{\mu}$ *versus* sensitivity β. Obtained from simulation of $L = 50$ system at averaged density $\bar{\rho} = 1.6$, after $t = 1000$ time steps (after [3]).

To implement the local alignment interaction we define

$$\mathbf{D}(\mathbf{r}, t) = \sum_{p=1}^{4} \sum_{i=1}^{4} \mathbf{c}_i s_i(\mathbf{r} + \mathbf{c}_p, t), \tag{10}$$

specifying the average momentum/flux of cells at the nearest neighbors of node $\mathbf{r}$. Mere cell alignment requires that the number of cells at each node, $\rho(\mathbf{r}, t) = \rho[s(\mathbf{r}, t)] \equiv \sum_{i=1}^{4} s_i(\mathbf{r}, t)$, is conserved during interaction; this implies that the spatially averaged density of particles per node $\bar{\rho}$ is constant in time.

The transition probability from $s(\mathbf{r}, t)$ to $\sigma(\mathbf{r}, t)$ in the presence of $\mathbf{D}(\mathbf{r}, t)$ is given by

$$A[s \to \sigma | \mathbf{D}, \beta] = \frac{1}{Z} \delta[\rho(\sigma), \rho(s)] \exp\left[\beta \mathbf{D} \cdot \mathbf{J}(\sigma)\right], \tag{11}$$

where the normalization factor $Z(\rho(s), \mathbf{D})$ is chosen such that

$$\sum_{\sigma} A[s \to \sigma | \mathbf{D}, \beta] = 1 \quad \forall s. \tag{12}$$

The interaction rules are designed to minimize the angle between the director field $\mathbf{D}$ and the post-interaction flux $\mathbf{J}(\sigma)$. The sensitivity parameter β, playing the role of an inverse temperature, controls the degree of local alignment: for $\beta = 0$ there is no alignment at all because the dynamics are diffusive; for $\beta \to \infty$ the two-dimensional inner product $\mathbf{D} \cdot \mathbf{J}(\sigma)$ – and therefore the local alignment – is maximized. It will turn out that a dynamical phase transition occurs at a critical value β_c of the sensitivity. Figure 7 shows the time evolution of an initially random distribution for $\beta > \beta_c$. The formation of locally aligned patches can clearly be observed.

An appropriate order parameter for the statistical characterization was already introduced in Eq. (4). It is the spatially averaged momentum

$$\bar{\mu}(t) = \frac{1}{L^2} \left| \sum_{\mathbf{r}} \sum_{i=1}^{4} \mathbf{c}_i s_i(\mathbf{r}, t) \right|, \tag{13}$$

which takes values between 0 and 1. For the random disordered initial configuration, $\bar{\mu}(t = 0) \approx 0$. After a relaxation time the system reaches an ordered state with $\bar{\mu} \gg 0$

Figure 9. Cohort formation in simulations of the lattice Boltzmann equation (Eq. (14)). Shown are snapshots of the systems after 0, 200, and 1000 time steps. The length of a line at each node in every direction indicates the probability of finding a cell in that direction. In the first time step instead of the local alignment interaction, a random perturbation (max $\pm 5\%$ for each channel) that keeps the local density constant is used to perturb the initially spatial uniform state. Parameters are: sensitivity $\beta = 1.5$, local density $\bar{\rho} = 0.8$, and system size $L = 50$.

(Fig. 8). The order appears in a continuous manner typical for second-order phase transitions (as $L \to \infty$). The precise character of the phase transition will be further investigated below.

3.2. Boltzmann Equation

To analyze the behavior of our cellular automaton model, we consider the time evolution of a statistical ensemble of systems. For technical details, we refer to [3] and [7]. In a mean-field description a central role is played by the average occupation numbers $f_i(\mathbf{r}, t) \equiv \langle s_i(\mathbf{r}, t) \rangle$. It is assumed that at each time step just before interaction, the probability distribution is completely factorized over channels $(\mathbf{r}, \mathbf{c}_i)$, so that the probability to find a microstate $\{s_i(\mathbf{r})\}$ at time t is given by $\prod_{\mathbf{r}} \prod_{i=1}^{4} [f_i(\mathbf{r}, t)]^{s_i(\mathbf{r})} [1 - f_i(\mathbf{r}, t)]^{1-s_i(\mathbf{r})}$. We denote the factorized average by $\langle \cdots \rangle_{\mathrm{MF}}$. Replacing $\langle \cdots \rangle_{\mathrm{MF}}$ with $\langle \cdots \rangle$, i.e., neglecting all correlations between occupation numbers, we obtain a closed evolution equation for $f_i(\mathbf{r}, t)$: the nonlinear Boltzmann equation

$$f_i(\mathbf{r} + \mathbf{c}_i, t + 1) = f_i(\mathbf{r}, t) + I_i(\mathbf{r}, t). \tag{14}$$

Here the term $I_i(\mathbf{r}, t) \equiv \langle \sigma_i(\mathbf{r}, t) - s_i(\mathbf{r}, t) \rangle_{\mathrm{MF}}$, taking values between -1 and 1, equals the average change in the occupation number of channel $(\mathbf{r}, \mathbf{c}_i)$ during interaction. Figure 9 shows snapshots of simulations of this (deterministic) equation starting from various initial conditions which provide the only sources for random effects. The simulations nicely mimic the automaton snapshots. Thus, it seems reasonable to expect valuable insights into the automaton dynamics from the stability analysis of the nonlinear Boltzmann equation.

3.3. Linear Stability Analysis

We focus on the linear stability of the spatially homogeneous and stationary solution $f_i(\mathbf{r}, t) = \bar{f} = \bar{\rho}/4$ to Eq. (14) with respect to fluctuations $\delta f_i(\mathbf{r}, t) = f_i(\mathbf{r}, t) - \bar{f}$.

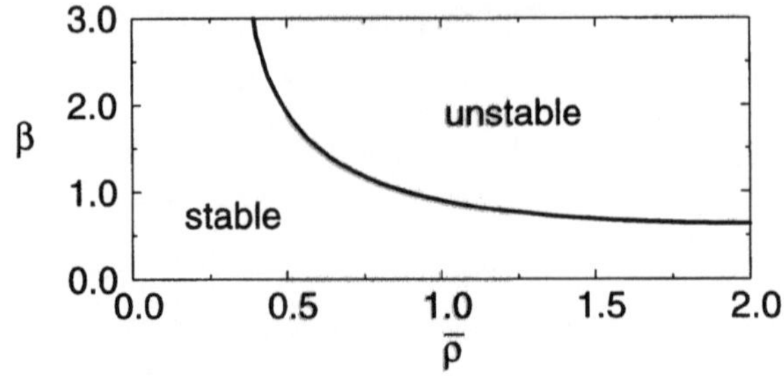

Figure 10. Phase diagram for alignment pattern formation in the cellular automaton model. Shown are the regions of stable and unstable behavior, as a function of sensitivity β and average density $\bar{\rho}$ (after [3]).

Therefore, we linearize Eq. (14), perform a Fourier transformation,

$$\delta f_i(\mathbf{k}, t) = \sum_{\mathbf{r}} e^{-i\mathbf{k}\cdot\mathbf{r}}\, \delta f_i(\mathbf{r}, t),$$

and obtain a linear system

$$\delta f_i(\mathbf{k}, t+1) \simeq \sum_{j=1}^{4} \Gamma_{ij}(\mathbf{k}) \delta f_j(\mathbf{k}, t). \tag{15}$$

The mean-field or Boltzmann propagator $\Gamma(\mathbf{k})$ describes how a small perturbation around a spatially uniform state evolves in time. It is given by

$$\Gamma_{ij}(\mathbf{k}) = e^{-i\mathbf{k}\cdot\mathbf{c}_i} \left[\delta_{ij} + \sum_{p=0}^{4} e^{i\mathbf{k}\cdot\mathbf{c}_p}\, \Omega_{ij}^{p} \right], \tag{16}$$

with $\mathbf{c}_0 \equiv 0$ and $\Omega_{ij}^{p} = \partial I_i(\mathbf{r}, t)/\partial f_j(\mathbf{r} + \mathbf{c}_p, t)|_{\bar{f}}$. Statements about stability are now reduced to statements about the matrices $\Gamma(\mathbf{k}, \beta, \bar{\rho})$. A detailed stability analysis of the model can be found in [3]. The results lead to the eigenvalue spectrum shown in Figure 11.

The spectrum indicates that the fastest growth occurs for wave number $\mathbf{k} = 0$. The corresponding maximum eigenvalue $\lambda_{x,y}$ possesses an eigenspace spanned by $e_x = (1, 0, -1, 0)$ and $e_y = (0, 1, 0, -1)$, referring to the x and y components of the total particle momentum. The onset of instability of the homogeneous state is determined by the condition $\lambda_{x,y} = 1$. The location of the critical line in the $(\beta, \bar{\rho})$ parameter plane is shown in Figure 10, which was obtained by numerically solving the equation $\lambda_{x,y}(\beta, \bar{\rho}) = 1$.

3.4. Phase Transition

The mean-field stability analysis illustrates the nature of the observed phase transition in cellular automaton simulations (Fig. 8). We have compared the results of our stability analysis with the computer simulations. Simulations exhibit an abrupt change in $\bar{\mu}$ at $\beta \simeq 0.7$ (Fig. 8), which agrees well with the prediction $\beta_c = 0.67$ obtained from our stability analysis (cf. Fig. 10). Note that the mean-field approximation obtained by neglecting all correlations, is sufficient to explain the essential patterns in the automaton

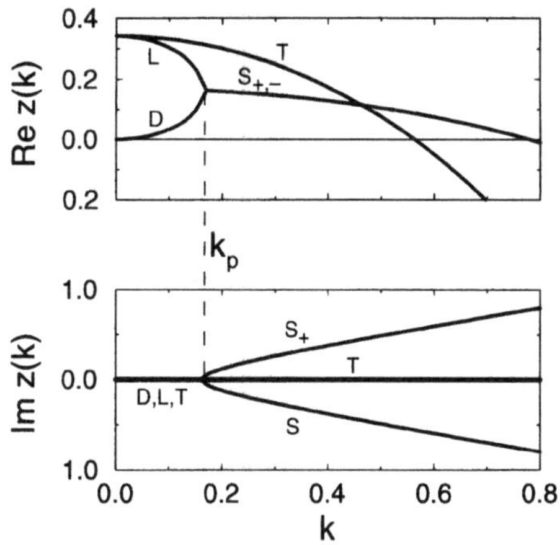

Figure 11. Eigenvalue spectrum for $\bar{\rho} = 1.6$, $\beta = 1.5$, and $k = k_x$ where $\mathbf{k} = (k_x, 0)$. Density (D), longitudinal (L) and transverse (T) momentum, as well as sound ($S_\pm$) modes are indicated. The stable mode that has eigenvector $e_{x^2-y^2}$ at $\mathbf{k} = 0$ is not shown (after [3]).

simulations because all correlations that are produced in the interaction step are immediately destroyed by the deterministic migration.

Expansion of the nonlinear Boltzmann equation around the critical point $(\bar{\rho}, \beta_c)$ allows us to determine critical exponents b', ρ' defined by

$$\bar{\mu} \sim (\beta - \beta_c)^{b'}, \quad \bar{\mu} \sim (\rho - \rho_c)^{\rho'}. \tag{17}$$

Our analysis yields $b' = \rho' = \frac{1}{2}$ (see [3]), which is in very good agreement with the exponents obtained in the off-lattice model (cf. section 2).

4. Discussion

The reviewed individual-based models of cohort formation have interesting biological interpretations. The structure of the phase space (Fig. 10) of the self-propelled systems we have studied suggests two possible scenarios for a change from non-cooperative to cooperative motility. First, relatively minor changes in cell behavior influencing their alignment or motility can have severe macroscopic manifestations with respect to swarming if they occur in a system close to criticality (cf. Fig. 3b). Second, a similar transition can also be achieved by simply increasing the local cell density. This result provides a possible clue to explain the behavioral change between non-cooperative and cooperative stages in life cycles of some bacteria or amoebae in which a reproductive feeding phase of individually moving cells is followed by social (coordinated) aggregation.

Note that the formation of patches in our models is due to the periodicity of the boundary conditions. There is no intrinsic aggregation tendency in the models *per se*. To arrive at realistic swarms in confined geometries "attracting interactions" have to be considered (an example can be found in [14]). Other, more specific models for complex bacterial pattern formation (including branching, chirality or vortices) based on long-range interactions (particularly chemotaxis) have also been studied (see [1] and references therein).

Finally, we stress that the methods employed here for the case of alignment interaction can easily be adapted to gain theoretical insight into the behavior of a wide range of other biologically motivated cellular interactions (e.g., adhesion, cell reversal), that so far have mainly been analyzed by observation of simulation outcomes [2, 7]. It turns out that the coarse-grained perspective of the cellular automaton model covers the essential aspects of the cell interaction behavior.

Acknowledgments

A. C. is grateful to Tamás Vicsek for many discussions. This work was supported by grants OTKA T019299, F026645, A. D. appreciates fruitful cooperation with Harmen Bussemaker.

References

[1] E. Ben-Jacob (1997) From snowflake formation to growth of bacterial colonies II: Cooperative formation of complex colonial patterns. *Contemporary physics* **38**(3), 205

[2] U. Börner, A. Deutsch and M. Bär (2002) Rippling patterns in aggregates of Myxobacteria arise from cell-cell collisions. *Phys Rev Lett* **89** (7), 078101

[3] H. Bussemaker, A. Deutsch, E. Geigant (1997) Mean-field analysis of a dynamical phase transition in a cellular automaton model for collective motion, *Phys Rev Lett* **78**, 5018–5021

[4] B. Chopard and M. Droz (1998) *Cellular automata modeling of physical systems* Cambridge University Press, Boston

[5] A. Czirók, H. E. Stanley, and T. Vicsek (1997) Spontaneously ordered motion of self-propelled particles. *J Phys A* **30**, 1375–1385

[6] A. Deutsch (1999) Principles of biological pattern formation: swarming and aggregation viewed as self-organization phenomena. *J Biosc* **24**(1), 115–120

[7] A. Deutsch and S. Dormann (2003) *Cellular automaton modeling of biological pattern formation.* Birkhäuser, Boston

[8] G. D. Doolen et al., eds (1990) *Lattice gas methods for partial differential equations.* Addison-Wesley, New York

[9] L. Edelstein-Keshet and B. Ermentrout (1990) Models for contact-mediated pattern formation: Cells that form parallel arrays. *J Math Biol* **29**, 33–58

[10] B. Ermentrout, L. Edelstein-Keshet (1993) Cellular automata approaches to biological modeling. *J Theor Biol* **160**, 97–133

[11] B. Hegedűs et al (2000) Locomotion and proliferation of glioblastoma cells in vitro: statistical evaluation of videomicroscopic observations. *J Neurosurg* **92**, 428–434

[12] J. M. Kosterlitz and D. J. Thouless (1973) Ordering, metastability and phase transitions in two-dimensional systems. *J Phys C* **6**, 1181

[13] A. T. Lawniczak and R. Kapral, eds (1995) *Pattern formation and lattice gas automata, fields institute communications.* Vol. 6 (American Mathematical Society, Providence, RI)

[14] H. Levine and W.-J. Rappel (2000) Self-organization in systems of self-propelled particles. *Phys Rev E* **63**, 017101

[15] J. v. Neumann (1966) *The theory of self-reproducing automata.* (A. W. Burks, ed.) University of Illinois Press, Urbana, IL

[16] L. Olsen, P. K. Maini, J. A. Sherratt, J. Dallon (1999) Mathematical modelling of anisotropy in fibrous connective tissue. *Mathem Biosc* **158**, 145–170

[17] H. Reichenbach and H. Kühlwein (1964) Myxobakterien. Film archive material, produced by *Institute for Scientific Films, Göttingen.* Botanical Institute, Karlsruhe

[18] B. Schönfisch (2001) Simple individual based models of movement, alignment and schooling behaviour. *Future Gen Comp Sys* **17**, 873–882

[19] A. Stevens and F. Schweitzer (1997) Aggregation induced by diffusing and nondiffusing media. In: *Dynamics of cell and tissue motion* (W. Alt, A. Deutsch, G. Dunn, eds.) Birkhäuser, Basel, 183–192,

[20] R. B. Stinchcombe (1983) Dilute magnetism. In: *Phase transitions and critical phenomena (C. Domb, J.L. Lebowitz, eds.)*, volume 7, Academic Press, New York

[21] J. Toner and Y. Tu (1995) Long-range order in a two-dimensional dynamical XY model: how birds fly together. *Phys Rev Lett* **75**, 4326

[22] J. Toner and Y. Tu (1998) Flocks, herds, and schools: a quantitative theory of flocking. *Phys Rev E* **58**(4), 4828–4858

[23] Tamás Vicsek, András Czirók, Eshel Ben-Jacob, Inon Cohen, and Ofer Shochet (1995) Novel type of phase transition in a system of self-driven particles. *Phys Rev Lett* **75**, 1226–1229

[24] S. Wolfram (1986) *Theory and applications of cellular automata.* World Publishing, Singapore

Polymer and Cell Dynamics – Multiscale Modeling
and Numerical Simulations, 221–241
ed. by W. Alt, M. Chaplain, M. Griebel und J. Lenz

Interactive Movement, Aggregation, and Swarm Dynamics

Wolfgang Alt[1], Till Bretschneider[2] and Ralf Müller[1]

1) Division of Theoretical Biology, Institute for Cellular and Molecular Botanics, University of Bonn, Kirschallee 1, D-53115 Bonn, Germany
2) Max-Planck Institute for Biochemistry, Martinsried, Germany

Summary. We consider a general class of models for biological interactive movement describing collective phenomena such as chemotaxis and tensiotaxis of cells or swarming of birds. Using the joint name **'bimes'** for such **bi**ological **i**nteracting and **m**oving **e**ntities, we suppose that communication between 'bimes' is performed *via* a specific medium by establishing therein a dynamic *interaction field*, whose perception stimulates the dynamic response and motile activity of each single 'bime'. In some particular cases, sample simulations are shown for one and two spatial dimensions. We discuss how a proposed general modeling approach using Voronoi tessellation and Delaunay triangulation might help to formulate and simulate collective cell movement with direct cell-cell contact, as occurs in models for wound healing and other cell tissue dynamics.

1. Introduction

Biological **i**nteracting and **m**oving **e**ntities, henceforth called 'bimes', occur with sizes on various spatial scales, performing dynamic features on characteristic spatiotemporal scales. Typical examples are

- *Myxobacteria* ($\sim \mu m$), gliding in self-produced slime and forming expanding 2D swarms or 3D 'fruiting bodies' ($\sim 100\ \mu m$ within tens of minutes); see [12, 26];
- *Amoeba cells* ($\sim 10\ \mu m$), crawling on surfaces or on each other, forming 2D streaming aggregations, 3D slugs, and, eventually, also 'fruiting bodies' ($\sim mm$ within hours); see [6, 9];
- *Epidermal cells* ($\sim 50\ \mu m$), crawling on surfaces or along/through (partially) self-produced filamentous polymer matrices and forming 2D areas of dense monolayers with the ability to close 'skin wounds' ($\sim cm$ within days); see [1];
- *Birds*, for example, *starlings* ($\sim 10\ cm$), flying in visual (or auditory) contact and producing variable shapes of dynamic 'swarms'; see [15, 22].

What distinguishes these biological examples of interactive dynamics from physical multiparticle systems such as polymer fluids or granular sand dunes? First of all, each 'bime' actively moves on its own energy supply and motor apparatus, thereby often tending to persist in its current *moving direction* or to maintain a favored *forward locomotion speed*, even if only relative to a substratum that is used or modified by other 'bimes'. Second, interactions between 'bimes' are primarily not induced by long-range 'potentials' (as in gravity or electromagnetic fields) but rather by short-range 'contact response' and specific *perception* of chemical, mechanical, or other signals (e.g., light, sound, or heat), which each 'bime' receives or collects from its 'neighborhood'. Finally, 'bimes' have a more-or-less expressed degree of freedom to choose between arbitrary (self-determined or environmentally induced) motile activity or response (and adaptation) to perceived signals.

In order to reveal such common principles of biological interactive movement and contrast them with purely physical dynamics, we propose a general formalism for modeling the stochastic motion of N 'bimes' in m dimensions, communicating through a specific medium by means of a so-called *interaction field*, which they themselves produce and perceive, thereby determining changes in their otherwise autonomous (stochastic) motile behavior.

For reference to further literature, see Section 6, as well as the review article by Dirk Drasdo (this volume, Part III).

2. Model Formulation for the Movement Dynamics

Let the movement of each bime be described by (stochastic) variables for *position* $X_i(t) \in \mathfrak{R}^m$ and *velocity* $V_i(t) \in \mathfrak{R}^m$, of the ith 'bime's body, $i = 1, \ldots, N$. Changes per time in position and velocity are supposed to be determined by the following simple system of stochastic ordinary differential equations of *Ornstein-Uhlenbeck* type,

$$dX_i = V_i \cdot dt \tag{1}$$

$$dV_i = (\Phi(V_i) + A_i) \cdot dt + \beta_i \cdot d\zeta \tag{2}$$

where β_i measures the amplitude of stochastic perturbations in the ith bime's velocity, e.g., due to environmental noise ('thermodynamics') or induced by individual arbitrariness, here modeled as mutually independent Brownian increments $d\zeta \in \mathfrak{R}^m$ during infinitesimally small time steps dt. The first deterministic *acceleration* term in Eq. (2) might describe a self-induced adjustment to a given *preferred speed* S^*, common for all bimes, say

$$\Phi(V_i) = \gamma(\| V_i \|) \cdot (S^* / \| V_i \| - 1) \cdot V_i(t) \tag{3}$$

with an *adjustment rate* $\gamma = \gamma(S)$ that could depend on the bime's actual *speed* $S_i(t) = \|V_i(t)\|$. The second *acceleration* term in Eq. (2), namely, the vector $A_i(t)$, comprises all influences on the ith bime's velocity by other 'bimes' $j \neq i$ and represents the central term for modeling biological interaction. Introducing a constant *body mass* m_i for each bime, the term $F_i = m_i A_i$ might then model the *interaction force* that it experiences. For the case $S^* = 0$, however, both acceleration (resp. force) terms could be summarized and written as

$$\Phi(V_i) + A_i = \gamma \cdot \{W_i - V_i\} \tag{4}$$

with now $W_i := A_i/\gamma$ representing the *preferred velocity* that the ith bime tries to attain as result of interactions with other 'bimes' $j \neq i$. Then a positive adjustment rate

γ determines the mean response time $1/\gamma$, which the ith bime needs to adjust its current velocity to the preferred one. In the case of 2D cell motion on a planar substratum, such an adjustment could be due, for instance, to a rearrangement of the contractile intracellular actin-myosin polymer system constituting the 'locomotive machinery' of migrating polarized cells (see [1, 20]).

For the limiting case $\gamma \to 0$, while $m_i \to \infty$ such that $\gamma m_i \simeq \gamma_0$, we would obtain an instantaneous adjustment $V_i = W_i$ or, written as $\gamma_0 V_i = m_i A_i = F_i$, a simple force balance equation between a *friction force* $\gamma_0 V_i$ and the interaction force F_i. In this case motion dynamics is reduced to the simple *Stokes' approximation* of Eq. (1)–(2), namely

$$dX_i = \frac{1}{\gamma_0} F_i \cdot dt \tag{5}$$

where stochastic perturbations could now be implemented into the *friction coefficient* γ_0 or into model expressions for the *interaction forces* F_i.

In each of these cases, model expressions for the acceleration term A_i, i.e., the *dynamic response* resulting from biological interactions between bimes, have not yet been specified.

3. General Model Formulation for Biological Interaction and Dynamic Response

3.1. The Medium

Interactions between bimes need a medium $\mathfrak{M}_t$ for transmission of signals between them. It could be the whole space $\mathfrak{R}^m$ or a certain subset of $\mathfrak{R}^m$ defining, for example, a 'container' to which the bimes movement is restricted. The medium might also be confined to the space *outside* the bimes. If, for instance, a cell body is represented by a simple ball $B_t^i = \mathfrak{B}_\varepsilon\{X_i(t)\} \subset \mathfrak{R}^m$ of radius ε, then in this case the medium would be the *extracellular space* $\mathfrak{M}_t = \mathfrak{R}^m \setminus \cup_{i=1}^N B_t^i$. For modeling we could use $m = 1$, 2, or 3.

3.2. The Interaction Field

Interaction can be thought to take place through *signal transduction*, whereby some quantity of a spatially located or expanded *interaction field* in the medium is produced or modified by the bimes themselves, while, simultaneously, this or some other field quantity is perceived by them. We represent such a (time-dependent) interaction field by a *supporting measure*, $d\mu_t$, which could describe a volume density, $d\mu_t(x) = \rho(t, x)\, dx$, or some lower dimensional measure, and by a *mediating tensor field*, $\mathfrak{T}_t$, which carries the field quantities that might be transmitted between bimes, typically represented by a symmetric matrix $\mathfrak{T}_t(x) \in \mathfrak{R}^{m,m}$.

Example I: Chemotaxis

In this classical example of biological interaction leading to the well-known model for amoeboid aggregation of *Dictyostelium* cells [6], $\rho(t, x)$ denotes the volume concentration of a *chemoattractant*, namely *cAMP*, which is released by the cells into the extracellular space $\mathfrak{M}_t = \mathfrak{R}^m \setminus \cup_{i=1}^N B_t^i$, diffuses therein, and is perceived by specialized receptors at the cell surfaces ∂B_t^i. Thus, the transmitted tensor field would be $\mathfrak{T}_t(x) = \rho(t, x)\, \mathfrak{I}$, where $\mathfrak{I} \in \mathfrak{R}^{m,m}$ denotes the unit matrix, since binding of chemoattractant molecules to surface receptors does not carry any further anisotropic spatial information.

3.3. Dynamics of the Interaction Field

Constitutive *field dynamics* are typically formulated as evolution or balance equations for certain mean field quantities, e.g., for *field concentration* $\rho(t,x)$ and *field velocity* $\mathbf{v}(t,x)$. First, the standard *mass-balance* equation

$$\partial_t \rho + \nabla_x \cdot (\rho \mathbf{v}) = \Pi \tag{6}$$

holds with Π denoting field production or degradation in the medium $\mathfrak{M}_t$. Second, the field velocity has to be determined.

In **Example I** above, diffusion is classically described by Fick's law

$$\mathbf{v} = -\frac{D}{\rho}\nabla_x \rho \tag{7}$$

and, for a bounded medium, the resulting parabolic Eq. (6) has to be complemented by suitable boundary conditions, which in this case specify the efflux of *cAMP* through the cell surfaces ∂B_t^i, namely,

$$\theta \cdot \rho \mathbf{v} = \lambda_i - \beta_i \, \rho \tag{8}$$

with a production rate λ_i that might be constant or depend on the temporal state of the ith cell (see below) and a degradation rate β_i due to phosphodiesterase fixed at the outer cell surface, (cf. [6]). Here $\theta \in \mathfrak{S}^{m-1}$ represents the normal direction at the cell surface $\partial B_t^i = \{x = X_i(t) + \varepsilon \cdot \theta \,|\, \theta \in \mathfrak{S}^{m-1}\}$.

In this chemotaxis example, the boundary conditions (8) simultaneously provide (generally needed) *effector equations* describing how the ith bime affects the interaction field.

Example II: Tensiotaxis

In this second prototype of biological interaction between cells, we are going to restrict our modeling to the 2D case, e.g., the crawling movement of *keratinocytes* (human epidermal skin cells) on a viscoelastic substratum, say a sheet of polyacrylamid. Thus, the medium is the whole plane, $\mathfrak{M}_t = \mathfrak{R}^2$, or a confined region.

Then again, concentration of the polymer susbstrate, $\rho(t,x)$, satisfies the mass balance equation (6) with $\Pi = 0$, while the viscoelastic property of the medium can be modeled by an explicit expression for the *tension field*, (see [7]), namely,

$$\mathfrak{T}_t(x) = \eta_s \left(\nabla_x \mathbf{v}(t,x)\right)_{sym} + (\eta_s - \eta_d)\left(\nabla_x \cdot \mathbf{v}(t,x)\right)\mathfrak{I} - \alpha\left(\rho(t,x) - \rho_0\right)\mathfrak{I} \tag{9}$$

describing shear viscosity (η_s), dilatational viscosity (η_d) and elasticity (α) in a most simple, linear way. Assuming that, on the spatiotemporal scale of creeping cell movement (typicall μm within *min*), inertial forces can be neglected, and the constitutive *force balance* laws reduce to the standard pseudo-stationary Stokes equations

$$\nabla_x \cdot \mathfrak{T}_t - \Phi_{\mathfrak{M}} \rho \mathbf{v} + F_t = 0 \tag{10}$$

yielding an elliptic equation for the velocity field $\mathbf{v}$, eventually to be supplemented by suitable boundary conditions. Here the second term describes eventual friction forces due to translocation of the viscoelastic medium ($\Phi_{\mathfrak{M}}$), whereas now the 'outer' force density field $F_t = \sum_{i=1}^{N} F_t^i$ represents the *effector* sum of forces that are exerted to the medium by the N bimes. Model expressions for them will be given below, since in this example they are directy coupled with *perception* of tension and subsequent *dynamic response.*

3.4. Perception of the Interaction Field

Each bime, represented by the ball B_t^i, is supposed to perceive the interaction field in a so-called *perception scope* $\mathfrak{P}_t^i \subset \mathfrak{M}_t$ and to collect, from the field quantities $d\mu_t$ and $\mathfrak{T}_t$ there, information about its environment by 'looking around', thus forming a *perceived measure* $dG_t^i(\theta)$ distributed over possible 'sensing directions' $\theta \in \mathfrak{S}^{m-1}$, i.e., all normal directions on the cell boundary ∂B_t^i.

In **Example I** (chemotaxis), where perception is induced by specific chemotactic receptors on each cell periphery, we set $\mathfrak{P}_t^i = \partial B_t^i$ and, by simply assuming a uniform distribution of receptors, we suppose an instantaneously perceived 'measure' of ligand concentration ρ at the cell boundary, thus $dG_t^i(\theta) = \rho\,(t, X_i(t) + \varepsilon \cdot \theta\,)\,d\theta\,.$

In **Example II** (tensiotaxis) perception is performed by specific substrate adhesion receptors underneath the cell, i.e., in a certain subset $K_t^i \subset B_t^i \subset \mathfrak{R}^2$, and they bind there, for instance, only to certain *binding sites* located at discrete points $Z_k(t)$, $k = 1, \ldots, K$, which are imbedded in the substratum and are moving with it. Thus, in our general terminology,

$$\mathfrak{P}_t^i = \bigcup_{Z_k(t) \in K_t^i} \{Z_k(t)\}$$

where, for example,

$$K_t^i = \left\{ z \in \mathfrak{R}^2 : 0 < \varepsilon_1 < |z - X_i(t)| < \varepsilon \right\}$$

describes a peripheral ring of lamellipodial activity surrounding an adhering cell (see Figure 4 below and [16]). Now, let us assume that at so-called *focal adhesion complexes* associated with such binding sites (*via* integrin adhesion receptors), intracellular actin filaments can 'anchor' and 'pull' the integrin receptors towards the cell center, for simplicity, resulting in a slight deformation or translocation of the binding site $Z_k(t)$ into this direction. Then, in the opposing direction $\theta_{ik}(t) := \frac{Z_k(t) - X_i(t)}{|Z_k(t) - X_i(t)|}$ the adhesion site could 'experience' a counter-force by the viscoelastic substratum that is proportional to substrate tension $\tau_{ik}(t) := \mathfrak{T}_t(Z_k(t) + \tilde{\varepsilon}\,\theta_{ik}(t))\,[\theta_{ik}(t)]$, where $\tilde{\varepsilon} \ll \varepsilon$. Thus, in our terminology, $dG_t^i(\theta) = \sum_{Z_k(t) \in K_t^i} \mathfrak{T}_t(Z_k(t) + \tilde{\varepsilon}\,\theta)\,[\theta]\;\delta_{\theta_{ik}(t)}(\theta)$ would define a 'perceived' measure, being the discrete sum of Dirac measures evaluated in certain (temporally varying) directions $\theta_{ik}(t)$. Notice, however, that we do not specify (because one does not know) in which manner this perceived quantity at any of the 'stressed' focal adhesion complexes is stored and transduced, i.e. whether the integrin molecule itself or an integrin-binding protein undergoes a conformational change or whether the attached actin filament bundle is rearranged (e.g., more bundled) under increased stress (cf. [7, 14]).

3.5. The Dynamic Response

Within each bime the perceived measure dG_t^i will, over certain internal transduction steps, induce a *dynamic response*, which finally might be expressed in the *acceleration term*

$$A_i = \int_{\mathfrak{S}^{m-1}} \theta \; d\alpha_i(\cdot, \theta) \tag{11}$$

as a (summation) integral over all 'sensing direction' vectors θ with certain time dependent weights $d\alpha_i(t, \theta)$. In general, these could be nonlinear functionals of the perceived measure dG_t^i (see Section 3.4), or even derived from it by differential equations describing internal *transduction kinetics*.

For **Example I** (chemotaxis) these usually are kinetic equations for the peripheral distribution of free and bound receptors, $f_i(t, \theta)$ and $b_i(t, \theta)$, depending on the sensed substrate concentration $\rho\,(t, X_i(t) + \varepsilon \cdot \theta\,)$ as well as kinetic equations for secondary messenger concentrations in a subsequently induced signaling cascade. These then have an effect, for instance, on the *motile activity* of cortical actin filaments within peripheral lamellipods. If such actin filament concentration is described by $a_i(t, \theta)$, the motile force output in direction θ could be a nonlinear function of it thus, $d\alpha_i(\cdot, \theta) = f\,(a_i(t, \theta))\; d\theta$.

Analogously, in **Example II** (tensiotaxis) the sensed substrate tension $\tau_{ik}(t) = \mathfrak{T}_t(Z_k(t) + \tilde{\varepsilon}\,\theta_{ik}(t))\,[\theta_{ik}(t)]$ at any adhesion site could be kinetically or, by fast kinetics, instantaneously transformed into a resulting force $F_{ik}(t) = f\,(\tau_{ik}(t))\,\theta_{ik}(t)$, by which anchoring actin filaments pull the cell body (center) into direction $\theta = \theta_{ik}(t)$. Thus, in this case we have $d\alpha_i(\cdot, \theta) = \sum_{Z_k(t) \in K_t^i} f\,(\tau_{ik}(t))\,\delta_{\theta_{ik}(t)}(\theta)$. Clearly, the counter-force that the ith cell exerts onto the substrate would then be $F_t^i = -\sum_{Z_k(t) \in K_t^i} F_{ik}(t)$, the term appearing in Eq. (10) above and representing the effect of the ith cell onto the substratum.

4. Aggregation Dynamics

It is widely known that mutual chemotaxis of motile cells leads to their aggregation, as in *Dictyostelium* mophogenesis and slug formation. However, most of the older models use either continuum 'chemotaxis' equations of diffusion-drift type for cell concentration or stochastic models with cells simply represented by moving dots (see, e.g., [9, 17]. Only some recent models for cell aggregation also use 'balls' (even of deforming shape), but they do not resolve detailed response dynamics at the cell periphery (see [6, 24] and the review by Dirk Drasdo, this volume, Part III).

4.1. Example I: Aggregation by Chemotaxis

To illustrate how the model derived above in Section 3 gives rise to mutual aggregation of cells, let us, for simplicity, assume that the dynamic reponse to receptor sensing is very fast and happens only locally at the cell surface. Then the resulting motile activity $a_i(t, \theta)$ of the ith cell in direction $\theta \in \mathfrak{S}^{m-1}$ could be a direct function of the sensed substrate concentrations $\rho\,(t, X_i(t) + \varepsilon \cdot \theta\,)$ such that, finally, the acceleration vector of the dynamic response (11) becomes

$$A_i = \int_{\mathfrak{S}^{m-1}} \theta \, \widetilde{f}\left(\rho\left(t, X_i(t) + \varepsilon \cdot \theta\right)\right) d\theta. \tag{12}$$

The response function $\widetilde{f}$ might be linear, but in realistic cases it usually reflects saturated kinetics for chemotactic sensing and signal transduction, e.g., in the simplest Michaelis-Menten case,

$$\widetilde{f}(\rho) = \chi \frac{\rho}{K + \rho}. \tag{13}$$

Now, let us assume that in Eq. (6) the source term is $\Pi = -\gamma\rho$ due to cAMP degredation with rate γ, e.g., by phosphodiesterase uniformly distributed in the medium. Then the parabolic boundary value problem to determine $\rho(t, x)$ consists in the simple linear diffusion equation

$$\partial_t \rho = \nabla_x \cdot (D \, \nabla_x \rho) - \gamma \, \rho \tag{14}$$

on the exterior domain $\mathfrak{M}_t = \mathfrak{R}^m \setminus \cup_{i=1}^N B_t^i$ with Robin (mixed) boundary conditions (8), thus

$$\theta \cdot \nabla_x \rho = (-\lambda + \beta \, \rho)/D \tag{15}$$

on ∂B_t^i, provided that cAMP production and degradation rates λ and β are constant and equal for all cells. Furthermore, when degradation and diffusion of cAMP are both relatively fast compared to the resulting cell acceleration, then in a *pseudo-steady-state approximation* for each fixed time instant t, the substrate concentration $\rho = \rho(t, \cdot)$ is the unique solution of $D\Delta_x \rho = \gamma\rho$ on $\mathfrak{M}_t$ satisfying boundary conditions (15). In order to show how this standard linear elliptic boundary value problem can be solved such that an explicit expression for the integrand in (12) is obtained, let us restrict the problem to the simple one-dimensional case. The following formulas can be proven in a straight-forward way:

Gradient perception in the limit of vanishing cell diameter $\varepsilon \to 0$
In space dimension $m = 1$, the unique bounded pseudo-steady-state solution $\rho = \rho(t, \cdot)$ of (14)-(15) satisfies, for $\varepsilon \to 0$, the following differential equation with Dirac measures

$$-D\Delta_x \rho + \gamma \, \rho = \sum_{i=1}^{N} (\lambda - \beta \, \rho) \, \delta_{X_i(t)} \tag{16}$$

and can be represented as a linear superposition of exponential functions

$$\rho(t, x) = \frac{\lambda}{2\gamma\overline{r} + \beta} \sum_{i=1}^{N} c_i \, \exp\left(-\left|x - X_i(t)\right| / \overline{r}\right), \tag{17}$$

with the vector $c = (c_1, \ldots, c_N)$ obtained as the unique positive solution of the matrix equation $c \cdot E = \mathbf{1}$, where $\mathbf{1} = (1, \ldots, 1)$, $E_{i,i} = 1$, and $E_{i,j} = \frac{1}{1+2\gamma\overline{r}/\beta} \exp\left(-r_{ij}/\overline{r}\right)$ for $i \neq j$, with cell-cell distance denoted by $r_{ij} := |X_i(t) - X_j(t)|$. The notation $\overline{r} = \sqrt{\frac{D}{\gamma}}$ defines a so-called *mean diffusional range distance* within the interaction field. For just a pair of cells, i.e., $N = 2$, the interaction field concentration can be easily computed and

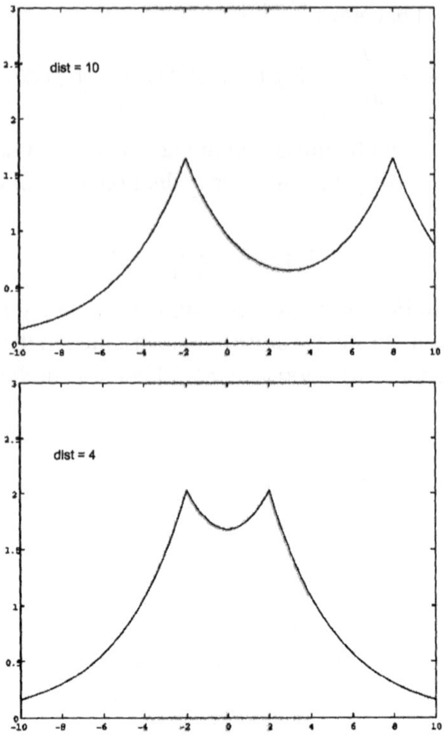

Figure 1. Concentration plot, Eq. (17), for the cAMP interaction field in the 1D chemotaxis model (Example I): Case of only two cells having a distance of 10 (upper picture) and 4 (lower picture; non-dimensional units). Parameters: $D = \gamma = 1$, $\lambda = 10$, $\beta = 0.001$.

estimated from above by the uniform bound $\frac{\lambda}{\gamma\bar{r}+\beta}$. For two different values of cell-cell distance the concentration profiles obtained are plotted in Figure 1.

In the general m-dimensional case, a corresponding localization limit for vanishing cell diameter $\varepsilon \to 0$ would require a suitable scaling, e.g., in the boundary condition (15), in order to guarantee uniform boundedness of the approximating cAMP densities. However, then the gradient $\nabla_x \rho(t,x)$ could become singular as $x \to X_i(t)$, and the response functional (12) has to be appropriately rescaled with $\varepsilon \to 0$. In any case, one would obtain the following approximation

$$\begin{aligned} A_i(t) &= \int \theta\, \tilde{f}(\rho(t, X_i(t) + \varepsilon \cdot \theta))\, d\theta \\ &= \tilde{f}'(\rho(t, X_i(t))) \int \theta \cdot \nabla_x \rho(t, X_i(t) + \varepsilon \cdot \theta)\,|_{\varepsilon \to 0}\, \theta d\theta + O(\varepsilon). \end{aligned} \quad (18)$$

Thus, the approximate dynamic response of the ith cell in such a simplified chemotaxis model is finally respresented by a *mean gradient perception*, detecting direction and amount of steepest ascent of the surrounding cAMP concentration field. In realistic

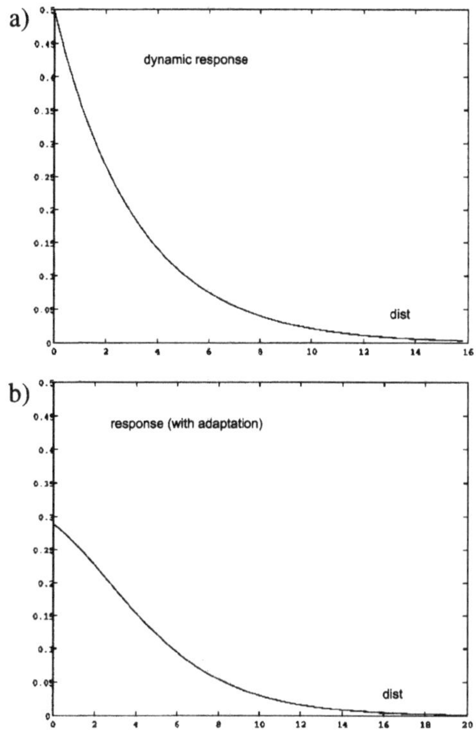

Figure 2. Dependence of the dynamic response, Eq. (18), on distance r_{ij} between two cells (a) for a linear response functional, (b) for an adapting response, model Eq. (13); parameters: $K = 1$, $\chi = 10$.

cases with adaptation as modeled by (13), for instance, the linear 'differential' perception is multiplied by a factor decreasing with the perceived 'absolute' cAMP concentration $\rho = \rho(t, X_i(t))$, namely, $\widetilde{f}'(\rho) = \frac{\chi K}{(K+\rho)^2}$.

In the simple 1D situation of just two cells, the integral in (18) represents the *mean gradient* perceived by a cell and is explicitely given (e.g., for the left-hand cell) by

$$\frac{1}{2}\left(\partial_x^+ \rho(t, X_1(t)) + \partial_x^- \rho(t, X_1(t))\right) = \frac{\lambda/\overline{r}\ \exp\left(-r_{ij}/\overline{r}\right)}{2\gamma\overline{r} + \beta\left[1 + \exp\left(-r_{ij}/\overline{r}\right)\right]}. \tag{19}$$

Depending on distance, it exponentially decays to 0 for $r_{ij} \to \infty$ with a rate scaled by the diffusional range distance $\overline{r}$ (see Fig. 2a). By Eq. (2) this induces aggregation of the two cells, which accelerate more the nearer they approach each other. For an 'adapting response' functional (13), there is a slight reduction in the slope for small distances (see Fig. 2b).

Clearly, similar results would be obtained in higher dimensional situations. For more than two cells, however, the chemotactic perception of a common interaction field generally would lead to more intricate response functionals, which may still be computed by using analogous linear superposition formulas as given above for the 1D case (cf. also [25, 26]).

4.2. Example II: Aggregation by Tensiotaxis

Here we briefly present some analytic results in the 1D case as well as numerical simulations for $m = 1$ and $m = 2$ (see [7]). Let us first consider only one single cell, one-domensionally represented by an interval $[X_i(t) - \epsilon, X_i(t) + \epsilon]$, and its effect on the 1D viscoelastic substratum, which then obeys the following nonlinear hyperbolic-elliptic system (on a fixed interval with appropriate boundary conditions)

$$\partial_t \rho + \nabla_x \cdot (\rho \mathbf{v}) = 0 \tag{20}$$

$$\partial_x \cdot (\eta \partial_x \mathbf{v} - \alpha(\rho - \rho_0)) - \Phi_{\mathfrak{M}} \rho \mathbf{v} = \sum_k f\left(\tau_{ik}(t)\right) \theta_{ik}(t)\, \delta_{Z_k(t)}\,, \tag{21}$$

where the sum runs over those binding sites whose position occupies the interval $|Z_k(t) - X_i(t)| < \varepsilon$. There the perceived tension is given by $\tau_{ik}(t) = \tau(t, Z_k(t) + \tilde{\varepsilon}\theta_{ik}(t))$ with $\tau = \eta\partial_x \mathbf{v} - \alpha(\rho - \rho_0)$, as in eq. (9), and the sign of the force term is determined by $\theta_{ik}(t) = sign\,(Z_k(t) - X_i(t))$.

We have to consider the regularity of the solution functions above. Since the right-hand side of (21) is a sum of Dirac-measures, at each such singular point $Z_k(t)$ the tension τ and, in general, $\partial_x \mathbf{v}$ too, will perform a jump in space. But then, due to (20), also the density ρ will jump across corresponding characteristics. However, this does not matter, since according to the sign $\theta_{ik}(t)$, the tension is evaluated either to the right or to the left, so that for negligibly small $\tilde{\varepsilon} \ll \varepsilon$ we can write $\tau_{ik}(t) = \tau(t, Z_k(t)\pm)$. For two and more cells, the same arguments hold, so that we are dealing with a well-posed hyperbolic-elliptic boundary value problem. Simulating the full system (5), (20)-(21) with an initially given homogeneous substrate density $\rho(0, x) = \rho_0$ and uniform stochastic distribution of adhesion sites $Z_k(0)$, we obtain, for a pair of cells, a continuing aggregation dynamics whose plots at two selected times can be viewed in Figure 3.

The numerical algorithm for these simulations is conceived for general 2D situations, where discretization of (20, 21) uses an adapted flux corrected transport scheme for 2D triangulated grids based on [4].

Pictures of corresponding simulations with two cells approaching each other *via* tensiotaxis on a circular substratum ('Petri dish') are shown in Figure 4.

5. Swarm Dynamics by Nearest-neighbor Interaction

Let us consider a flock (or 'swarm') of N 'birds', for example, starlings in a 3D 'swarm cloud' (often resembling a dynamic 'fluid drop' in air) or wild geese in 1D 'skeins' (resembling viscoelastic threads) or, in general, biological entities that perform *collective movement* by keeping a more or less fluctuating but, on the average, constant *mean neighbor distance*, say δ.

Such a collective movement can be achieved, for instance, by visual interaction in a manner that each 'bird' selects in its *perception scope* only those neighbors that are closest in distance, measuring certain 'relations' to them and simultaneously suppressing

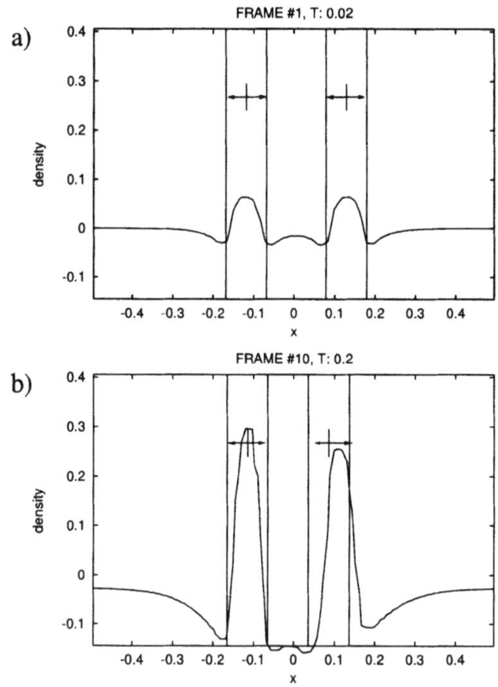

Figure 3. Simulation of the 1D tensiotaxis model (Example II) for two cells migrating on a viscoelastic polymer substrate as the interaction field. Substrate concentration is plotted for successive times, (a) $t = 0.02$, (b) $t = 0.2$, with the substrate being compressed underneath each cell. The low concentration region in between the two cells corresponds to maximum tension. Positions of cell centers are given by short vertical strokes, where arrows indicate the net force transmitted to the substrate on either side. Proximal vertical lines indicate cell boundaries.

all other sensory inputs. Assuming, for simplicity, that 'birds' are able to look around in any direction without difference, then the *nearest neighbors* of the ith 'bird' might be defined by the well-known *Voronoi tessellation*, namely, as those j for which the $(m-1)$-dimensional *contact surface* (or *Voronoi surface*)

$$\Delta_{ij}(t) := \{x \epsilon \mathfrak{R}^m : |x - X_i(t)| = |x - X_j(t)| < |x - X_k(t)| \, \forall k \neq i, j\} \qquad (22)$$

is not empty. Observing that this nearest-neighbor relation, iNj, is commutative, we will use the notation $j \epsilon N_i$ or, equivalently, $i \epsilon N_j$. The contact surface Δ_{ij} (more precisely, the hypersurface spanned by it) orthogonally cuts the connecting vector $R_{ij} := X_j - X_i$ between two 'birds' exactly at half their *neighbor distance* $r_{ij} := |R_{ij}|$. The vector R_{ij} represents the visual interaction line, along which signals between two neighbors are exchanged, in a commutable way (according to our assumption above) through the whole space as medium, $\mathfrak{M}_t = \mathfrak{R}^m$. Thus, we might consider the *interaction field* as represented by a $(m-1)$-dimensional Hausdorff measure $d\mu_t = \sum_{iNj} d\sigma_{\Delta_{ij}(t)}$, supported by all contact surfaces; furthermore, the *interaction tensor* $\mathfrak{T}_t|_{\Delta_{ij}(t)}$ could be defined by

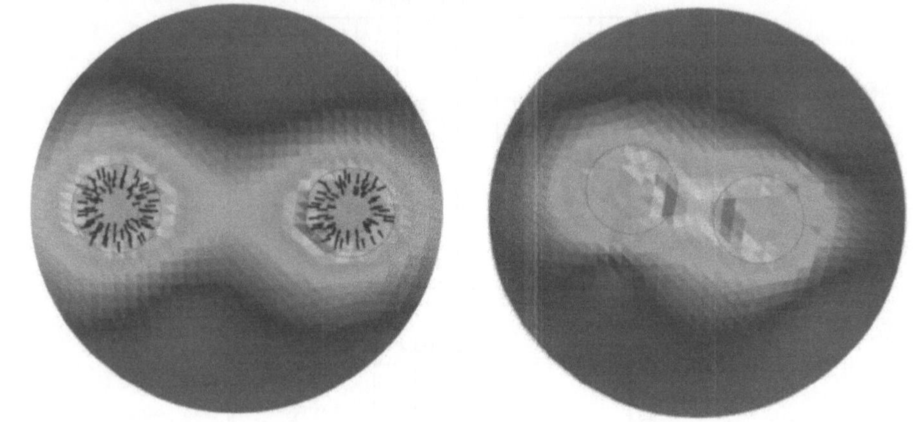

Figure 4. Simulation snapshots of two cells migrating and interacting on a viscoelastic polymer substrate in the 2D tensiotaxis model (Example II). Arrows indicate forces acting on the substrate at adhesion sites. They are confined to the outer rim (lamella region) of the cell. Colors code tension in the polymer matrix (see Eq. (9) and color plate 13).

the ellipse (or m-dimensional ellipsoid) whose 'main axis' is R_{ij} and whose orthogonal cross-section has size $|\Delta_{ij}(t)|$. (See the model below for a justification.) Thus, for each 'bird' position $X_i(t)$ we have an open *Voronoi neighborhood* $\mathfrak{V}_i(t)$ whose boundary, the so-called *Voronoi polyhedron* $\partial\mathfrak{V}_i(t) = \bigcup_{j\epsilon N_i} \overline{\Delta_{ij}(t)}$, is just a concatenation of the related contact surfaces to its nearest neighbors. See the 2D example in Figure 5, which also shows the so-called dual *Delaunay triangulation* generated by the connected network of edges R_{ij} between nearest neighbors.

5.1. Differential Perception and 'Visco-elastic' Response

By using the formalism of Section 3, let us assume that each 'bird', represented by a ball B^i_t around $X_i(t)$ with (negligibly small) radius ε, has a 'round about' perception scope that is defined by the radial projection $\mathfrak{P}^i_t$: $\partial B^i_t \simeq \mathfrak{S}^{m-1} \to \partial\mathfrak{V}_i(t)$ onto the interaction field, such that for any sensing direction $\theta \in \mathfrak{S}^{m-1}$ of its 'spherical vision field' the optical input induces a *perceived measure* $dG^i_t(\theta) = \mathfrak{G}\left[\mathfrak{T}_t\left(\mathfrak{P}^i_t(\theta)\right), d_t\mathfrak{T}_t\left(\mathfrak{P}^i_t(\theta)\right)\right] d\theta$. This measure is constant for each angular 'visual cone' $\Theta^j_i(t) := \left\{\theta : \mathfrak{P}^i_t(\theta) \subset \Delta_{ij}(t)\right\}$ towards a nearest neighbor, as it takes values from the interaction tensor $\mathfrak{T}_t|_{\Delta_{ij}(t)}$ and its temporal derivative. The first model assumption of constancy (as well as neglecting more distant birds) idealizes the concept of contrast enhancement frequently found in biological sensing processes. The second assumption – that a bird perceives not only instantaneously measured values, $\mathfrak{T}_t$, but also their change in time, $d_t\mathfrak{T}_t$, due to certain sensory adaptation mechanisms or memory effects, namely *differential perception* – is another typical biological feature, also occuring in cell-cell interactions but not present in simple physicochemical interactions.

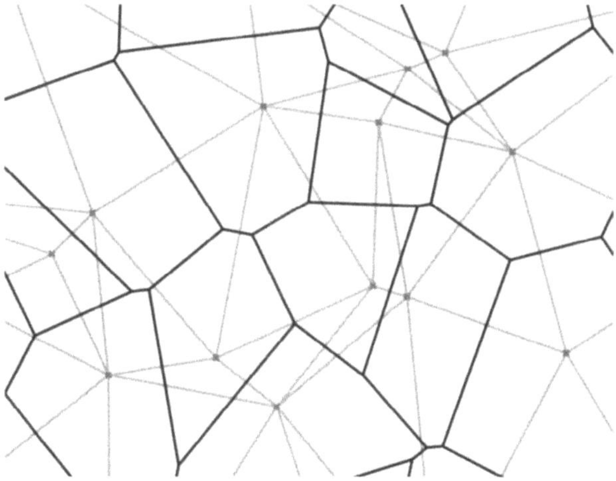

Figure 5. Part of a 2D Voronoi tessellation (black lines) and Delaunay triangulation (gray lines) induced by a set of points X_i, representing the network of nearest neighbors in a simulated bird swarm (from [22]).

Notice that the *'space angle'* $\left|\Theta_i^j\right|$ measuring the size of visual contact with the j-neighbor is related to the 'vision scope size' of the contact surface $|\Delta_{ij}(t)|$ *via* trigonometric formulas and the scaling factor $\left(r_{ij}/2\right)^{m-1}$. On the other hand, the neighbor body B_t^j will induce an 'angular spot' $\widetilde{\Theta}_i^j$ on the 'vision field', with mean angular direction $\theta_{ij} := R_{ij}/r_{ij}$ and size $\left|\widetilde{\Theta}_i^j\right| \sim \varepsilon^{m-1} \cdot r_{ij}$, so that the ith bird can obtain all defining quantities of the interaction tensor $\mathfrak{T}_t|_{\Delta_{ij}(t)}$, namely, $R_{ij} = r_{ij}\theta_{ij}$ and $|\Delta_{ij}(t)|$, by elementary neuro-physiological calculations from sensory inputs in its angular vision field.

Example III: Visco-elastic nearest-neighbor response

If we assume that the ith bird perceives the quantities $R_{ij}, d_t R_{ij}$, and $|\Delta_{ij}|$ with respect to each of its nearest neighbors, $j \epsilon N_i$, then the induced acceleration in (11) might be given as a quasi-linear combination, $d\alpha_i(\theta) = (\widetilde{\alpha}_{ij}\, R_{ij} + \widetilde{\mu}_{ij}\, d_t R_{ij})\, d\theta$, for $\theta \in \Theta_i^j$, with 'elasticity' coefficient, $\widetilde{\alpha}_{ij}$, and 'viscosity' coefficient, $\widetilde{\mu}_{ij}$, non-linearly depending on nearest neighbor distance r_{ij} and 'vision scope' size $|\Delta_{ij}|$. Thus, the resulting dynamic acceleration response (11) could be expressed as

$$\begin{aligned} A_i &= \sum_{j \epsilon N_i} \left|\Theta_i^j\right| \left\{\widetilde{\alpha}_{ij}\left(r_{ij}, |\Delta_{ij}|\right) R_{ij} + \widetilde{\mu}_{ij}\left(r_{ij}, |\Delta_{ij}|\right) d_t R_{ij}\right\} \\ &= \sum_{j \epsilon N_i} \left\{\alpha\left(r_{ij}, |\Delta_{ij}|\right) R_{ij} + \mu\left(r_{ij}, |\Delta_{ij}|\right) d_t R_{ij}\right\} . \end{aligned} \tag{23}$$

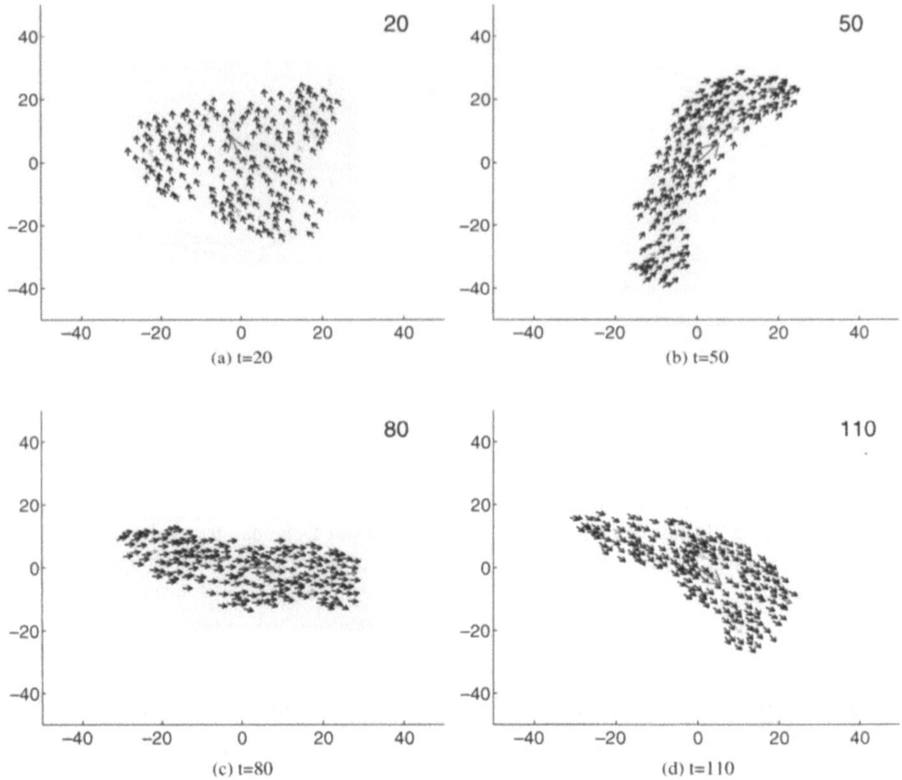

Figure 6. Simulation snapshots for the 2D bird swarm multiparticle model with monotone increasing 'elasticity' and decreasing 'viscosity' depending on nearest-neighbor distance (positions of 200 birds with fixed swarm center, at times $t = 20, 50, 80$, and 110 s). Initial configuration was a stochastically perturbed hexagonal arrangement. Further stochastic noise as well as sudden changes in nearest-neighbor relations induce fluctuations, such that major changes in size and shape of the swarm occur. Notice quite smooth free boundaries of the dynamic swarm. For further discussion, see text. From [22]

5.2. Numerical Simulations of Bird Swarms

In space dimension $m = 2$ this model has been used for extensive simulations under special assumptions on only distance dependent coefficients: $\alpha(r)$ is monotone increasing with a unique zero r^* defining a preferred equilibrium distance, with neighboring birds *repelling* each other at distances $r < r^*$ and *attracting* at $r > r^*$; $\mu(r)$ is monotone decreasing, modeling *mutual velocity adjustment* between nearest neighbors, becoming weaker for larger distances. Figure 6 plots the configuration of a simulated swarm with $N = 200$ birds at four consecutive times, with center of mass fixed. It is obvious that the geometric swarm shape varies but keeps a well-defined and quite smooth boundary. Moreover, mean swarm orientation changes considerably during the time intervals of 30 s

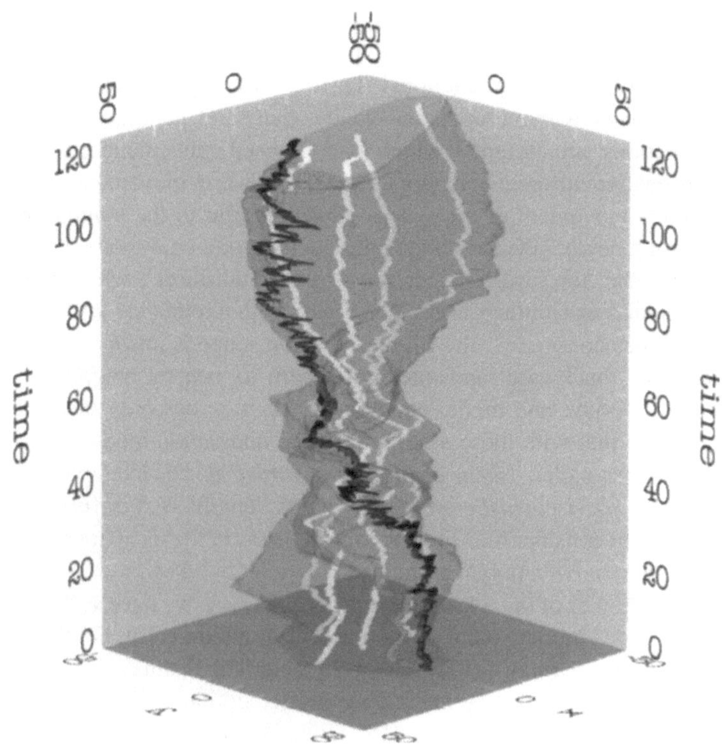

Figure 7. Time-space plot of a complete bird swarm simulation as in Figure 6 (with fixed swarm center, time $t = 0 - 120$ s, from bottom to top). The surface depicts the swarm boundary in the x-y plane changing in time; the black curve on the boundary indicates the mean moving direction of the swarm (viewed from the center); other curves represent the relative positions of five selected birds. For further discussion see text and [22] (see color plate 14).

between successive pictures. A corresponding 3D time-space plot of the same swarm simulation (Figure 7, again with fixed swarm center) shows continuous dynamics for a whole time span of 120 s. Observe the fluctuating swarm size with two more-or-less expressed maxima, suggesting a possible temporal fluctuation period of about 60 s. Moreover, the mean moving direction of the swarm, indicated by the black curve along the boundary of the 'swarm hull', reveals periods of quite strong directional persistence interrupted by sudden directional changes. Finally, with the chosen parameters in Figure 7, one might detect viscous fluid-like behavior in the relative motion of 'particles' with respect to each other: Single birds marked by different colors only slightly shift their relative position within the swarm, while their (joint) mean moving direction can turn even more than 180 degrees, thus leading to a characteristic type of stochastically perturbed *shearing deformation* of the whole swarm.

Two-dimensional simulations as well as certain analytical calculations have been performed in a similar model for swarms of 'particle agents' with 'harmonic interactions' [13]. There, the only interaction functional is $\alpha(r) = ar$, representing pure linear attraction between 'particles' without repulsion. On the other hand, general monotone nonlinear 'dissipation functions' are assumed, which lead to speed adjustment terms as in Eq. (3). The authors reveal two different behavioral modes for swarm movement. One mode is the usual translational movement of the swarm center, similar to the simulations described before. The second newly detected mode consists in a rotational movement of elongated swarms, whereby the 'top birds' are attracted by the 'tail birds', which they observe at one side. Corresponding simulations are performed with a relatively low degree of noise.

In space dimension $m = 1$ the ith bird's vision scope is just $\theta = \pm 1$, i.e., looking either 'forward' or 'backward' and interacting with its nearest neighbors $j = i \pm 1$. Then Δ_{ij} is just a point, and model equation (23) *per se* has only distance dependent coefficients. Notice that with these assumptions the interaction model (1)–(3) describes the dynamics of a so-called *'skein of migrating birds'* as, for example, wild geese or cranes, generally a *1D viscoelastic ordered chain* of 'length' N. For further applications in biology, one might consider files of migrating cells (as in the morphogenetic movement of embryonic tissue) or even polymer chains in filament bundles.

For various choices of the 'elasticity' coefficient $\alpha(r)$, we have systematically studied the behavior of this model, mainly by numerical simulations [2]. For the simplest case of a *linear viscoelastic chain*, we observe two characteristic features of 1D swarm dynamics (see Fig. 8): Longterm fluctuations in swarm size, similar to the 2D case (Fig. 7) but more regular, with a period of about $T = 18$ s (Fig. 8a), and so-called *compression waves* (like in traffic flow) with a persistent wave speed of about $c = 1\,\mathrm{m/s}$ (Fig. 8b). Notice that the mean swarm diameter is 10m, so that a compression wave passes through the swarm from one boundary to the other in about half the time of a size fluctuation period. These waves, emerging by stochastic perturbations, can arise at both the front and the rear of the swarm. We remark that a uniform swarm translation to the right according to an assumed preferred moving speed $S^* > 0$ in (3) was subtracted in the plots. In more realistic asymmetric models, where the bird's vision scope is biased in forward direction, compression waves appear only at the swarm front (for more details, see [2]).

Moreover, for a non-monotone 'elasticity' coefficient $\alpha(r)$ starting to decrease for distances r larger than two times the equilibrium distance r^*, meaning that mutual attraction decays for larger distance between neighboring birds, we observe swarm separation (see Fig. 9). For the chosen parameters a simulated swarm of 18 birds separates by chance into a bigger and a smaller part (Fig. 9a), where within the bigger part transient seperation can be observed (Fig. 9b). Thus, compared to the simulations in Figure 8, the overall swarm dynamics is destabilized due to the chosen nonlinearity $\alpha(r)$.

5.3. Scaling and Continuum Limit Equations for $\delta \to 0$

Let us consider large swarms with infinitely increasing bird number, $N \to \infty$, and with vanishing mean neighbor distance, $\delta \to 0$, such that the mean swarm diameter $N \cdot \delta \sim 1$ stays bounded. If then the spatial coordinates r are replaced with r/δ and the acceleration term in (23) is properly rescaled by a factor, namely $1/\delta^m$, the suggestive question

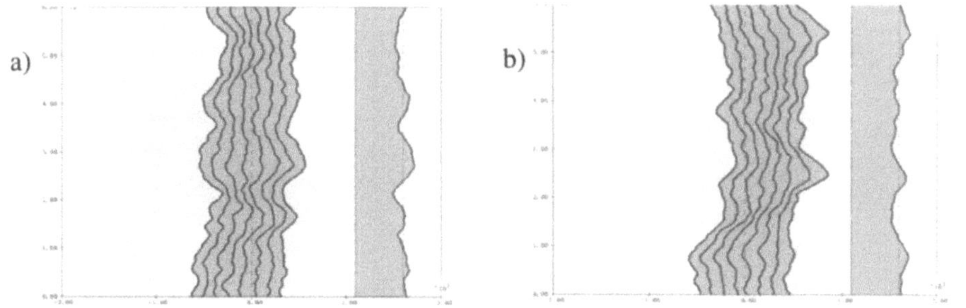

Figure 8. Two consecutive plots (twice 60 s, running from bottom to top) for the stochastic 1D viscoelastic swarm model with $\alpha(r)$ linearily increasing, equilibrium distance $r^* = 1.25$ m. Simulated positions (left plot) of a skein of 8 birds showing fluctuations in size (plot to the right) with a mean period of about 18 s (best seen in a) and compression waves with constant speed of about 1 m/s (best seen in b).

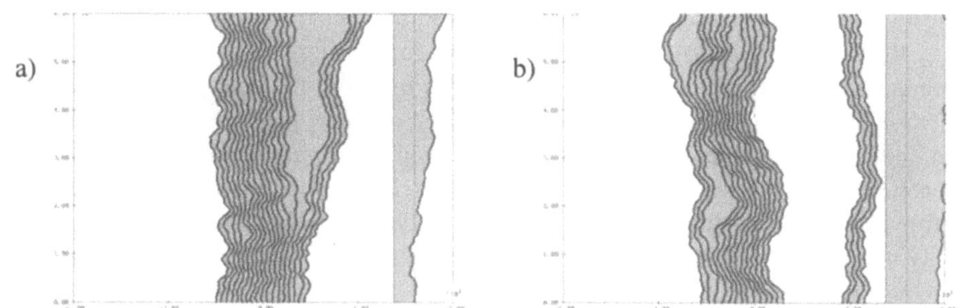

Figure 9. Plots similar to Fig. 8 with $\alpha(r)$ increasing for lower $r \sim r^*$ but decreasing to zero for $r \to \infty$. Simulated skein of 18 birds showing swarm separation (a) and ongoing swarm instability (b).

arises whether there are limiting differential equations determining the resulting continuum dynamics. In the 1D case ($m = 1$), this could be positively answered [2]. Suitable definitions of 'densities' (as step functions) and straightforward formal approximation procedures (using difference quotient arguments) yield as limiting equations the standard *compressible Navier-Stokes equations* with proper nonlinearities and natural Neumann conditions at the two *moving free boundary* points. Moreover, estimations for the naturally defined local *viscoelastic swarm tension* can be achieved, such that suitable scaling of the additive Gaussian noise in the stochastic equation (2), namely $\beta_i \sim \sqrt{\delta}$, guarantees the preservation of additive stochastic perturbations in the continuum limit. Discretization and numerical solution of this *stochastic hyperbolic-parabolic free boundary problem* reproduce the two characteristic features of the stochastic multiparticle system (presented in Section 5.2) also in the continuum limit. Typical swarm size fluctuations with recognizable period as well as compression waves with remarkably constant speed (e.g., seen

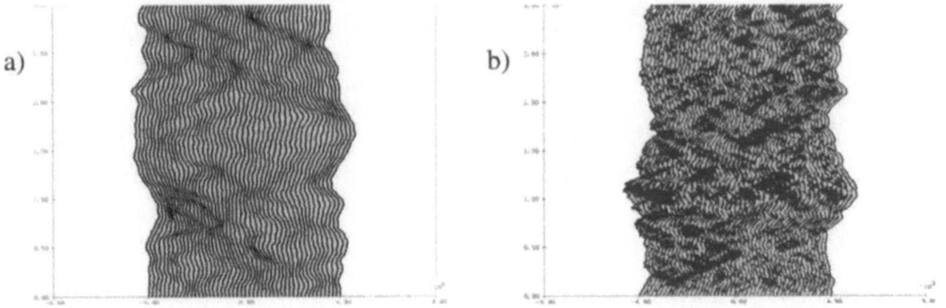

Figure 10. Comparison of stochastic multiparticle simulations (left, $N = 50$ birds) with discretized numerical solutions of the stochastic Navier-Stokes continuum limit (right, $N = 50$ discretization intervals)

in Fig. 10a for 50 birds) are again induced, in the continuum model, by stochastic perturbations of the otherwise stable trivial situation of a steadily moving swarm with uniform density (Fig. 10b, using discretization by 50 uniform compartments).

Recently, the formal approximation procedure leading to a continuum limit has been generalized to the m-dimensional situation by using the dual pair of *Voronoi tessellation* and *Delaunay triangulation* [2]. On each of these, piecewise constant densities and linearily interpolated velocities are defined, and *uniformity criteria* for the geometric nearest-neighbor constellation in the moving 'particle network' are formulated that would enable convergence estimates and guarantee that the two definitions for density and velocity merge in the limit $\delta \to 0$. Under further restrictive assumptions on the parameter functions in (23) the outcome would again be the *compressible Navier-Stokes equations* (cf. [8]) for model application in myxobacteria swarming. However, there remain several analytic and algorithmic problems to be resolved. One is the complex problem of appropriate 'boundary closure' of the Voronoi-Delaunay network and 'boundary conditions' for the suitably defined 'normal' and 'tangential components' of the velocity (see also [5]). Another problem is a request for the derivation of proper differential equation systems in the general modeling case (23), leading possibly to anisotropic fluid dynamics with extra equations for the stress tensors in correspondence to the interaction tensors $\mathfrak{T}_t$ defined above and appearing in (23). For further details and discussions see [2].

6. Dynamics of Cell Monolayers and Voronoi-Delaunay Neighborhoods

Two-dimensional dynamic cell monolayers appear in many biological tissues as well as in corresponding experimental assays. There, motile cells adhere to a substratum, perform autonomous migration and, when contacting other cells, mutually exert forces or communicate by further mechanical or chemical transmissions, and, eventually, concert their mechanical activities into cooperative movement of the monolayer or into morphogenetic tissue arrangement. The most prominent example is epithelial cell migration during early embryogenesis, where a whole sheet of cells cooperatively moves towards determined regions and undergoes 3D folding [28]. A further important area is cell arrangement in endothelial or epithelial monolayers, e.g., at the inner surface of blood vessels [23] or intestinal crypts [21]. In recent years techniques of image processing have been improved so that *quantitative morphodynamics* can be performed, both *in vivo* and *in vitro*, e.g., statistical analysis of translocations and shape deformations of single cells within confluent monolayer *in vitro* [10]. Tracing of individual cells is, in the mean time, even possible for 3D tissue movement, e.g., in *Dictyostelium* slugs [3].

As a matter of fact, the idea of using the geometric *Voronoi tessellation* induced by 'cell centers' in order to represent their shapes and mutual contact zones in a simplified way dates back about 20 years to Honda's geometrical article [18] and a more histological one [19] which also considers 3D situations. Thereafter, this model idea had been used only sporadically, e.g., for modeling and simulating cell sorting based on minimization of contact surface energy [27]. However, recently it has been picked up again for modeling confluent 2D cell layers in [5] and [21].

For a survey of stochastic, multiparticle simulation models including cell-cell contacts, see the review article by Drasdo (this volume, Part III). In order to model some kind of 'cell integrity', the most frequently used procedure is to replace the 'elastic' repulsion function (see $\alpha(r)$ in the section above) with a 'hard core potential' or some approximation, simultaneously maintaining cell-cell attraction by either keeping the attractive 'elastic' part or introducing an exterior field, e.g., chemotactic [6].

However, tissue-forming cells have a tendency to keep mutual adherence, at least transiently under suitable chemical and mechanical conditions. Therefore, modeling direct mechanical transmissions over contact surfaces is required. (For discussion of the most detailed model by Weliky & Oster [28], see again Drasdo, this volume.)

The Voronoi-Delaunay approach, which was used by [21] for modeling the arrangement of newly produced cells in a cylindrical representation of intestinal crypts, brought us a major step forward in reproducing experimental findings. These authors use a *simplified elastic model* without viscosity, i.e., $\alpha(r)\,R = \mu \cdot (r - r_0) \cdot R/r$, though with an equilibrium distance (rest length of a model 'elastic spring') that is linearily increasing with time after cell division.

On the other hand, Bottino [5] generalizes his Voronoi-Delaunay simulation approach, which he had used for modeling intracellular cytoplasm polymer fluid dynamics, to cell monolayer dynamics. Hereby he applies a simple *incompressibility* condition to the case of cells. By introducing a suitable penalty term in the force balance equations, he keeps all cells at an almost fixed 'volume'.

Both papers do not model autonomous cell migration forces (as in the tensiotaxis model above) but rather bring about cell movement from 'pushing forces' onto contact surfaces with nearest neighbors. Moreover, both papers use the 'over-damped' Stokes' approximation (5) so that they assume relatively fast accomodation of the cell's migratory apparatus to incoming stimuli. Finally, 'viscosity' is not modeled explicitely as in our 'differential response' model above; it is only induced by passive flow of cells which thereby change their *neighborhood relations* – meaning that 'contact bonds' are eventually broken.

What should be considered in the future are more detailed models for mechanical interaction *via Voronoi contact sufaces*, e.g., by using the modeling ideas presented in this article – combination of ideas from Section 5.1 on bird swarms with previous ideas on tensiotaxis – at least in the following ways (cf. also the recent approach in [11]):

1. Perception of tension at the Voronoi surfaces, i.e., the cell-cell contacts, could be transduced (under consideration of additional forces such as adhesion or cortical contraction) into a resulting dynamic response of mutual force application at the cell-cell contact acreas, i.e., again *via* the Voronoi surfaces.
2. Alternatively or complementarily, a more detailed model could introduce interaction fields that 'pass through' the cells. The simplest example would be a pressure field. More intricate could be the idea of regarding the 'synthethium' of all intracellular actin filament networks, mutually connected *via* cadherine receptors on contact surfaces, as a global mechanical interaction field.

Acknowledgments

Most of the presented results were obtained within the DFG research program (SFB 256) Nonlinear Partial Differential Equations, University Bonn. We thank our colleagues at the Institute for Applied Mathematics for many helpful hints and discussions, in particular about stochastics and numerics.

References

[1] W. Alt, O. Brosteanu, B. Hinz and H.W. Kaiser (1995) Patterns of spontaneous motility in videomicrographs of human epidermal keratinocytes (HEK). *Biochemistry and Cell Biology* **73**, 441–459

[2] W. Alt (2002) Nonlinear hyperbolic systems of generalized Navier-Stokes type for interactive motion in biology. In: *Geometric Analysis and Nonlinear Partial Differential Equations* (S. Hildebrandt, H. Karcher eds.) Springer-Verlag, Berlin, 431–459

[3] J.T. Bonner (1998) A way of following individual cells in the migrating slugs of *Dictyostelium discoideum*. *Proc. Nat. Acad. Sci.* **95**, 9355–9359

[4] J.P. Boris, A.M. Landsberg, E.S. Oran, J.H. Gardner (1993) LCPFCT – A flux-corrected transport algorithm for solving generalized continuity equations. *NRL Memorandum Report* 93–7192

[5] D.C. Bottino (2000) Computer simulations of mechanochemical coupling in a deforming domain: Application to cell motion. In: (Ph. K. Maini and H. Othmer, eds.) *Mathematical Models for Biological Pattern Formation*, Springer, New York, 295–314

[6] T. Bretschneider, B. Vasiev, C.J. Weijer (1997) A model for cell movement during *Discyostelium* mound formation. *J. Theor. Biol.* **189**, 41–51

[7] T. Bretschneider (2003) Reinforcement of cytoskeleton-matrix bounds and tensiotaxis: A cell based model. In: (A. Deutsch, J. Howard, M. Falcke, W. Zimmermann eds.) *Function and Regulation of Cellular Systems: Experiments and Models.* To appear in Birkhäuser-Verlag, Basel

[8] A. Czirok, T. Vicsek (2000) Collective behaviour of interacting self-propelled particles. *J. Phys. A* **281**, 17–29

[9] J.C. Dallon, H.G. Othmer (1997) A discrete cell model with adaptive signaling for aggregation of *Dictyostelium discoideum. Phil. Trans. Royal Soc. London: Ser. B* **352**, 391–417

[10] P. Dieterich, M. Odenthal-Schnittler, C. Mrowietz, M. Krämer, L. Sasse, H, Oberleitner, H.-J. Schnittler (2000) Quantitative morphodynamics of endothelial cells within confluent cultures in reponse to fluid shear stress. *Biophys. J.* **79**, 1285–1297

[11] P. Dieterich, J. Seebach, H.-J. Schnittler (2003) Quantification of shear stress induced cell migration in endothelial cultures. In: *Function and Regulation of Cellular Systems: Experiments and Models.* (A. Deutsch, J. Howard, M. Falcke, W. Zimmermann eds.) To appear in Birkhäuser-Verlag, Basel

[12] M. Dworkin, D. Kaiser, eds. (1993) *Myxobacteria II.* Amer. Society of Microbiology, Washington

[13] W. Ebeling, F. Schweitzer (2001) Swarms of particle agents with harmonic interactions. *Theory Biosciences* **120**, 207–224

[14] C.G. Galbraith, M.P. Sheetz (1998) Forces on adhesive contacts affect cell function. *Curr. Opin. Cell Biol.* **10**, 566–571

[15] F.H. Heppner (1997) Three-dimensional structure and dynamics of bird flocks. In: (J.K. Parrish, W.M. Hammer eds.) *Animal Groups in Three Dimensions.* Cambridge University Press, 68–89

[16] B. Hinz, W. Alt, C. Johnen, V. Herzog, H.W. Kaiser (1999) Quantifying lamella dynamics of cultured cells by SACED, a new computer assisted motion analysis. *Exp. Cell Research* **251**, 234–243

[17] T. Höfer, J.A. Sherratt, P.K. Maini (1995) Cellular pattern formation during *Dictyostelium* aggregation. *Physica D* **85**, 425–444

[18] H. Honda (1978) Description of cellular patterns by Dirichlet domains: The two-dimensional case. *J. Theor. Biol.* **203**, 317–333

[19] H. Honda (1983) Geometrical models for cells in tissues. *Int. Rev. Cytology* **81**, 191–248

[20] Th. Libotte, H.-W. Kaiser, W. Alt, T. Bretschneider (2001) Polarity, protrusion-retraction dynamics and their interplay during keratinocyte cell migration. *Exp. Cell Research* **270**, 129–137

[21] F.A. Meinicke, S.C. Potten, M. Loeffler (2001) Cell migration and organization in the intestinal crypt using a lattice-free model. *Cell Prolif.* **34**, 253–266

[22] R. Müller (2002) Dynamik in stochastischen Vielteilchensystemen zur Modellierung von Vogelschwärmen. *Diploma thesis* Univ. Bonn

[23] R.M. Nerem (1993) Hemodynamics and the vascular endothelium. *J. Biomech. Engin.* **115**, 510–514

[24] E. Palsson (2001) A three-dimensional model of cell movement in multicellular systems. *Future Gen. Comp. Systems* **17**, 835–852

[25] A. Stevens (2000) The derivation of chemotaxis equations as limit dynamics of moderately interacting stochastic many-particle systems. *SIAM J. Appl. Math.* **61**, 183–212

[26] A. Stevens, F. Schweitzer (1997) Aggregation induced by diffusing and nondiffusing media. In: *Dynamics of Cell and Tissue Motion* (W. Alt, A. Deutsch, G. Dunn eds.) Birkhäuser, Basel, 183–192

[27] D. Sulsky, S. Childress, J.K. Percus (1986) A model for cell sorting. *J. Theor. Biol.* **106**, 275–301

[28] M. Weliky, G. Oster (1990) The mechanical basis of cell rearrangement. I. Epithelial morphogenesis during Fundulus epiboly. *Development* **109**, 373–386

Polymer and Cell Dynamics – Multiscale Modeling
and Numerical Simulations, 243–250
ed. by W. Alt, M. Chaplain, M. Griebel und J. Lenz

A Discrete-cell Model of Tissue-equivalent Compaction

R. Brent Rice[1] and Victor H. Barocas[2]

1) Department of Chemical Engineering, University of Colorado
2) Department of Biomedical Engineering, University of Minnesota, 7-106 BSBE, 312 Church St. SE, Minneapolis, MN 55455, USA

Summary. We (and others) have spent considerable effort over the last decade to develop continuum models partial differential equations (PDEs) of cell-populated collagen gels. These models have a number of advantages: they are relatively simple and easy to solve, they resemble transport equations familiar to engineers, and they are a natural description of the material on the scale of interest. There have, however, been recent advances in understanding the behavior of individual cells or fibrils. We are thus exploring ways to incorporate discrete cell/fibril information into our continuum model in a logical and tractable way.

1. Introduction

The ability of cells to generate contractile stresses is of great importance to bodily function. In addition to the obvious contractile role of muscle cells, fibroblast-driven contraction is significant in such processes as morphogenesis [1] and wound healing [2]. Because of the importance of this phenomenon, many scientists have undertaken different studies to quantify the force (or stress) that cells generate and to compare force generation under different conditions.

Our goal is to develop a framework within which to compare different measurements of cell contraction (N.B. – we do not distinguish between traction and contraction forces [3] but rather refer to all such forces as "contraction" forces because they lead to contraction of the surrounding tissue) and to enable use of single-cell data in population-scale simulations. Before discussing mathematical models, we briefly discuss the types of experiments done to measure cell contraction force. Although many cell types have been studied by such techniques, we focus our attention on fibroblasts, which are commonly studied and provide representative examples of the different classes of experiments.

Cell contraction experiments can be classified broadly as *single cell* or *multicell*. This distinction is quite simple, but the experiments in the two classes are very different. Single-cell measurements are in general difficult to do, and they often involve non-physiological environments. The experiments of [4] and [5], for example, were performed on silastic sheets, and those of [6] involved a cell interacting with a micromechanical device. Multicell experiments, in contrast, are often easier to perform, but the number of viable cells present may be inconvenient to measure. Furthermore, since multicell experiments generally involve cells and a non-cellular support (such as a collagen gel [7]), the interaction between the cells and the support can make interpretation of the measurement difficult. Finally, the multicell experiments often involve alignment of the cell population, which can have significant influence on the overall behavior of the system [8].

The different experiments all have the necessary features to provide qualitative comparison between two cases. One could, for example, add a suspected cytoactive agent to the medium in any of the experiments described above, and a cellular response would be measurable (e.g., [9, 10]). If direct comparison were one's only interest, there would be no need for further study. There are, however, important challenges that transcend a single experiment. One would like to be able to compare different studies of cell behavior in a rational way, requiring a method to determine, for example, whether the cell traction parameter reported by [11] is equivalent to the cell traction forces reported by [5]. These properties should be equivalent in some sense, since they are both measures of the contractility of the cell, but they arise from very different descriptions of cell behavior, so the relation among them is not clear.

An additional challenge arises from the use of cell contractility data to predict *in vivo* behavior. Sophisticated models of wound healing [12] and morphogenesis [1] are able to make excellent qualitative predictions of physiological phenomena, but quantitative prediction remains out of reach due in part to lack of accurate parameter values. Similarly, design of bioartificial tissues [13] would be dramatically improved by good, readily available cell contractility measurements. Because of the sheer number of cells involved in physiological processes, continuum models of cell behavior, based on cell concentration rather than individual cells, have generally been preferred. In order for such a model to make use of single-cell data, a multiscale approach is needed.

In this article, we derive a simplified discrete-cell form of our continuum-based anisotropic biphasic theory (ABT, [14]) of tissue-equivalent mechanics. The discrete-cell equations are solved by the finite element method, and the results are compared with an analytical solution for a test case.

2. Simplified Continuum Model

We begin by introducing the following simplifications to the ABT:

- The problem is treated as two-dimensional (plane stress) and isotropic.
- The linear viscoelastic fluid model of the ABT is replaced with the Newtonian fluid model (N.B. – this is mathematically equivalent to using a linear elastic model for the test problems solved herein).
- Only the initial stages of compaction are considered, so there is no change in sample volume.
- The sample is taken to be sufficiently thin that resistance to permeation is negligible. This assumption implies that the compaction of the gel is not limited by interstitial drag on the exuded fluid.

With this set of simplifications, the ABT stress balance reduces to the following simple form:

$$\mu\nabla\cdot\left[\nabla\mathbf{v}+(\nabla\mathbf{v})^T+\frac{2\nu}{1-\nu}(\nabla\cdot\mathbf{v})\,\mathbf{I}\right]+\tau_0\nabla c=0\,, \tag{1}$$

where μ is the viscosity of the matrix, $\mathbf{v}$ is the velocity, ν is Poisson's ratio (N.B. – the second coefficient of viscosity could equivalently be used, but we prefer ν for consistency with previous presentation of the model and to facilitate comparison with linear elasticity below), $\mathbf{I}$ is the identity tensor, τ_0 is the cell traction parameter, and c is the cell concentration.

3. Discrete-cell Model

We now modify the model from the previous section to include cells as discrete, force-generating inclusions rather than as a continuum. The τ_0 term must be removed from Eq. (1) and replaced with a discrete-cell form:

$$\mu\nabla\cdot\left[\nabla\mathbf{v}+(\nabla\mathbf{v})^T+\frac{2\nu}{1-\nu}(\nabla\cdot\mathbf{v})\,\mathbf{I}\right]+\sum_{i=1}^{N_{cells}}\nabla\cdot\sigma_i=0\,, \tag{2}$$

where σ_i is the stress due to an individual cell. Figure 1 shows an idealized cell of length L and width $L\epsilon$ aligned at some angle α from the horizontal axis. In this model, we assume that the cell generates force *only* along its polarity axis α. The cell generates a force per unit thickness F_0 and thus a stress $F_0/L\epsilon$ over its entire area of $L^2\epsilon$. Assuming that L and $L\epsilon$ are small compared to the length scale of interest, we conclude that

$$\sigma_i=F_0L\delta(x-x_i)\delta(y-y_i)\alpha\alpha\,, \tag{3}$$

where the cell is located at position (x_i, y_i) and δ is the Dirac delta function (with units of inverse length). Substituting Eq. (3) into Eq. (2) yields the desired expression:

$$\mu\nabla\cdot\left[\nabla\mathbf{v}+(\nabla\mathbf{v})^T+\frac{2\nu}{1-\nu}(\nabla\cdot\mathbf{v})\,\mathbf{I}\right]+\sum_{i=1}^{N_{cells}}F_0L\mathbf{s_i}=0\,, \tag{4}$$

where

$$\mathbf{s_i}\equiv\left[\begin{array}{c}\alpha_x\alpha_x\delta'(x-x_i)\delta(y-y_i)+\alpha_x\alpha_y\delta(x-x_i)\delta'(y-y_i)\\ \alpha_x\alpha_y\delta'(x-x_i)\delta(y-y_i)+\alpha_y\alpha_y\delta(x-x_i)\delta'(y-y_i)\end{array}\right], \tag{5}$$

and $\delta'(x)$ is the dipole that arises from differentiation of the delta function. Eq. (4) represents the discrete-cell equivalent of Eq. (1).

4. Numerical Solution of the Discrete-cell Equation

Eq. (4) is solved by the standard Galerkin finite element method (SGFEM) with bilinear basis functions. Although SGFEM is not well suited to the discontinuous velocity profile that arises from Equation 4, the simplicity and flexibility of the method make it sufficiently

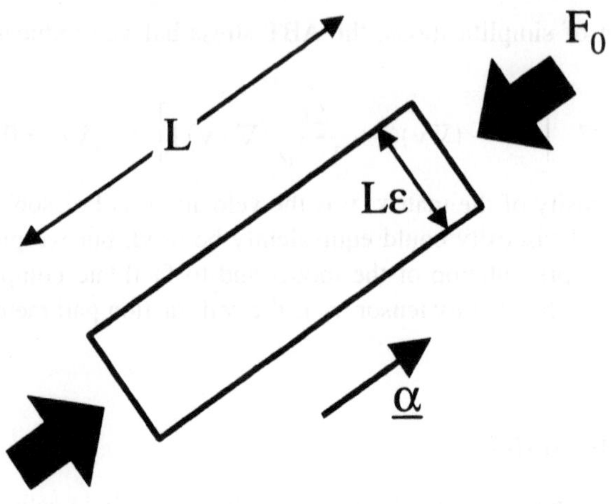

Figure 1. Schematic of cell force generation

attractive to explore. The model equation is multiplied by a test function ϕ and integrated over the domain Ω:

$$\int_\Omega \phi \cdot \left[\mu\nabla \cdot \left[\nabla\mathbf{v} + (\nabla\mathbf{v})^T + \frac{2\nu}{1-\nu}(\nabla\cdot\mathbf{v})\,\mathbf{I}\right]\right] + \int_\Omega \sum_{i=1}^{N_{cells}} F_0 L \phi \cdot \mathbf{s_i} = 0 \quad (6)$$

The first term is integrated by parts, and the second term is integrated by taking advantage of the equalities $\int \delta(x-a)f(x)dx = f(a)$ and $\int \delta'(x-a)f(x)dx = -f'(a)$. Since we are considering only cases with Dirichlet boundary conditions in this work, we omit the boundary term for brevity and write

$$\begin{aligned} &\textstyle\int_\Omega \mu\nabla\phi : \left[\nabla\mathbf{v} + (\nabla\mathbf{v})^T + \frac{2\nu}{1-\nu}(\nabla\cdot\mathbf{v})\,\mathbf{I}\right] + \\ &\textstyle\sum_{i=1}^{N_{cells}} F_0 L \left[\alpha_{i,x}\alpha_{i,x}\frac{\partial\phi_x}{\partial x} + \alpha_{i,x}\alpha_{i,x}\left(\frac{\partial\phi_x}{\partial y} + \frac{\partial\phi_y}{\partial x}\right) + \alpha_{i,y}\alpha_{i,y}\frac{\partial\phi_y}{\partial y}\right]_{x_i,y_i} = 0 \end{aligned} \quad (7)$$

The resulting problem is now solved by decomposing $\mathbf{v}$ into basis functions ϕ. Equation 7 can also be derived by treating the cells as an internal boundary, substituting the cellular stress into the weak form, and letting the cell dimensions shrink to zero.

5. Analytical Solution of the Discrete-cell Equation

Because of the linearity of the problem, Eq. (6) can be solved analytically under certain conditions. In particular, if we consider the case of an infinite domain with displacement and tension going to zero at infinity, we observe (substituting a Young's modulus $E = 2\mu(1+\nu)$) that Eq. (6) is equivalent to the plane stress equation for linear elasticity, which has been solved *via* a Green's function [5, 15]. Although a general solution is possible, we will analyze here only the simplest case of a single cell at the origin. The solution to this problem is given by

$$\mathbf{v} = \left[\begin{array}{c} \frac{1}{2\mu(1+\nu)} \int_{-\infty}^{+\infty} \int_{-\infty}^{+\infty} G_{11}T_1 + G_{12}T_2 d\xi d\eta \\ \frac{1}{2\mu(1+\nu)} \int_{-\infty}^{+\infty} \int_{-\infty}^{+\infty} G_{12}T_1 + G_{22}T_2 d\xi d\eta \end{array} \right], \quad (8)$$

where the components of G and T are defined as follows:

$$\begin{aligned} G_{11} &= \tfrac{(1+\nu)^2}{4\pi} \left[\tfrac{(x-\xi)^2}{(x-\xi)^2+(y-\eta)^2} - \tfrac{3-\nu}{2(1+\nu)} \ln[(x-\xi)^2 + (y-\eta)^2] \right] \\ G_{12} = G_{21} &= \tfrac{(1+\nu)^2}{4\pi} \tfrac{(x-\xi)(y-\eta)}{(x-\xi)^2+(y-\eta)^2} \\ G_{22} &= \tfrac{(1+\nu)^2}{4\pi} \left[\tfrac{(y-\eta)^2}{(x-\xi)^2+(y-\eta)^2} - \tfrac{3-\nu}{2(1+\nu)} \ln[(x-\xi)^2 + (y-\eta)^2] \right] \end{aligned} \quad (9)$$

and

$$\begin{aligned} T_1 &= F_0 L\alpha_x\alpha_x\delta'(\xi)\delta(\eta) + F_0 L\alpha_x\alpha_y\delta(\xi)\delta'(\eta) \\ T_2 &= F_0 L\alpha_x\alpha_y\delta'(\xi)\delta(\eta) + F_0 L\alpha_y\alpha_y\delta(\xi)\delta'(\eta) \end{aligned} . \quad (10)$$

Evaluating the integrals yields the following velocity profile for a single cell in an infinite viscous domain:

$$\mathbf{v} = -\frac{F_0 L(1+\nu)}{8\pi\mu} \left[\begin{array}{c} \frac{2x^3}{r^4} - \frac{2x}{r^2} + \frac{3-\nu}{1+\nu}\frac{x}{r^2} \\ \frac{y(x^2-y^2)}{r^4} \end{array} \right], \quad (11)$$

where $r^2 = x^2 + y^2$.

6. Comparison of Analytical and Numerical Results

The numerical model was tested by solving the problem of contractile stress in a unit square centered at the origin with a single cell at the origin aligned with the x-axis. At the boundaries of the square, the velocity was set equal to the infinite-domain analytical solution from Eq. (11). Thus, the exact solution to the finite-domain problem is given by Eq. (11).

The SGFEM solution, as expected, showed severe oscillations, so four-point smoothing was employed. Figure 2 shows vector plots of three cases: (i) the analytical solution, (ii) the (smoothed) numerical solution, and (iii) the numerical solution reduced by a factor of five. The same arrow sizing is used in each plot, so it is readily apparent that although the numerical solution is of similar structure to the analytical solution (Figs. 2a, 2c), the velocity is badly overestimated (Figs. 2a, 2b). We suspect that the overestimation is the result of the inability of the continuous model to describe the singularity and that the solver is attempting to capture the infinite slope at the origin by introducing large (and alternately signed) values at nodes near the origin.

To demonstrate the versatility of the numerical technique, we considered a unit square containing cells distributed in a preset pattern, as seen in Figure 3. The fluid velocity vectors show the complex pattern generated by cellular traction. Since this problem is linear, and since the boundaries of the problem are far enough from the cells that the assumption of infinite domain is fairly accurate, the same analytical method described above could also have been used to attain the results of Figure 3.

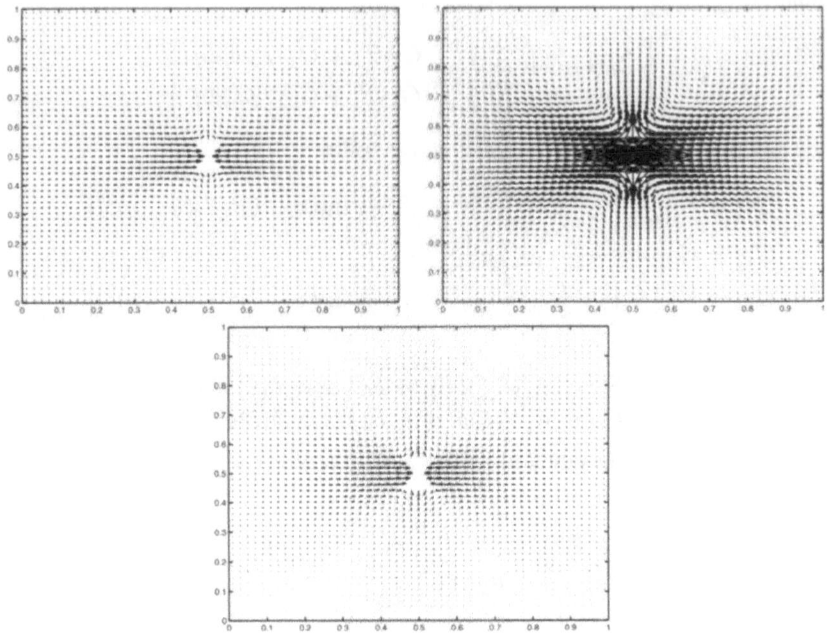

Figure 2. Velocity field for a single cell within a unit square of gel. (a) Analytical solution. (b) Numerical solution. (c) Numerical solution divided by 5. The center of the gel is not shown because the very high velocities associated with the singularity generate arrows that obscure the rest of the profile.

7. Discussion

We have presented a discrete-cell model of cell traction in two dimensions. In the simplest, linear cases, a Green's function solution is available by analogy with classical linear elasticity. For more general problems – especially those in which cells move, the gel properties change as collagen density changes, or the cell traction force generation depends on the local stress state – a numerical solution is needed. Our preliminary numerical studies using standard Galerkin finite elements do a good job on the qualitative behavior of the solution at moderate distance from a cell, but they are not able to capture the singularity that arises near each individual cell.

The availability of an analytical solution to the linear problem near the cell suggests that a hybrid method could be employed. In addition to the standard basis functions, a special *singular basis function* could be introduced based on the Green's function solution for the infinite problem (cf. [16] for the reentrant corner singularity). Although our problem is slightly different in that the singularity arises from the underlying behavior rather than from the presence of a corner, the concept should be applicable to our problem.

Finally, we express our hope that the discrete-cell model described herein will eventually serve as a tool to connect single-cell data with multicell data and to allow design of

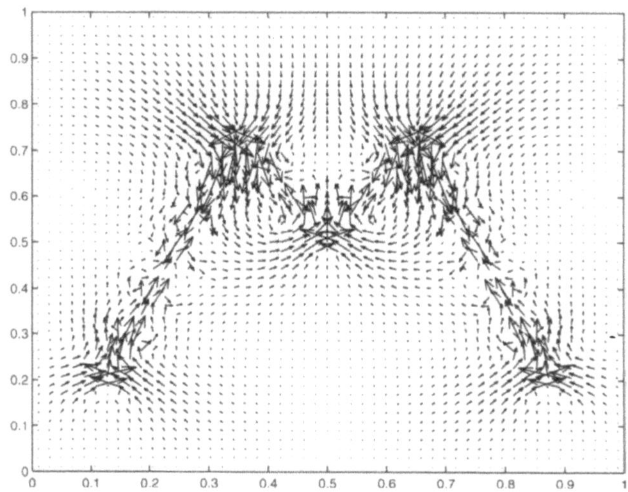

Figure 3. Simulated compaction of a unit square of gel with cells arranged in an "M" pattern.

cell-populated systems. By performing test studies using both Eqs. (1) and (4), we will be able to assess the validity of our traction parameter τ_0 and explore other possible forms for the relationship between contractile stress and cell concentration in terms of force generation by individual cells.

Acknowledgments

This work was made possible by supercomputing and workstation resource grants from the University of Minnesota Supercomputing Institute for Digital Simulation and Advanced Computation.

References

[1] G. F. Oster, J. D. Murray and A. K. Harris (1983) Mechanical aspects of mesenchymal morphogenesis. *J Embryol Exp Res* **78**, 83–125

[2] R. A. F. Clark and P. M. Henson (1988) *The Molecular and Cellular Biology of Wound Repair.* Plenum Press

[3] G. Gabbiani (1994) Modulation of fibroblastic cytoskeletal features during wound healing and fibrosis. *Pathol Res Pract* **190**, 851–853

[4] T. R. Fray, J. E. Molloy, M. P. Armitage and J. C. Sparrow (1998) Quantification of single human dermal fibroblast contraction. *Tissue Engineering* **4**, 281–291

[5] M. Dembo and Y.-L. Wang (1999) Stresses at the cell-to-substrate interface during locomotion of fibroblasts. *Biophysical Journal* **76**, 2307–2316

[6] C. G. Galbraith and M. P. Sheetz (1997) A micromachined device provides a new bend on fibroblast traction forces. *Proceedings of the National Academy of Sciences of the United States of America* **94**, 9114–9118

[7] E. Bell, B. Ivarsson and C. Merrill (1979) Production of a tissue-like structure by contraction of collagen lattices by human fibroblasts of different proliferative potential in vivo. *Proceedings of the National Academy of Sciences of the USA* **76**, 1274–8

[8] V. H. Barocas and R. T. Tranquillo (1997) A finite element solution for the anisotropic biphasic theory of tissue-equivalent mechanics: the effect of contact guidance on isometric cell traction measurement *Journal of Biomechanical Engineering* **119**, 261–269

[9] H. F. Yee, A. C. Melton and B. N. Tran (2001) RhoA/Rho-associated kinase mediates fibroblast contractile force generation. *Biochemical and Biophysical Research Communications* **280**, 1340–1345

[10] T. Wakatsuki, B. Schwab, N. C. Thompson and E. L. Elson (2001) Effects of cytochalasin D and latrunculin B on mechanical properties of cells. *Journal of Cell Science* **114**, 1025–1036

[11] V. H. Barocas, A. G. Moon and R. T. Tranquillo. (1994) The fibroblast-populated microsphere assay of cell traction force – Part 2. Measurement of the cell traction coefficient. *Journal of Biomechanical Engineering* **117**, 161–170

[12] L. Olsen, J. A. Sherratt and P. K. Maini (1995) A mechanochemical model for adult dermal wound contraction and the permanence of the contracted tissue displacement profile. *Journal of Theoretical Biology* **177**, 113–128

[13] V. H. Barocas, T. S. Girton and R. T. Tranquillo (1998) Engineered alignment in media-equivalents: consequences of cell induced compaction on magnetic prealignment. *Journal of Biomechanical Engineering* **120**, 660–666

[14] V. H. Barocas and R. T. Tranquillo (1997) An anisotropic biphasic theory of tissue-equivalent mechanics: the interplay among cell traction, fibrillar network deformation, fibril alignment, and cell contact guidance. *Journal of Biomechanical Engineering* **119**, 137–145

[15] S. Timoshenko (1934) *Theory of Elasticity.* McGraw-Hill

[16] Z. Cai and S. Kim (2001) A finite element method using singular functions for the Poisson equation: Corner singularities. *SIAM Journal of Numerical Analysis* **39**, 286–299

Polymer and Cell Dynamics – Multiscale Modeling
and Numerical Simulations, 251–259
ed. by W. Alt, M. Chaplain, M. Griebel und J. Lenz

A Hybrid Discrete-continuum Technique for Individual-based Migration Models

Alexander R.A. Anderson

Department of Mathematics, University of Dundee, Dundee DD1 4HN, Scotland

Summary. In this chapter we present a modeling technique that is used to simulate individual-based migration models that are very appropriate for the modeling of cell responses to external signaling cues. The technique, referred to here as the hybrid discrete-continuum (HDC) technique, initially formulates a general problem of cell migration in response to external signaling cues in terms of system partial differential equations (PDEs) describing the evolution of mean-field quantities such as cell density and chemical concentrations. Using a discretized form of the PDEs, it is shown that one can generate a (discrete, stochastic) biased random walk model that enables one to track the migration paths of individual cells. The technique provides a simple but effective way of linking macro- and microscale events.

1. Introduction

Many mathematical models for organism migration exist that have been developed for many different areas of biology and applied at a variety of different spatial scales. However, most of these models that consider space explicitly fall into one of two categories: (i) continuum population models, e.g., systems of differential equations such as reaction-diffusion equations [6, 13, 14, 17, 18, 25] and (ii) discrete individual models, e.g., cellular automata [11, 12, 22, 26], interacting particle systems [10], Monte Carlo models [9], and random walk models [1, 15, 20]. (Note that random walk models are often used to derive continuum diffusion approximations.)

When deciding which model should be used, the number and scale of the organisms being modeled is important, and the manner in which the organisms interact with their environment and each other is also important. Discrete, stochastic interactions between organisms cannot be captured by the continuum approach, and, likewise, global population interactions cannot be captured by the discrete approach. Therefore the most appropriate modeling technique depends on both the number of organisms and the scale at which they are being studied. However, in recent years a third category of models has emerged: hybrid models that allow modelers to exploit the advantages of both continuum and discrete models, e.g., a combination of discrete cellular automata that also employ reaction-diffusion equations [5, 23, 24], and, *vice versa*, partial differential equation models that use individual-based components [2, 3, 4, 7, 21]. The advantages of using such hybrid approaches are clear when dealing with organisms that involve processes at different scales, e.g., a cell (microscale) migrating up a chemical gradient produced by a population of cells (macroscale). One can model the chemical as a continuous concentration (with, e.g., reaction-diffusion equations) and the organism as a discrete individual (with, e.g., a cellular automata or a biased random walk).

Our modeling technique falls into this third, hybrid category. In this chapter we shall discuss the technique itself, the range of problems for which this technique is applicable, and its advantages and limitations.

2. The Master Partial Differential Equation

We now define the generic structure of the problems for which our hybrid technique is applicable. This includes a reasonably broad class of partial differential equations (PDEs). In particular, we are interested in modeling organism migration in response to some form of external stimulus, e.g., with diffusion-taxis models. For ease of exposition, we shall consider models posed in a two-dimensional spatial domain. However, this technique easily extends to three dimensions and indeed to N dimensions (although this is not physically relevant).

Let Ω be the two-dimensional domain for the problem, where $\Omega = \{(x, y) : a_1 < x < b_1, a_2 < y < b_2\}$. We shall assume that the organism motion is driven by both unbiased and biased migration (in response to various external stimuli). We denote by $n(x, y, t)$ the density of the organism at the point (x, y) at time t and define the PDE modeling its migration as

$$\frac{\partial n}{\partial t} = \frac{\partial}{\partial x}\left(D_n \frac{\partial n}{\partial x} - n \sum_{i=1}^{w} \chi_i \frac{\partial m_i}{\partial x}\right)$$
$$+\frac{\partial}{\partial y}\left(D_n \frac{\partial n}{\partial y} - n \sum_{i=1}^{w} \chi_i \frac{\partial m_i}{\partial y}\right) + f\left(\frac{\partial n}{\partial x}, \frac{\partial m_i}{\partial x}, \frac{\partial n}{\partial y}, \frac{\partial m_i}{\partial y}\right), \qquad (1)$$

where D_n is the diffusion coefficient and represents the magnitude of the unbiased motion, $\chi_i \geq 0$ are the taxis coefficients and represent the magnitude of the biased migration in response to the gradient of m_i (with $i = 1$ to w), representing one or more external stimuli. In addition to the taxis term, the function f represents other directed (biased) types of migration, e.g., convection. Since most of the external stimuli we consider are chemicals, it is sufficient to define each m_i as satisfying a general reaction-diffusion equation, leading to the system

$$\frac{\partial \underline{m}}{\partial t} = \mathbf{D}\Delta \underline{m} + \underline{g}(\underline{m}, n), \qquad (2)$$

where $\underline{m} = [m_1\ m_2\ m_3 \ldots\ m_w]^T$ is a w dimensional vector representing the number of different external stimuli, $\underline{g}$ represents the reaction kinetics, and $\mathbf{D}$ is the diagonal matrix of positive, constant diffusion coefficients.

3. The Hybrid Discrete-continuum Technique

The hybrid discrete-continuum technique (hereafter referred to as the HDC technique) that we will use is a generalization of the method developed by Anderson et al. [2] and first of all involves discretizing (using standard finite-difference methods) the partial differential equation governing the migration of the organism (Eq. (1)). The resulting coefficients of the five-point finite-difference stencil are then used to generate the probabilities of movement of an individual organism in response to its local milieu. In effect, we will derive a biased random walk governing the motion of an organism based on the system of partial differential equations (1) and (2) of the previous section. In this sense, our discrete model is most similar in formulation to the reinforced random walk models of Othmer and Stevens [19], hereafter referred to as OS. In particular, the form of the transition probabilities for the gradient model of OS is very similar to the probabilities of movement that will be derived from our discrete model (see also [1, 8]). In fact, as we will see below we can use the techniques in OS to obtain the continuum equation (1) from the transition probabilities.

We now set about formulating the discrete model and deriving the movement probabilities for an organism. We first discretize (1) using the Euler finite central difference approximation [16]. This involves approximating the continuous two-dimensional domain $(a_1, b_1) \times (a_2, b_2)$ as a grid of discrete points of mesh size h and time t by discrete increments of magnitude k. This results in the discrete organism equation

$$n_{i,j}^{q+1} = n_{i,j}^{q} P_0 + n_{i+1,j}^{q} P_1 + n_{i-1,j}^{q} P_2 + n_{i,j+1}^{q} P_3 + n_{i,j-1}^{q} P_4, \tag{3}$$

where the subscripts specify the location on the grid and the superscripts the time steps. That is, $x = ih$, $y = jh$, and $t = qk$ where i, j, and q are positive integers.

In a numerical simulation of the continuous model (1), the purpose of the discrete equation (3) is to determine the organism density at grid position (i, j) and time $q + 1$ by averaging the density of the four surrounding neighbors at the previous time step q. For our discrete model, we will use the five coefficients P_0 to P_4 from (3) to generate the stochastic motion of an individual organism. The central assumption of the HDC technique is that these five coefficients can be thought of as being proportional to the probabilities of the organism being stationary (P_0) or moving left (P_1), right (P_2), up (P_3), or down (P_4) one grid distance (h) at each time step (k).

The exact form of the coefficients P_0 to P_4 will depend upon the precise forms of (1) and (2), but they will certainly be functions of the external stimuli. In order to understand more fully the way in which the discrete model is developed and how it relates to the work in OS, we shall use a simpler form of Eq. (1). We now consider migration to be driven only by diffusion and chemotaxis in response to a single external chemical stimulus (m), as defined by

$$\frac{\partial n}{\partial t} = D_n \nabla^2 n - \chi \nabla . (n \nabla m), \tag{4}$$

where D_n and χ and m are as above. If we now discretize (4) in the manner discussed above, then upon comparing with (3), we see that the coefficient P_0, which is proportional to the probability of no movement, has the form

$$P_0 = 1 - \frac{4kD_n}{h^2} - \frac{k\gamma}{h^2}\left(m^q_{i+1,j} + m^q_{i-1,j} - 4m^q_{i,j} + m^q_{i,j+1} + m^q_{i,j-1}\right), \tag{5}$$

and that the coefficients P_1, P_2, P_3, and P_4, which are proportional to the probabilities of moving left, right, down, and up, respectively, have the forms

$$\begin{aligned} P_1 &= \frac{kD}{h^2} - \frac{k\gamma}{4h^2}\left[m^q_{i+1,j} - m^q_{i-1,j}\right], \\ P_2 &= \frac{kD}{h^2} + \frac{k\gamma}{4h^2}\left[m^q_{i+1,j} - m^q_{i-1,j}\right], \\ P_3 &= \frac{kD}{h^2} - \frac{k\gamma}{4h^2}\left[m^q_{i,j+1} - m^q_{i,j-1}\right], \\ P_4 &= \frac{kD}{h^2} + \frac{k\gamma}{4h^2}\left[m^q_{i,j+1} - m^q_{i,j-1}\right]. \end{aligned} \tag{6}$$

Clearly, when the concentration of m is zero around the organism, P_1 to P_4 are equal. Also when there is an equal concentration gradient of m on either side of the organism, the values $m_{i,j-1}$ and $m_{i,j+1}$ cancel each other out, as do $m_{i-1,j}$ and $m_{i+1,j}$, and thus P_1 to P_4 are again equal. Therefore, in both these circumstances unbiased random movement will be produced. Otherwise, the appearing m-terms are small differences of the concentration values, at the order of h. Thus, for sufficiently small grid size h, all defined probabilities stay positive. If there is a higher concentration of m on one side of the organism than the other, the probabilities (P_1 to P_4) will no longer be equal, and hence directed movement, towards the higher m concentration, will result. The value of P_0 will also change depending on the local concentration of m, e.g., if the surrounding grid points have a higher concentration of m, then P_0 will be smaller and therefore the organism will be less likely to be stationary. Notice that P_0 stays positive, provided that the standard discretization condition $\frac{4kD_n}{h^2} < 1$ is fulfilled.

Eq. (6) is very similar to the transition probabilities of the reinforced random walk model in OS. In particular, its gradient models have a "random" component and a "taxis" component. OS used its discrete transition probabilities to then derive a partial differential equation in the continuous limit. It is possible to do this for our model, as we now show. In line with OS, we concentrate on the semi-discrete (i.e., discrete in space but not time), one-dimensional form of (4). Using central differences, the semi-discrete, one-dimensional form of (4) is given by

$$\frac{\partial n_i}{\partial t} = n_{i+1}P_1 + n_{i-1}P_2 - n_i P_0, \tag{7}$$

where

$$P_0 = 2\mu + 4\eta\left[m_{i+1} + m_{i-1} - 2m_i\right] \tag{8}$$

and

$$\begin{aligned} P_1 &= \mu - \eta\left[m_{i+1} - m_{i-1}\right] \\ P_2 &= \mu + \eta\left[m_{i+1} - m_{i-1}\right] \end{aligned} \tag{9}$$

with $x = ih$, $\mu = \frac{D}{h^2}$, and $\eta = \frac{\gamma}{4h^2}$. In OS, Eq. (7) is called the *continuous time master equation* and it defines the conditional probability $n_i(t)$ that an organism is at position $i \in \mathcal{Z}$ at time t, conditioned on the fact that it begins at $i = 0$ when $t = 0$. The transition probabilities P_1, P_2 defined by (9) represent the probability of a one-step jump to $i \pm 1$ per unit time and P_0, defined by (8), represents the mean time spent at location i.

We now have the complete one-dimensional, biased random walk model defined by (7)–(9). The basis for our HDC technique is the assumption that the probabilities (8) and (9) represent the probability of an organism being stationary, moving one step left or right each time step. To validate this assumption, we shall use (7)–(9) to try to recover (4) in the continuum limit. Upon substituting (8) and (9) into (7) and rearranging, we obtain

$$\begin{aligned}\frac{\partial n_i}{\partial t} = \mu[n_{i+1} + n_{i-1} - 2n_i] \quad &- \quad \eta n_{i+1}[m_{i+1} - m_{i-1}] \; + \\ \eta n_{i-1}[m_{i+1} - m_{i-1}] \quad &- \quad 4\eta n_i[m_{i+1} - m_{i-1} - 2m_i].\end{aligned} \tag{10}$$

We shall consider a lattice of mesh size h and therefore let $x = ih$ (as before). Expanding the r.h.s. of (10) as a function of x in a Taylor series to second order in h, we obtain

$$\begin{aligned}\frac{\partial n}{\partial t} = \mu(h^2 n'') - 2\eta h m' \left(n + hn' + \frac{h^2}{2}n''\right) + 2\eta h m' \left(n - hn' + \frac{h^2}{2}n''\right) \\ -4\eta h^2 m'' + O(h^4),\end{aligned}$$

where $'$ denotes $\frac{\partial}{\partial x}$, which upon simplifying gives

$$\frac{\partial n}{\partial t} = \mu h^2 n'' - 4\eta h^2 (n'm' + nm'') + O(h^4). \tag{11}$$

Suppose there is a scaling of the transition rates such that $m = \sigma\tilde{m}$, $\mu = \sigma\tilde{\mu}$ and such that

$$\lim_{\substack{h \to 0 \\ \sigma \to \infty}} (\sigma h^2) = D \tag{12}$$

exists for some positive constant D. Then taking $D_n = D\tilde{\mu}$ and $\chi = 4D\eta$, we find the "diffusion limit" of Eq. (11) becomes

$$\frac{\partial n}{\partial t} = D_n \frac{\partial n}{\partial x} - \chi \left(\frac{\partial n}{\partial x}\frac{\partial \tilde{m}}{\partial x} + n\frac{\partial^2 \tilde{m}}{\partial x^2}\right). \tag{13}$$

This is clearly equivalent to the one-dimensional form of (4). However, it is also possible to do the same analysis with the two- or even three-dimensional form of (4).

These results therefore show that by using simple finite differences to discretize a continuous partial differential equation, modeling a population, of the form (1), we can obtain an equivalent biased random walk model for an individual in that population. However, please note that the biased random walk derivation here should be taken as more of an analogy than a true definition. This is because the biased random walk governs the migration probabilities that might subsequently be changed *via* interactions with other organisms or individual-based processes.

4. Individual-based Processes

When modeling migration of an organism in this manner, once the biased random walk model has been derived, the organism is now effectively discrete but the external stimuli remain continuous (i.e., a hybrid of discrete and continuum modeling). The advantage of working at the scale of the organism is that we can now consider the inclusion of individual-based processes, e.g., proliferation, mutation, birth, and death. The inclusion of such processes can allow us to model the migration and interaction of many organisms where the former is driven by biased random walks (derived from the PDEs) and the latter is driven by individual-based rules implemented at the scale of the organism.

In the following chapter we discuss three different examples where the HDC technique has been applied. We shall consider (i) tumor induced angiogenesis, where cell migration is driven by a diffusion/chemotaxis /haptotaxis model and the individual-based processes are proliferation, branching, and anastomosis; (ii) tumor invasion, where cell migration is modeled by diffusion/chemotaxis and the individual-based processes are proliferation, degradation, and cell-cell adhesion; and (iii) *Dictyostelium discoideum* aggregation, where migration of the amoebae is driven by diffusion/chemotaxis and the individual processes are cAMP production and amoeba layering.

5. HDC Simulation Process

To predict an organism's migration using the five coefficients in Eq. (3), we use the following process for each time step of the simulation. For each grid point we numerically solve the discrete form of the stimuli equation (2), taking into account whether an organism is occupying the current grid point or not (e.g., $n = 0$ in the stimuli equations if an organism does not occupy the current grid point). This allows us to generate the five coefficients P_0 to P_4 (e.g., Eqs. (5) and (6)) of movement for an organism located at the current grid point. Probability ranges are then computed by summing the coefficients to produce five ranges; $R_0 = 0$ to P_0 and $R_j = \sum_{i=0}^{j-1} P_i$ to $\sum_{i=0}^{j} P_i$, where $j = 1$ to 4. We then generate a random number between 0 and 1, and depending on the range into which this number falls, the current organism under consideration will remain stationary (R_0) or move left (R_1), right (R_2), down (R_3) or up (R_4). The larger a particular range, the greater the probability that the corresponding coefficient will be selected. Therefore, each organism is restricted to move to one of its four orthogonal neighboring grid points or remain stationary at each time step of the simulation.

6. Discussion

One question that might arise from using the HDC technique is: Why derive a discrete model from the continuum PDE model instead of starting with a discrete model in the first place? There are two main reasons for this: (i) when modeling the migration of organisms that are influenced by stimuli at different scales, it is easier to model these initially at the continuum level; (ii) Because the biased random walk, in the HDC technique, has been derived from a numerical discretization of the PDE (governing organism migration), the space and time steps for the biased walk already match those for the numerical solution of the "continuous" variables (the external stimuli) and therefore make interactions between these discrete and continuous variables much more straightforward.

One of the main advantages of the HDC technique is that individual-based process are modeled at the individual level. This can clearly be seen by the fact that we are required only to solve the stimuli equation (2) numerically and use the five movement probabilities (e.g., (6)) generated from the discrete organism equation (3) as the basis for the biased random walk model. This of course leads to the question of how to deal with "organism density" in the stimuli equations (2), since a discrete organism has no "density". For example, say the organism (n) reacts chemotactically to and produces a diffusable stimulus (m). Then we must calculate how much stimulus an individual organism can produce in a given space (h) and time step (k), and that is how much must be added to the stimulus equation for that specific grid location.

Another advantage of the HDC technique is the stochastic element in the organism's migration, i.e., even if one probability range is larger than all the others, due to the random choice, there is still no certainty that this range will be selected. This implies that no two organisms movements will be identical, as would be the case in reality.

Due to the derivation of the discrete model being based on a finite difference approximation of a PDE, there is some restriction in the choice of the space (h) and time (k) steps in order to ensure stability. A basic rule for stability of a diffusion-driven model requires that the constants h and k should satisfy the condition $\frac{4kD}{h^2} < 1$ where D is the diffusion coefficient. Since in general the spatial grid size (h) will be fixed by the scale of the organism (this need not always be the case), the time step (k) must be chosen sufficiently small in order to satisfy the above condition.

In conclusion, while the HDC technique will not be applicable to every migration system, the idea of using standard numerical techniques for solving partial differential equations to derive biased random walk models is clearly a very simple way of obtaining a hybrid model of individual organism migration. Indeed, the HDC technique provides a powerful means of linking microscale events to macroscale events, individual behavior to population behavior, and, as we shall see in the next chapter, has potential application to a wide range of problems in mathematical biology.

Acknowledgments

Dr. A.R.A. Anderson is supported by a Personal Research Fellowship from the Royal Society of Edinburgh and an equipment grant from the Royal Society.

References

[1] Alt, W. (1980) Biased random walk models for chemotaxis and related diffusion approximations. *J Math Biol* **9**, 147–177

[2] Anderson, A.R.A., B.D.S. Sleeman, I.M. Young and B.S. Griffiths (1997) Nematode movement along a chemical gradient in a structurally heterogeneous environment: II. Theory. *Fundam Appl Nematol* **20**, 165–172

[3] Anderson, A.R.A. and M.A.J. Chaplain (1998) Continuous and Discrete Mathematical Models of Tumor-Induced Angiogenesis Angiogenesis. *Bull Math Biol* **60**, 857–899

[4] Anderson, A.R.A., M.A.J. Chaplain, E.L. Newman, R.J.C. Steele and A.M. Thompson (2000) Mathematical Modelling of Tumor Invasion and Metastasis. *J Theoret Med* **2**, 129–154

[5] Bretschneider, T., B. Vasiev and C.J. Weijer (1999) A model for Dictyostelium slug movement. *J Theor Biol* **199**, 125–136

[6] Chaplain, M.A.J. (1996) Avascular growth, angiogenesis and vascular growth in solid tumors: the mathematical modelling of the stages of tumor development. *Math Comput Modelling* **23**, 47–87

[7] Dallon, J.C. and H.G. Othmer (1997) A discrete cell model with adaptive signaling for aggregation of *Dictyostelium discoideum*. *Phil Trans R Soc Lond* **B352**, 391–417

[8] Davis, B (1990) Reinforced Random Walk. *Probab Th Rel Fields* **84**, 203–229

[9] Drasdo, D (1996) Different growth regimes found in a Monte Carlo model of growing tissue cell populations. In: (F. Schweitzer ed.) *Self-Organization of complex structures: From individual to collective dynamics.* London: Gordon and Breach

[10] Durret, R., S.A. Levin (1994) Stochastic spatial models: a user's guide to ecological applications. *Phil Trans R Soc Lond B* **343**, 329–350

[11] Ermentrout, G.B. and L. Edelstein-Keshet (1993) Cellular automata approaches to biological modelling. *J Theor Biol* **160**, 97–133

[12] Hassell, M.P., H.N. Comins and R.M. May (1991) Spatial structure and chaos in insect population dynamis. *Nature* **353**, 255–258

[13] Höfer, T., J.A. Sherratt and P.K. Maini (1995) Cellular pattern formation during Dictyostelium aggregation. *Physica D* **85**, 425–444

[14] Kareiva, P.M (1983) Local movements in herbivorous insects: Applying a passive diffusion model to mark-recapture field experiments. *Oecologia (Berlin)* **57**, 322–327

[15] Kareiva, P.M. and N. Shigesada (1983) Analyzing Insect Movement Data as a Correlated Random Walk. *Oecologia* **56**, 234–238

[16] Mitchell, A.R. and D.F. Griffiths (1980) *The finite difference method in partial differential equations.* Chichester: Wiley

[17] Okubo, A (1980) *Diffusion and Ecological Problems : Mathematical Models.* Berlin: Springer-Verlag

[18] Okubo, A (1986) Dynamical aspects of animal grouping: swarms, schools, flocks, and herds. *Adv Biophys* **22**, 1–94

[19] Othmer, H. and A. Stevens (1997) Aggregation, blowup and collapse: The ABCs of taxis and reinforced random walks. *SIAM J Appl Math* **57**, 1044–1081

[20] Patlak, C.S (1953) Random walk with persistence and external bias. *Bull. of Math, Biophys.* **15**, 311–338

[21] Pitcairn, A. W., M. A. J. Chaplain, C. J. Weijer and A. R. A. Anderson (2000) A Discrete-Continuum Mathematical Model of *Dictyostelium* Aggregation. *European Communication in Mathematical and Theoretical Biology* **2**, 6–11

[22] Qi, A., X. C. Du Zheng and B. An (1993) A Cellular Automaton Model of Cancerous Growth. *J Theor Biol* **161**, 1–12

[23] Savill, N.J. and P. Hogeweg (1997) Modelling morphogenesis: from single cells to crawling slugs. *J Theor Biol* **184**, 229–235

[24] Stott, E.L., N.F. Britton, J.A. Glazier and M. Zajac (1999) Stochastic simulation of benign avascular tumor growth using the Potts model. *Math Comput Modelling* **30**, 183–198

[25] Tracqui, P (1995) From passive diffusion to active cellular migration in mathematical models of tumor invasion. *Acta Biotheor* **43**, 443–464

[26] Vasieva, O.O., B.N. Vasiev, V.A. Kaprov and A.N. Zaikin (1994) A model of dictyostelium discoideum aggregation. *J Theor Biol* **171**, 361–367

Polymer and Cell Dynamics – Multiscale Modeling
and Numerical Simulations, 261–279
ed. by W. Alt, M. Chaplain, M. Griebel und J. Lenz

Application of the Hybrid Discrete-continuum Technique

Alexander R.A. Anderson and Alan W. Pitcairn

Department of Mathematics, University of Dundee, Dundee DD1 4HN, Scotland

Summary. Following on from the previous chapter, here we present a number of specific biological problems involving cell migration that can be modeled using the hybrid discrete-continuum (HDC) technique. The three problems considered are (i) angiogenesis (blood vessel formation) in response to a solid tumor, (ii) tissue invasion by cancer cells, and (iii) aggregation in the slime mold *Dictyostelium discoideum*. Tumor-induced angiogenesis is the formation of blood vessels (vasculature) from a preexisting vasculature in response to chemical stimuli (tumor angiogenic factors, TAF) produced and secreted by the solid tumor. The formation of the new vasculature facilitates further tumor growth and invasion. The mathematical model describes the formation of the blood-vessel network in response to chemical stimuli (TAF) supplied by a solid tumor. The model also takes into account essential cell-matrix interactions *via* the inclusion of the matrix macromolecule fibronectin. A discretized form of the partial differential equation model is then used to develop a biased random walk model that enables us to track individual endothelial cells at the blood vessel tips and incorporate anastomosis, mitosis, and branching explicitly into the model. A second important aspect of solid tumor growth is the invasion of the local tissue by the cancer cells. A crucial part of the this process is the ability of the cancer cells to degrade the surrounding tissue or extracellular matrix (ECM) through the secretion of matrix degrading enzymes (MDEs). As the ECM is degraded, the cancer cells actively migrate into the space created, thereby spreading locally into the healthy host tissue. If there are blood or lymph vessels in the vicinity, these can then provide a route for the cancer cells to spread widely throughout the host's body. An initial system of PDEs is formulated considering the evolution of cancer cell density, MDEs, and ECM density. An HDC model is then developed that enables us to track the migration and spread of indivual cancer cells into the host tissue. The results have implications for the surgical removal of solid tumors. The social amoeba *Dictyostelium discoideum* is a widely studied paradigm for biological pattern formation. Mathematical models can show how a collective pattern of cell communication and motion driven by the chemoattractant cyclic adenosine 3'5'-monophosphate (cAMP) arises from excitable local cAMP kinetics and cAMP diffusion. The master continuum PDE model describes the evolution of cAMP concentration, the fraction of active receptors on *Dictyostelium* cells, and cell density. The HDC technique is then used to develop a biased random walk model that enables us to track individual *Dictyostelium* amoeba as they move through the aggregation territory in response to the local cAMP gradients. The theoretical aggregation patterns generated by computer simulations of the discrete model are compared with the morphology of such patterns observed in *in vivo* experiments.

1. Introduction

In the previous chapter, we discussed the hybrid discrete-continuum (HDC) technique for organism migration. In this chapter, we consider three different applications of the HDC technique *via* three different cellular organisms: endothelial cell migration in tumor angiogenesis, tumor cell migration in tumor invasion, and amoebae migration in *Dictyostelium discoideum* aggregation. Through these three examples, we aim to show how the individual cellular processes can be included and how these HDC models can reproduce experimental observations.

2. Tumor-induced Angiogenesis

Angiogenesis, the formation of blood vessels from a preexisting vasculature, is a crucial component of many mammalian growth processes. Uncontrolled or excessive blood-vessel formation is essential for tumor genesis and is also observed in arthritis, abnormal neovascularization of the eye, duodenal ulcers, and following myocardial infarction [10]. The angiogenic response to a tumor is a critical part of the metastatic cascade, providing a crucial bridge between the avascular and vascular growth phases, leading to invasion and metastasis (see next section). Prevention of angiogenesis is now an important area of cancer research, and therefore the study of capillary network formation in response to tumors is very important.

The initiation of tumor-induced angiogenesis involves the tumor secreting a number of chemicals, collectively known as tumor angiogenic factors, or TAF [10], into the surrounding tissue. Once secreted, the TAF then diffuses through the tissue, creating a chemical gradient between the tumor and any existing vasculature. Upon reaching any neighboring blood vessels, endothelial cells (EC) lining these vessels are first induced to degrade the parent venule basement membranes and then migrate through the disrupted membrane towards the tumor through the extra cellular matrix (ECM). The initial response of the EC is a chemotactic one to the TAF concentration gradient [18]. Interactions between the EC and certain components of the ECM (e.g., fibronectin, laminin) have been shown to enhance cell adhesion to the matrix. In particular, cultured EC are known to synthesize and secrete cellular fibronectin [21] and respond haptotactically to gradients of fibronectin [16]. Therefore, in addition to the chemotactic response of the EC to the TAF, there is also a haptotactic response to gradients of certain matrix macromolecules present within the extracellular matrix [5].

The precise sequence of events involved in angiogenesis is now well known [18] thanks largely to the development and use of a wide range of angiogenesis assays, both *in vivo* and *in vitro*: (i) EC migration (ii) sprout formation (Fig. 1a), (iii) EC proliferation near sprout tips, (iv) branching and anastomosis (Fig. 1b, 1c), and (vi) vascularization (Fig. 1d).

2.1. The HDC Model of Angiogenesis

In this section we describe the continuum equations and subsequently the discrete equations that will be the basis of our HDC technique. In order to simplify the mathematical

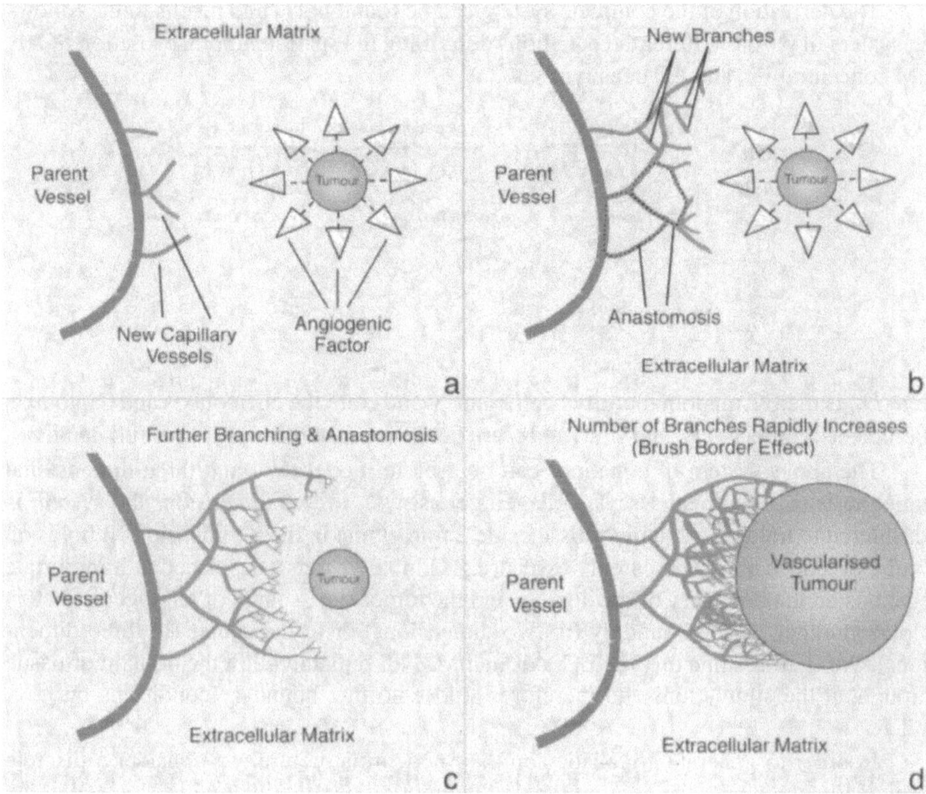

Figure 1. Schematic diagram of the sequence of key events of angiogenesis: (a) EC stimulated by the AF migrate from the parent vessel into the ECM and begin to form immature capillary vessels. (b) Capillary vessels grow further and form loops (anastomoses) and begin to branch. (c) Further growth of the capillary network towards the tumor results in more branches and anastomosis. (d) Vessel branching rapidly increases as the network approaches the tumor, eventually penetrating the tumor and thus completing angiogenesis and allowing the vascularized tumor to grow and invade the surrounding ECM.

modeling, we assume first of all that there is a net excess of angiogenic factors being produced (angiogenic inhibitors may easily be considered in a modified model). We also assume that the motion of the EC is influenced by three main factors: random motility (analogous to molecular diffusion); chemotaxis, i.e., directed motion in response to gradients of soluble angiogenic factors [24]; and haptotaxis, i.e., directed motion in response to gradients of bound molecules within the ECM (e.g., fibronectin, [16]). We consider a generic angiogenic factor (AF) and a generic matrix macromolecule (MM). We denote the EC density per unit volume by n, the AF concentration by c, and the MM concentration by f.

The derivation of the complete system can be found in [1] and results in the following system of partial differential equations describing the spatiotemporal evolution of AF, MM concentration, and EC density:

$$
\begin{aligned}
\frac{\partial n}{\partial t} &= \overbrace{D_n \nabla^2 n}^{\text{random motility}} \overbrace{- \chi \nabla \cdot (n \nabla c)}^{\text{chemotaxis}} \overbrace{- \rho \nabla \cdot (n \nabla f)}^{\text{haptotaxis}} , \\
\frac{\partial f}{\partial t} &= \overbrace{\beta n}^{\text{production}} - \overbrace{\gamma n f}^{\text{degradation}} , \\
\frac{\partial c}{\partial t} &= - \overbrace{\eta n c}^{\text{uptake}} .
\end{aligned} \tag{1}
$$

Here D_n is the EC random motility coefficient, χ and ρ are the chemotaxis and haptotaxis coefficients, respectively, and β, γ, and η are positive constants (see [1] for full details).

The above system of equations can be used to model two- and three-dimensional angiognesis (e.g., corneal assays, collagen gel assays). In two dimensions the system is considered to hold on a square of tissue side 2 mm, while in three dimensions it holds on a cube of side 2 mm. The parent vessels (e.g., limbal vessels) are placed at appropriate locations on the boundary of the domain and the tumor (i.e., source of angiogenic factor) is placed either on the boundary (in two dimensions) or in the center (in three dimensions). We shall assume that the EC, AF, and MM all remain within the domain of tissue throughout the simulations and therefore impose no-flux boundary conditions on n, c, and f.

In order to generate a realistic capillary network structure, we consider a discrete form of Eq. (1). Under the assumption that the motion of an individual EC located at the tip of a capillary vessel governs the growth of the whole vessel (in terms of its direction), we use the HDC method of Anderson [4]. This first of all involves discretizing (using standard finite-difference methods) the system of partial differential equations (1) using a grid of discrete points (mesh size h), at discrete time intervals k. The discrete endothelial cell equation in three-dimensions is

$$
\begin{aligned}
n_{l,m,w}^{q+1} = n_{l,m,w}^{q} P_0 &+ n_{l+1,m,w}^{q} P_1 + n_{l-1,m,w}^{q} P_2 + n_{l,m+1,w}^{q} P_3 \\
&+ n_{l,m-1,w}^{q} P_4 + n_{l,m,w+1}^{q} P_5 + n_{l,m,w-1}^{q} P_6,
\end{aligned} \tag{2}
$$

where the subscripts specify the location on the grid and the superscripts the time steps. That is, $x = lh$, $y = mh$, $z = wh$, and $t = qk$ where l, m, w, and q are positive integers.

For the HDC technique we use the seven coefficients P_0 to P_6 from Eq. (2) to generate the motion of an individual EC at the tip of a capillary vessel. These coefficients can be thought of as being proportional to the probabilities of the EC being stationary (P_0) or moving west (P_1), east (P_2), south (P_3), north (P_4), down (P_5), or up (P_6), i.e., the movement probabilities for an EC biased random walk. Note that the two-dimensional form of Eq. (2) is similar but does not contain the coefficients P_5 and P_6 (see [1, 9]).

The seven coefficients P_0 to P_6 from Eq. (2) are proportional to the probabilities of an individual EC being stationary or moving to one of the six neighboring grid points.

Each coefficient is actually a function of the local AF and MM concentrations, and therefore the motion of an individual cell at the vessel-tip is governed by its interactions with angiogenic factors and matrix macromolecules in its local environment. In fact, each of the coefficients P_1 to P_6 consists of three components:

$$P_n = \text{random movement} + \text{chemotactic} + \text{haptotactic},$$

thus showing how the discrete EC equation is linked to the continuous EC equation of system (1). The coefficient P_0 has a similar form.

2.2. Individual-based Processes: Branching, Anastomosis, and Proliferation

Since we are considering migration at the cellular level, we can incorporate microscale processes such as EC proliferation, the generation of new capillary vessels (branching), and the fusion of two vessels (anastomosis) explicitly into the model. While there is a good deal of information regarding the actual events of the generation of new sprouts (sprout branching) and the formation of loops (anastomosis), there is no explanation as to the precise mechanisms that cause them [18]. Figure 1 gives a schematic illustration of the processes that we will model explicitly using the HDC technique.

We will assume that the generation of new sprouts (branching) occurs only from existing sprout tips and only when sufficient space exists. It is also reasonable to assume that the newly formed sprouts are unlikely to branch immediately. From these assumptions, we obtain the following two conditions, which must be satisfied before a capillary sprout can branch at its tip and generate a new sprout: (i) the age of the current sprout is greater than some threshold branching age ψ, i.e., new sprouts must mature for a length of time at least equal to ψ before being able to branch; and (ii) there is sufficient space locally for a new sprout to form, i.e., branching into a space occupied by another sprout is not possible. Given that each of the above two conditions is satisfied, we assume that each sprout tip has a probability, P_b, of generating a new sprout (branching) and that this probability is directly proportional to the local TAF concentration. Therefore, very little branching occurs initially (near the parent vessel where the TAF concentration is low), but as the endothelial cells migrate closer to the tumor (the TAF concentration slowly increases), the number of new sprouts slowly increases. A short distance from the tumor, the frequency of branching dramatically increases (as does the TAF concentration), creating the "brush border" effect, i.e., a very dense network of interconnected vessels at the leading edge.

Anastomosis, the formation of loops by capillary sprouts, is another very important feature of angiogenesis that can be captured explicitly by the HDC technique. As the sprouts progress towards the tumor, driven by the movement probabilities of Eq. (2), at each time step of the simulation, the endothelial cells at the sprout-tips can move to any of the six orthogonal neighbors on the discrete grid. If upon one of these moves another sprout is encountered, then anastomosis can occur (see Fig. 1b). For simplicity, we assume that as a result of the anastomosis, only one of the original sprouts continues to grow (the choice of which is purely random).

During angiogenesis, EC are initially recruited from the parent vessel and migrate toward the tumor without proliferation. Approximately 36–48 hours into the process, cell mitosis is observed [18, 22] and is confined to a region just behind the sprout tip. Endothelial cell doubling time has been estimated at 18 hours [30], and we model the process of cell division in the discrete model by assuming that some of the cells behind the sprout tip divide (into two daughter cells) every 18 hours. We assume that this has the effect of increasing the length of a sprout by approximately one cell length every 18 hours. In terms of the HDC model, this is equivalent to the sprout length being increased by one grid point every half time unit. Since the age of each EC at a sprout tip is known, we can easily determine when proliferation should occur. Owing to the inherent randomness of the HDC technique, proliferation will occur asynchronously in separate sprouts, as is observed experimentally [18].

2.3. HDC Model Simulation Results

In the following simulations we use the experimentally measured parameter values $D_n = 0.00035$, $\chi = 0.38$, $\rho = 0.34$, $\beta = 0.05$, $\gamma = 0.1$ and $\eta = 0.1$ [1, 24]. Through trial and error, it was found that a threshold branching age of $\psi = 0.5$ (equivalent to a dimensional time of 0.75 days) produced simulated networks that were qualitatively similar in morphology to those networks observed *in vivo*.

We aim to predict the growth of a vascular network in two different experimental assays, i.e., (i) a 2D experiment, that of the animal corneal implant and (ii) a 3D experiment, that of a small implant in a collagen gel assay or modeling an actual *in vivo* scenario of the capillary response to a spherical tumor positioned at the center of the cube of tissue. In both simulations, we have assumed that the tumors initially secrete AF, which sets up a fixed spatial distribution of AF throughout the tissue. This AF profile is assumed to have a high concentration near the source (i.e., tumor) and decays in a linear fashion throughout the domain. A matrix macromolecule (such as fibronectin, which allows cells to adhere to the matrix) is initially distributed evenly throughout the domain in both simulations. The predicted growth of a capillary network in response to the AF and MM concentration fields is shown in Figures 2 and 3. The simulation results show that the EC migrate towards the angiogenic source (the tumor) under the influence of chemotaxis and haptotaxis and proliferation. New capillary vessels are formed *via* branching, and these vessels can also branch producing secondary vessels, and so on; finally, two vessels that are close to one another can fuse together to form a loop (anastomosis). These processes (vessel growth, cell proliferation, branching, and anastomosis) continue to occur until the tumor is eventually reached, i.e., the tumor becomes vascularized.

Figure 2 shows two sets of results from simulations of the model in a square domain of side 2 mm at times $t = 7$ and $t = 15$ days. Both simulations have the parent vessel located along the left-hand edge of the domain (y), while the AF source is located along the right-hand edge for one simulation (representing a large tumor) and in the center of the right-hand edge for the other (representing a small circular tumor). At $t = 7$ days, both simulations have a similar initial outgrowth of sprouts, but in the case of the circular tumor simulation, we clearly have more anastomosis occurring. By $t = 15$ days, both networks have developed into highly branched, connected structures that have vascularized the tumors. However, due to the two different underlying AF concentration fields, in the case of the small circular tumor the resulting network is more focused than that of

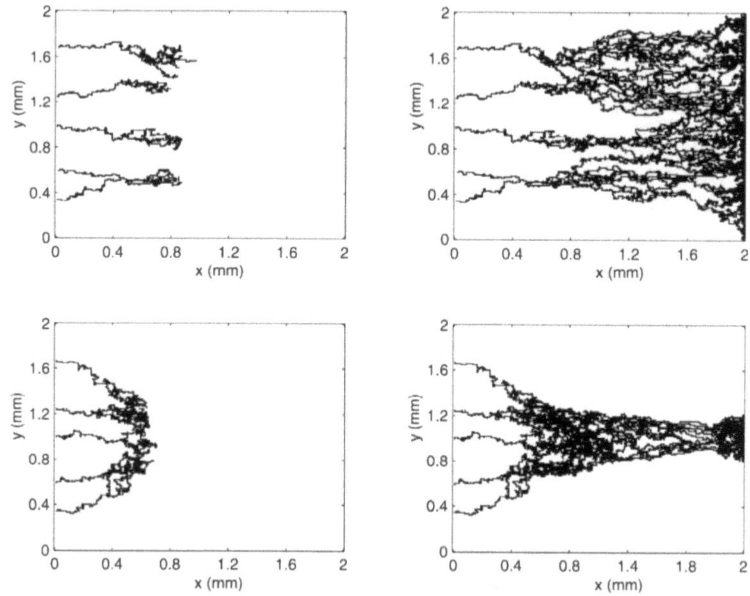

Figure 2. Spatiotemporal evolution of 2D capillary networks from numerical simulations with the HDC technique. The figure shows the evolution of the capillary network towards a large tumor (upper figures) and a smaller circular tumor (lower figures) for the times $t = 7$ days (left) and $t = 15$ days (right).

the large tumor. The latter has produced a broader network, which has more anastomosis occurring closer to the tumor. The "brush border effect" can be seen in the larger tumor simulation.

Figure 3 shows four snapshots in time of capillary network formation in a cube of tissue of side 2 mm, again in response to local gradients of angiogenic factors and matrix macromolecules ($t = 1, 3, 7, 10$ days). At $t = 1$ day we see that the initial vessels have grown a short distance towards the tumor but have branched only a little. As time progresses ($t = 3$ days), the vessels begin to branch more and migrate further towards the tumor. By $t = 7$ days there is a sufficient number of branches, which are close enough to one another for some anastomosis to occur, therefore leading to a better connected vasculature. Finally, at $t = 10$ days we have connection with the tumor and the completion of the angiogenic process. The number of branches rapidly increases between $t = 7$ and $t = 10$ days, creating a dense vasculature that almost engulfs the tumor.

By using the HDC technique, we have captured the structure and morphology of capillary networks, including the early formation of loops (anastomosis), the essential dendritic structure, and the formation of the experimentally observed "brush border". The HDC model also incorporates a realistic method of modeling mitosis and its effect on the

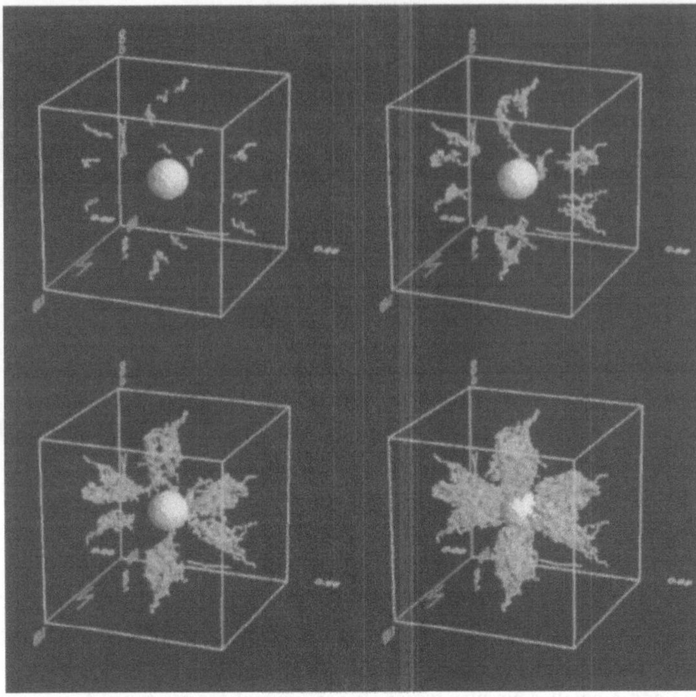

Figure 3. Spatiotemporal evolution of a 3D capillary network from numerical simulations with the HDC technique. The figure shows the evolution of the capillary network towards a spherical tumor for the times $t = 1, 3, 7$, and 10 days (see color plate 15).

sprouts, i.e., actual extension in the length of the sprout due to cell division. The above results show the importance of the tumor geometry, and consequently the AF concentration profile, in influencing EC migration and, therefore, the shape of the capillary network.

3. Invasion

The development of a primary solid tumor (e.g., a carcinoma) begins with a single normal cell becoming transformed as a result of mutations in certain key genes. This transformed cell differs from a normal one in several ways, one of the most notable being its escape from the body's homeostatic mechanisms, leading to inappropriate proliferation. An individual tumor cell has the potential, over successive divisions, to develop into a cluster (or nodule) of tumor cells. Further growth and proliferation leads to the development of an avascular tumor consisting of approximately 10^6 cells. This cannot grow any further, owing to its dependence on diffusion as the only means of receiving nutrients and removing waste products. For any further development to occur, the tumor must initiate angiogenesis – the recruitment of blood vessels (see previous section). Once angiogenesis is complete, the blood network can supply the tumor with the nutrients it needs to grow further. There is now also the possibility of tumor cells finding their way into the circulation and being deposited at distant sites in the body, resulting in metastasis.

The complete process of metastasis involves several sequential steps, each of which must be successfully completed by cells of the primary tumor before a secondary tumor (a metastasis) is formed. A summary of the key stages of the metastatic cascade is as follows: (i) cancer cells escape from the primary tumor; (ii) they locally degrade the surrounding tissue and continue migration; (iii) they enter the lymphatic or blood circulation system (*intravasation*); (iv) they must survive their journey in the circulation system; (v) they must escape from the blood circulation (*extravasation*); (vi) the cancer cells (from the primary tumor) must then establish a new colony in distant organs; (vii) the new colony of cells must then begin to grow to form a secondary tumor in the new organ.

A crucial part of the invasive/metastatic process is the ability of the cancer cells to degrade the surrounding tissue or *extracellular matrix* (ECM) [12, 13, 23]. This is a complex mixture of macromolecules (MM), some of which, like the collagens, are believed to play a structural role and others, such as laminin, fibronectin and vitronectin, which are important for cell adhesion, spreading, and motility. We note that all of these macromolecules are *bound* within the tissue, i.e., they are non-diffusible. The ECM can also sequester growth factors and itself be degraded to release fragments that can have growth-promoting activity. Thus, while ECM may have to be physically removed in order to allow a tumor to spread or intra- or extra-vasate, its degradation may, in addition, have biological effects on tumor cells.

A number of *matrix degradative enzymes* (MDEs) such as the *plasminogen activator* (PA) system and the large family of *matrix metalloproteinases* (MMPs) have been described [15, 17, 25], and both of these have been repeatedly implicated in all of the above steps of tumor invasion and metastasis.

3.1. The HDC Model of Invasion

We will base our mathematical model on generic solid tumor growth, which for simplicity we will assume is at the avascular stage. In principle, our model can be extended to include such interactions between the tumor and the vasculature, although the general form of our model will be the same for both invading vascular and avascular tumors. We choose to focus on three key variables involved in tumor cell invasion, thereby producing a minimal model, namely, tumor cell density (denoted by n), MDE concentration (denoted by m), and MM concentration (denoted by f). Initially, we derive a system of coupled nonlinear partial differential equations to model tumor invasion of surrounding tissue and use these as the basis for the HDC technique.

The complete system of equations describing the interactions of the tumor cells, MM, and MDEs is

$$
\begin{aligned}
\frac{\partial n}{\partial t} &= \overbrace{D_n \nabla^2 n}^{random\ motility} - \overbrace{\chi \nabla \cdot (n \nabla f)}^{haptotaxis} , \\
\frac{\partial f}{\partial t} &= - \overbrace{\delta m f}^{degradation} , \\
\frac{\partial m}{\partial t} &= \overbrace{D_m \nabla m}^{diffusion} + \overbrace{\mu n}^{production} - \overbrace{\lambda m}^{decay} ,
\end{aligned}
\tag{4}
$$

where D_n and D_m are the tumor cell and MDE diffusion coefficients, respectively, χ is the haptotaxis coefficient, and δ, μ, and λ are positive constants (see [2] for further details).

This system is considered to hold on some square spatial domain Ω (a region of tissue) with appropriate initial conditions for each variable. We assume that the MM, tumor cells, and consequently the MDEs remain within the domain of tissue under consideration, and therefore no-flux boundary conditions are imposed on $\partial\Omega$, the boundary of Ω. See [2] for full details of the model derivation.

The HDC technique will be used to follow the path of an individual tumor cell and first of all involves discretizing (using standard finite-difference methods) the system of partial differential equations (4). We then use the resulting coefficients of the five-point finite-difference stencil to generate the probabilities of movement of an individual cell in response to its local milieu (see [4] for details).

If there are gradients in the MM, through haptotaxis the defined probability coefficients will become biased towards the direction of increased MM concentration. The motion of an individual cell is therefore governed by its interactions with the matrix macromolecules in its local environment.

3.2. Indivdual-based Processes: Proliferation and MDE Production

In our model we assume that each individual cell has the capacity for proliferation and will produce two daughter cells provided that the following two conditions are satisfied: (i) the parent cell has reached maturity and (ii) there is sufficient space surrounding the parent cell for the two new daughter cells to occupy. We defined cell maturity to be 500 discrete time steps. While this time scale is arbitrary, with a precise estimate of parameter values in the original model, this maturity time can be made to correspond with an actual cell cycle time for specific cancer cells. In order to satisfy condition (ii), we assumed that one daughter cell replaces the parent cell and that the other daughter cell will move to any one of the parent cell's four orthogonal neighbors that is empty. If more than one of the neighboring grid points is empty, then the new cell position is chosen randomly from these points. If no empty neighbors exist, then proliferation is delayed until space becomes available.

Since we are modeling individual tumor cells, we must consider MDE production at the level of a single cell. In the continuum model (4), we have MDE production as being proportional to the tumor cell density; MDE is produced at a grid point only if a tumor cell is occupying that grid point. Since we have no precise parameter estimates for this production rate, we take $n = 0.1$ in the discrete form of the MDE equation when a tumor cell is occupying the current location and take $n = 0$ otherwise.

3.3. HDC Model Simulation Results

The following simulations were carried out on a 200×200 grid, which is a discretization of the unit square, $[0, 1] \times [0, 1]$, with a space step of $h = 0.005$ and a time step of $k = 0.001$. No-flux boundary conditions were imposed on the square grid, restricting the tumor cells, MDE and MM to within the grid. Initially, 500 tumor cells were centered around $(0.5, 0.5)$, and the MDE concentraion was zero throughout the domain ($m(x, y) = 0$). We consider the effects, upon tumor invasion, of three different MM initial distributions: (i) homogeneous ($f(x, y) = 1$), (ii) random ($0 \leq f(x, y) \leq 1$), and (iii) heterogeneous ($0 \leq f(x, y) \leq 1$). For clarity we shall label the resulting tumor cell distributions as (i) homogeneous tumor, (ii) random tumor, and (iii) heterogeneous tumor. The parameter values used in all the following simulations are $D_n = 0.0001$, $D_m = 0.001$, $\chi = 0.005$, $\delta = 10$, $\mu = 0.1$, and $\lambda = 0$ [2].

Figure 4 shows the results for each of the three variables for the three different initial MM distributions at $t = 4000$ time steps. In the first row we see that tumor cells invading through an initially homogeneous distribution of MM have produced the most invasive and symmetric tumor; this symmetry is also seen in both the MDE and MM distributions. The homogeneous tumor has also produced the largest number of individual cells. Clearly this is due to the combined effects of a faster invasion rate and, subsequently, access to empty space for proliferation leading to further invasion. In comparison with the random MM distribution results (in the second row), we see a smaller tumor with a slightly more ragged boundary. The random tumor has degraded less of the MM (i.e., invaded less of the ECM) and subsequently has produced fewer tumor cells and MDE. Finally, for the heterogenous MM distributions, we see quite a different tumor geometry (row three). In particular, the symmetry observed in the other two simulations is gone, and, perhaps more importantly, we see the appearance of a completely seperate cluster of tumor cells (bottom right-hand corner). These cells have the greatest potential to metastasize and may be difficult to detect clinically if they can invade further away from the main tumor mass.

Given that all three simulations use the same parameters, with the exception of the MM initial distributions, these results illustrate the importance of tumor cell/MM interactions and subsequently the ECM structure in aiding or hindering the migration of individual cells that may have the potential to metastasize.

4. Dictyostelium Discoideum Aggregation

Dictyostelium discoideum (Dd) cells proliferate as unicellular amoebae in soil where they consume bacteria for nourishment. When food becomes scarce, the amoebae enter the aggregation phase of their life cycle, which leads to the formation of a multicellular aggregate, or mound, of cells. During early aggregation, waves of a chemoattractant chemical, cAMP, emanate from an aggregation center. Dd amoebae detect the cAMP pulses

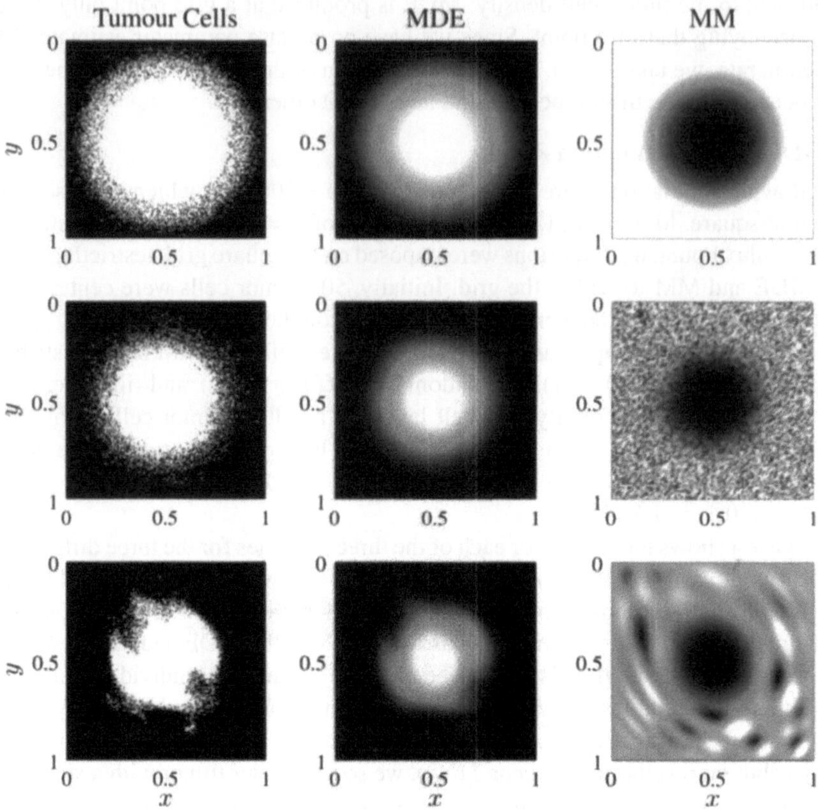

Figure 4. Spatial distribution at $t = 4000$ time steps, of a tumor (column one) producing MDE (column two) and subsequently invading three different MM distributions (column three): (i) homogeneous, first row; (ii) random, second row; (iii) heterogeneous, third row (see color plate 16).

and move chemotactically towards the aggregation center. The cells also help to generate further cAMP pulses through their autocatalytic response to cAMP. Receptors on the cell membranes respond to cAMP gradients, and these stimulated receptors initiate a process within the Dd amoebae which leads to their producing cAMP internally. This synthesized cAMP is then secreted by the stimulated Dd amoebae and it forms the basis for further cAMP pulses. Stimulated amoebae do not produce cAMP indefinitely because an (as yet unknown) element of the internal signal transduction pathway (that leads from cAMP receptor stimulation to cAMP synthesis) is a reaction-limiting step that causes cAMP synthesis to cease.

Previous authors [6–8, 11, 20, 26–28] have adopted different approaches to the mathematical modeling of the aggregation phase. A recent overview of this, as well as a comprehensive account of the Dd life cycle, can be found in [29].

Below we present a hybrid discrete-continuum mathematical model that describes aggregation in *Dictyostelium*. The model uses a discretized form of a system of reaction-diffusion equations to generate the motion of individual Dd cells. As a result, the individual cells move in response to local gradients in the continuous cAMP concentration. In addition to an equation for cell density, the continuum model we consider includes an equation for the chemo-attractant, cAMP, and an equation for the fraction of active receptors on a cell's surface. In the next section we present the mathematical model and describe the discrete-continuum technique used to simulate aggregation in Dd.

4.1. The HDC Model of Dd Aggregation

Previous work (e.g., [14, 29]) has shown that a minimal continuum model of Dd aggregation must contain three variables: the fraction of active (as opposed to desensitized) receptors, the concentration of extracellular cAMP and the density of *Dictyostelium* cells. We apply our technique to one particular model developed by Höfer et al. [11] using a simplified Martiel-Goldbeter model for the cAMP and receptor kinetics [14]. The Höfer continuum system of reaction-diffusion equations, with v denoting the fraction of receptors in the active state, n denoting the cell density, and u denoting the concentration of extracellular cAMP, is:

$$\begin{aligned}
\frac{\partial n}{\partial t} &= D_n \nabla^2 n - \nabla.(\chi(v) n \nabla u), \\
\frac{\partial v}{\partial t} &= -k_1 u v + k_2 (1 - v), \\
\frac{\partial u}{\partial t} &= \alpha(n) \rho (b v + v^2) \frac{a + u^2}{1 + u^2} - (d_1 \alpha(n) + d_2) u + \nabla^2 u,
\end{aligned} \tag{5}$$

where $k_1, k_2, k_v, \chi_0, \beta, a, b, d_1, d_2$ and ρ are all positive parameters, D_n is the cell random motility coefficient, $\chi(v) = \chi_0 \dfrac{v^\beta}{k_v^\beta + v^\beta}$ is the chemotactic response function and

$$\alpha(n) = \begin{cases} n/(1.2 - n), & \text{if } n < 1, \\ 5, & \text{if } n \geq 1. \end{cases} \tag{6}$$

In this model, cAMP is produced by the cells at a rate proportional to the amount of active cAMP receptors in the cell membranes. The number of active receptors is, in turn, a function of the cAMP concentration in the extracellular space. Dd cells have both an unbiased and a biased component to their motion, where the biased component is a chemotactic response to cAMP gradients (see [11, 19] for further detail).

We now use the HDC technique to generate the paths of individual Dd cells by first of all discretizing the system of partial differential equations (5) using standard finite-difference methods. As with the other two examples in this chapter, the central assumption of the HDC technique is that the five coefficients from the discretization of the continuum Dd cell equation (which are functions of the cAMP concentration near an individual Dd cell) are proportional to the Dd cell moving to one of its orthogonal neighbors or remaining stationary.

4.2. Individual-based Processes: cAMP Production, Degradation, and Cell Layering

To make our discrete-continuum model truly hybrid, both the generation and degradation (due to membrane-bound phosphodiesterase) of cAMP must be dependent on the presence of discrete cells rather than on the continuum cell density. To achieve this, $\alpha(n)$ is changed to $\alpha(cnum)$ with the following form:

$$\alpha(cnum) = \begin{cases} 2 \times cnum, & \text{if } 0 < cnum \leq 6, \\ 12, & \text{if } cnum > 6, \end{cases} \tag{7}$$

where $cnum$ is the integer number of discrete cells occupying a gridpoint at a particular time. Because Dd cells have a diameter of approximately 11μm [8] there will come a point during aggregation where the cell density in the aggregation plane ($x - y$ plane) is such that no more Dd cells can occupy a given area: a confluent cell-layer exists. Aggregation forces cause cells to pile on top of each other when cell density goes beyond that of the confluent cell layer and, as a result, a mound structure forms. Our model deals only with cells occupying the $x - y$ aggregation plane when we consider cellular cAMP production/degredation. We choose $cnum = 6$ to represent a confluent cell layer. As a result, although more than six cells can occupy a grid point (with cells 7, 8, 9 ... existing in a pseudo 3D level), we limit our cellular production/degredation term $\alpha(cnum)$ to a maximum value of $\alpha(6) = 12$. It can be seen that both production and degradation of cAMP will be proportional to the number of discrete cells occupying a gridpoint until the number of cells is greater than six. For reasons of computational simplicity and to allow the formation of a "mound", there is no limit placed on the number of cells that can occupy a gridpoint. Taking the model into three space dimensions would be the solution to this problem, but, at present, we believe that our approach is a reasonable compromise.

We solve the discretized form of these PDEs, which allows us to incorporate the discrete term $cnum \equiv cnum^q_{l,m}$ into what were originally PDEs dealing with continuum variables. Evolution of the $cnum$ values at each gridpoint is determined by the previously defined probabilities P_0 to P_4, which incorporate cellular the response to local cAMP gradients and unbiased cellular diffusion.

4.3. HDC Simulation Results

Zero-flux boundary conditions were imposed on the domain, meaning that Dd cells and cAMP were confined within the domain. In all simulations presented in the next section, individual cells are randomly placed at half of the grid points at the start of the simulation. Although simulations could be run with the total number of discrete cells between 1 and N (the total number of gridpoints), using N/2 cells (i.e., placing cells on half the gridpoints) provides a good balance between computational speed and the visual clarity of the data output. Further initial conditions are as specified in the captions below Figure 5.

The parameter values are taken from [11]: $D_n = 0.012$, $\chi_0 = 0.4$, $\rho = 100.0$, $a = 0.014$, $b = 0.2$, $d_1 = 23.5$, $d_2 = 0.9$, $k_1 = 2.5$, $k_2 = 2.5$, $K_v = 0.7$, $m = 10$. The simulation was carried out on a 200×200 grid. Results from the hybrid discrete-continuum model are presented in Figure 5 where we observe the initially randomly distributed cells forming aggregation streams and moving along these streams towards the aggregation center. Eventually, the vast majority of the cells have gathered into a circular group at the center, taking approximately four hours. This is in agreement with experimental observations of mound formation during Dd aggregation [29].

In line with the methodology used by Dallon and Othmer [8] and van Oss et al. [26], our discrete cells locally produce/degrade the cAMP. Like the van Oss model, the parameters we use are based on the values quoted by Martiel and Goldbeter [14]. However, the Dallon-Othmer model incorporates more biologically realistic G-protein cAMP kinetics and is therefore a better approximation of cellular behavior in this respect. Despite this, our results show both qualitative and quantitative agreement with experimental observations. In our model, when discrete cells move, they do so at a constant speed with only the direction of motion being influenced by local cAMP gradients. This is the case for some of the van Oss simulations, but others feature cellular motion with speed dependent on cAMP gradient. Furthermore we, like van Oss et al. [26], have a single grid for both the continuum and discrete components of the model, whereas Dallon and Othmer have separate grids, one for the discrete cells and another for continuum chemical concentrations, and interpolation between the two is required.

5. Discussion

In this chapter we have shown how the hybrid discrete-continuum technique of Anderson [4] can be applied to three different biological problems. In particular, we have shown the ability of the HDC model to incorporate both individual discrete processes and continuum processes. Clearly, the usefulness of the HDC technique depends on the situation being modeled and the adequacy of the continuum model in the first place. With these three examples, we have focused on a minimal modeling approach with the belief that if we can reproduce experimental observations with such a model, then it must be a reasonable approximation to the real system. It is then our goal to develop these models both at the continuum and individual levels to include further processes that will not only improve reproducibility of experimental work but also lead to more accurate predictability of possible outcomes. To this end, we now discuss briefly the possible application and development of each model in the next three paragraphs.

In the first model we have shown that we can reproduce detailed angiogenic structures in response to different tumor geometries (i.e., initial TAF concentrations). Both the number of vessels and connections between them (i.e., anastomosis) are strongly affected by the tumor geometry. The importance of these structural changes within the angiogenic network can only really be examined by considering the flow of blood though the network. In recent work, we have incorporated blood flow through our simulated angiogenic networks and have used this model to examine the uptake of a chemotherapy drug by the tumor. These results suggest that smaller tumors with more tightly connected networks will deliver less drug to the tumor than will a network resulting from a larger tumor. Another possible use of our model would be in anti-angiogenesis strategies [3] such as the preferential killing of endothelial cells, the inhibition of endothelial cell proliferation *via* a chemical such as angiostatin, the development of anti-chemotactic drugs, and

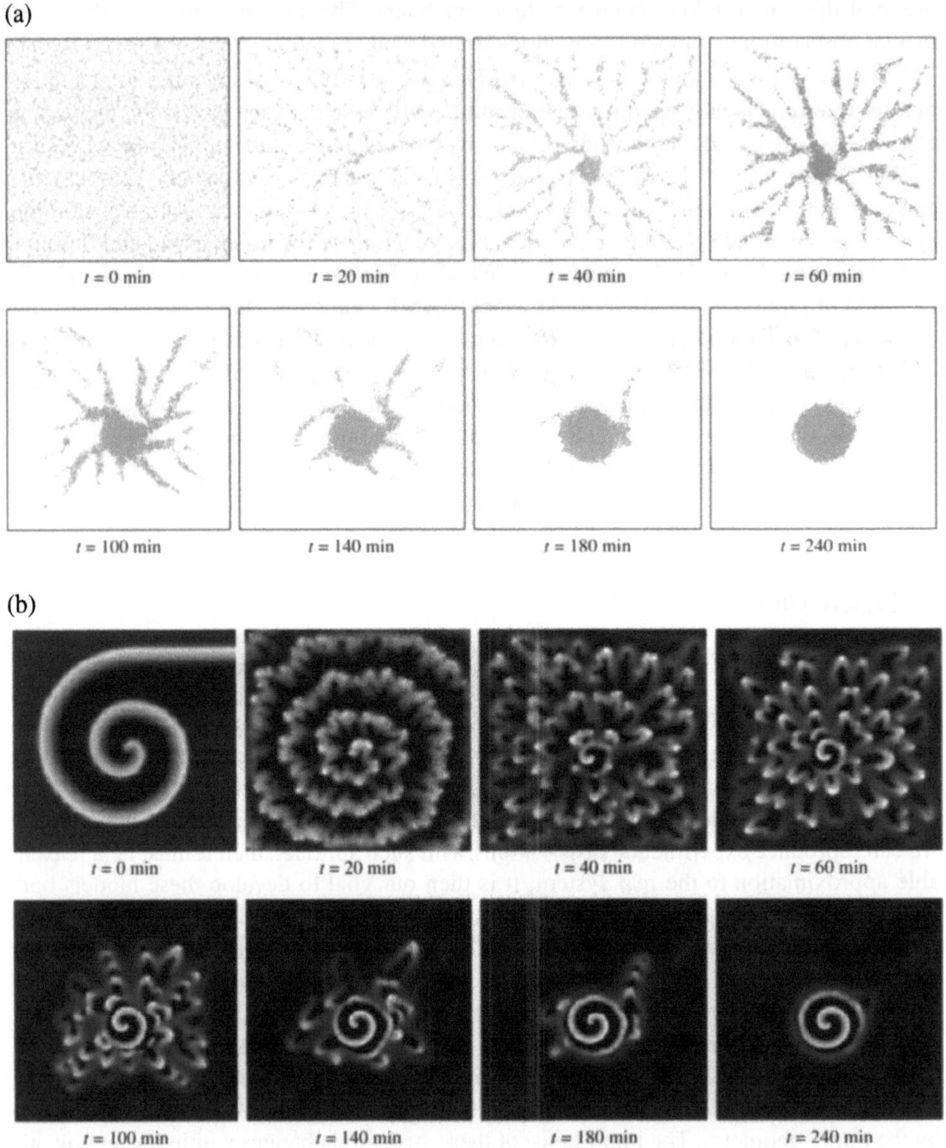

Figure 5. Spatiotemporal evolution of the discrete Dd cells (a) and extracellular cAMP (b) in a numerical simulation of the hybrid discrete-continuum model using the Höfer kinetics. The dimensional domain size is 7.0 mm $\times$ 7.0 mm (200 $\times$ 200 gridpoints), and snapshots are taken at the times specified. Initial conditions: a spiral wave of cAMP with corresponding spiral in the fraction of active receptors. Discrete cells occupy, randomly, half of the gridpoints. Boundary conditions are zero-flux.

the development of anti-haptotactic drugs. In particular, anti-angiogenic treatment is being recoginzed as an adjuvant chemotherapy in effective treatment of secondary tumors (metastases), and mathematical models of angiogenesis may have an increasingly important role to play in the development and testing of these therapies.

From a clinical point of view, the HDC model of tumor invasion has enormous potential. Even at this early stage in the model's development, the behavior of the simulated tumors closely parallels histological observations, especially when a heterogeneous MM distribution is introduced (cf. Fig. 4, row 3). It is therefore conceivable that measurement in tumors of some of the parameters used in these models will provide precise information on the invasive behavior of individual neoplasms. For example, it should then be possible to estimate the likely extent of local infiltration by a tumor and thereby tailor the radicality of surgical excision for that individual situation. It may also be possible to assess more accurately than at present the likelihood of metastatic disease, which will have important implications for adjuvant systemic therapy.

With regard to the HDC model of Dd aggregation, it is clear that the model can reproduce experimental results. However, to really exploit the potential of the HDC technique, many more properties need to be included at the cellular level. In particular, the modeling of individual receptor kinetics and cell-cell adhesion, the inclusion of mutant amoebae with different receptor affinities, and the extension to three space dimensions to more accurately model mound formation are needed.

Acknowledgments

Dr. A.R.A. Anderson is supported by a Personal Research Fellowship from the Royal Society of Edinburgh and an equipment grant from the Royal Society. Mr. A.W. Pitcairn is funded by an EPSRC Ph.D. studentship. The authors would like to thank Prof. M.A.J. Chaplain and Dr. F.A. Davidson for helpful comments and revisions.

References

[1] Anderson, A. R. A. and Chaplain, M. A. J. (1998). Continuous and discrete mathematical models of tumor-induced angiogenesis angiogenesis. *Bull. Math. Biol.*, **60**, 857–899

[2] Anderson, A. R. A., M. A. J. Chaplain, E. L. Newman, R. J. C. Steele and A. M. Thompson. (2000a) Mathematical modelling of tumor invasion and metastasis. *J. Theoret. Med.* **2**, 129–154

[3] Anderson A. R. A., M. A. J. Chaplain, C. Garcia-Reimbert and C. A. Vargas (2000b) A gradient-driven mathematical model of anti-angiogenesis. *Math. Comp. Mod.* **32**, 1141–1152

[4] Anderson, A. R. A. (2003) A hybrid discrete-continuum technique for individual based migration models. *This book, part III*

[5] Bowersox, J.C. and N. Sorgente (1982) Chemotaxis of aortic endothelial cells in response to fibronectin. *Cancer Res.* **42**, 2547–2551.

[6] Bretschneider, T., B. Vasiev and C. J. Weijer (1997) A model for cell movement during *Dictyostelium* mound formation. *J Theor Biol* **189**, 41–52

[7] Bretschneider, T., B. Vasiev and C. J. Weijer (1999) A model for *Dictyostelium* slug movement. *J Theor Biol* **199**, 125–136

[8] Dallon, J.C. and H.G. Othmer (1997) A discrete cell model with adaptive signaling for aggregation of *Dictyostelium discoideum. Phil Trans R Soc Lond* **B352**, 391–417

[9] Chaplain M. A. J. and Anderson A. R. A. (1999) Modelling the Growth and Form of Capillary Networks. In: *On Growth and Form. Spatio-Temporal Patterning in Biology* (eds. M.A.J. Chaplain, G.D. Singh, J. McLachlan) Chichester: Wiley, chapter 13, 225–250

[10] Folkman, J (1995) Angiogenesis in cancer, vascular, rheumatoid and other disease. *Nature Medicine* **1**, 21–31

[11] Höfer, T., J.A. Sherratt and P.K. Maini (1995) Cellular pattern formation during Dictyostelium aggregation. *Physica D* **85**, 425–444.

[12] Lawrence, J.A. and P.S. Steeg (1996) Mechanisms of tumor invasion and metastasis. *World J Urol* **14**, 124–130

[13] Liotta, L.A., C.N. Rao and S.H. Barsky (1983) Tumor invasion and the extracellular matrix. *Lab Invest* **49**, 636–649

[14] Martiel, J. and A. Goldbeter (1987) A model based on receptor desensitization for cyclic AMP signaling in *Dictyostelium* cells. *Biophysical Journal* **52**, 807–828

[15] Matrisian, L.M (1992) The matrix-degrading metalloproteinases. *Bioessays* **14**, 455–463

[16] McCarthy, J.B. and L.T. Furcht (1984) Laminin and fibronectin promote the directed migration of B16 melanoma cells in vitro. *J Cell Biol* **98**, 1474–1480

[17] Mignatti, P. and D.B. Rifkin (1993) Biology and biochemistry of proteinases in tumor invasion. *Physiol Rev* **73**, 161–195

[18] Paweletz, N. and M. Knierim (1989) Tumor-related angiogenesis. *Crit Rev Oncol Hematol* **9**, 197–242

[19] Pitcairn A. W., M. A. J. Chaplain , C. J. Weijer and A. R. A. Anderson (2000) A Discrete-Continuum Mathematical Model of *Dictyostelium* Aggregation. *ECMTB* **2**, 6–11

[20] Savill, N. J., and Hogeweg, P (1997) Modelling morphogenesis: From single cells to crawling slugs. *J Theor Biol* **184**, 229–235

[21] Sawada, H., H. Furthmayr, H. Konomi and Y. Nagai (1987) Immunoelectronmicroscopic localization of extracellular matrix components produced by bovine corneal endothelial cells in vitro. *Exp. Cell Res* **171**, 94–109

[22] Sholley, M.M., G. P. Ferguson, H.R. Seibel, J. L. Montour and J. D. Wilson (1984) Mechanisms of neovascularization. Vascular sprouting can occur without proliferation of endothelial cells. *Lab Invest* **51**, 624–634

[23] Stetler-Stevenson, W.G., S. Aznavoorian and L.A. Liotta (1993) Tumor cell interactions with the extracellular matrix during invasion and metastasis. *Ann Rev Cell Biol* **9**, 541

[24] Stokes, C.L., M.A. Rupnick, S.K. Williams and D.A. Lauffenburger (1990) Chemotaxis of human microvessel endothelial cells in response to acidic fibroblast growth factor. *Lab Invest* **63**, 657–668

[25] Thorgeirsson, U.P., C.K. Lindsay, D.W. Cottam and Daniel E. Gomez (1994) Tumor invasion, proteolysis, and angiogenesis. *J. Neuro-Oncology* **18**, 89–103

[26] van Oss, C., A. V. Panfilov, P. Hogeweg, F. Siegert and C. J. Weijer (1996) Spatial pattern formation during aggregation of the slime mould *Dictyostelium discoideum. J. theor. Biol.* **181**, 203–213

[27] Vasiev, B. N., P. Hogeweg and A. V. Panfilov (1994) Simulation of *Dictyostelium discoideum* Aggregation via Reaction-Diffusion Model. *Physical Review Letters* **73**, 3173–3176

[28] Vasiev, B. N., F. Siegert and C. J. Weijer (1997) A hydrodynamic model for *Dictyostelium discoideum* mound formation. *J. Theor. Biol.* **184**, 441–455

[29] Weijer, C. J (1999) The Role of Chemotactic Cell Movement in *Dictyostelium* Morphogenesis In Chaplain, M. A. J., Singh, G. D., and McLachlan, J. C. (Editors), 173–199, *On Growth and Form: Spatio-temporal Pattern Formation in Biology*, Chichester: Wiley.

[30] Williams, S.K (1987) Isolation and culture of microvessel and large-vessel endothelial cells; their use in transport and clinical studies. In: *Microvascular Perfusion and Transport in Health and Disease* (McDonagh, P. ed) Basel: Karger, 204–245

List of contributors and workshop participants

Wolfgang Alt, Abteilung Theoretische Biologie, Botanisches Institut der Universität Bonn, Kirschallee 1, 53115 Bonn, Germany; e-mail: wolf.alt@uni-bonn.de

Alexander R.A. Anderson, Department of Mathematics, University of Dundee, Dundee DD1 4HN, Scotland; e-mail: anderson@maths.dundee.ac.uk

Andreas Arend, Theoretische Physik, Universität des Saarlandes, 66041 Saarbrücken, Germany; e-mail: andreas@lusi.uni-sb.de

Victor H. Barocas, Assistant Professor of Biomedical Engineering, University of Minnesota, 7-106 BSBE, 312 Church St. SE, Minneapolis, MN 55455, USA; e-mail: baroc001@tc.umn.edu

Till Bretschneider, Max-Planck Institute of Biochemistry, Am Klopferspitz 18A, 82152 Martinsried,Germany; e-mail: bretschn@biochem.mpg.de

Hans-Joachim Bungartz, Institut für parallele und verteilte Systeme, Universität Stuttgart, Universitätsstraße 38, 70569 Stuttgart, Germany; e-mail: bungartz@informatik.uni-stuttgart.de

Mark A.J. Chaplain, The SIMBIOS Centre, Division of Mathematics, University of Dundee, Dundee DD1 4HN; e-mail: chaplain@maths.dundee.ac.uk

András Czirók, Dept. of Biological Physics, Eötvös University, Pázmány P. stny. 1A, H-1117 Budapest, Hungary; e-mail: derenyi@angel.elte.hu; e-mail: andras@angel.elte.hu

Micah Dembo, Department of Biomedical Engineering, Boston University, 44 Cummington Street, Boston MA 02215-2407, USA; e-mail: mxd@bu.edu

Imre Derényi, Dept. of Biological Physics, Eötvös University, Pázmány P. stny. 1A, 1117 Budapest, Hungary; e-mail: derenyi@angel.elte.hu

Andreas Deutsch, Center for High Performance Computing, Technical University Dresden, Zellescher Weg 12, 01069 Dresden, Germany; e-mail: deutsch@zhr.th-dresden.de

Dirk Drasdo, Max-Planck-Institute for Mathematics in the Science, Inselstr. 22-26, D-04103 Leipzig, Germany; e-mail: drasdo@mis.mpg.de

Jeanie L. Drury, 12505 100th Ave NE #5, Kirkland, WA 98034, USA; e-mail: jldrury@att.net

Ralf Everaers, Max-Planck-Institute for Polymer Research, Ackermannweg 10, P.O.Box 3148, 55128 Mainz, Germany; e-mail: everaers@mpip-mainz.mpg.de

Zénó Farkas, Dept. of Biological Physics, Eötvös University, Pázmány P. stny. 1A, H-1117 Budapest, Hungary, e-mail: zeno@angel.elte.hu

Dieter Felix, Abteilung Theoretische Biologie, Botanisches Institut der Universität Bonn, Kirschallee 1, 53115 Bonn, Germany; e-mail: dfelix@uni-bonn.de

Aaron Fogelson, Department of Mathematics, University of Utah, 155 South 1400 East, JWB 233, Salt Lake City, Utah 84112, USA; e-mail: fogelson@math.utah.edu

Mark Gerstein, Department of Molecular Biophysics & Biochemistry, Yale University, P.O. Box 208114, New Haven, CT 06520-8114, USA

Michael Griebel, Abteilung Wissenschaftliches Rechnen und Numerische Simulation, Institut für Angewandte Mathematik, Wegelerstr. 6, 53115 Bonn, Germany; e-mail: griebel@iam.uni-bonn.de

Helmut Grubmüller, Max-Planck Institute for Biophysical Chemistry, Karl-Friedrich-Bonhoeffer-Institut, Arbeitsgruppe für theoretische molekulare Biophysik, Am Faßberg 11, 37077 Göttingen, Germany, e-mail: hgrubmu@gwdg.de

Daniel Hoffmann, Center of Advanced European Studies and Research (caesar), Ludwig-Erhard-Allee 2, 53175 Bonn, Germany; e-mail: Daniel.Hoffmann@caesar.de

Josef Käs, Abteilung für die Physik weicher Materie, Institut für Experimentelle Physik I, Universität Leipzig, Linnéstr. 5, 04103 Leipzig, Germany; e-mail: jkaes@physik.uni-leipzig.de

Diego Kienle, Institut für Festkörperphysik, Forschungszentrum Jülich, 52425 Jülich, Germany; e-mail: d.kienle@fz-juelich.de

Werner G. Krebs, San Diego Supercomputer Center, Dept. 0505, University of California, San Diego, 9500 Gilman Drive, La Jolla, CA 92093-0505, USA; e-mail: werner.krebs@aya.yale.edu

Andrew Kuharsky, Quantlab Financial, 4200 Montrose Blvd., Houston 77006, Texas, USA

Martin Kühn, Lehrstuhl für Wassergüte und Abfallwirtschaft, Technische Universität München, Boltzmannstr. 3, 85748 Garching, Germany; e-mail: m.kuehn@bv.tum.de

Jürgen Lenz, Bioreact GmbH, Botanisches Institut, Universität Bonn, Kirschallee 1, 53115 Bonn, Germany; e-mail: lenz@bioreact.de

Siewert-Jan Marrink, Laboratory of Biophysical Chemistry, University of Groningen, Nijenborgh 4, NL-9747 AG Groningen, The Netherlands; e-mail: S.J.Marrink@chem.rug.nl

Jean-Louis Martiel, Laboratoire TIMC-IMAG, UMR CNRS 5525, Faculté de Médecine de Grenoble, 38706 La Tronche Cedex, France; e-mail: Jean-Louis.Martiel@imag.fr

Miriam Mehl, Institut für Informatik, TU München, Boltzmannstr. 3, 85748 Garching, Germany; e-mail: mehl@in.tum.de

Ralf Müller, Abteilung Theoretische Biologie, Botanisches Institut der Universität Bonn, Kirschallee 1, 53115 Bonn, Germany; e-mail: rmueller@uni-bonn.de

Eirikur Palsson, Department of Biological Sciences, Simon Fraser University, 8888 University Drive, V5A 1S6 Burnaby, Canada; e-mail: epalsson@alumni.princeton.edu

Alan W. Pitcairn, Department of Mathematics, University of Dundee, Dundee DD1 4HN, Scotland; e-mail: pitcairn@maths.dundee.ac.uk

Emmanuel Promayon, Laboratoire TIMC-IMAG, UMR CNRS 5525, Institut Albert Bonniot, 38706 La Tronche Cedex, France; e-mail: Emmanuel.Promayon@imag.fr

Dennis C. Rapaport, Physics Department, Bar-Ilan University, Ramat-Gan 52900, Israel; e-mail: rapaport@mail.biu.ac.il

R. Brent Rice, Department of Chemical Engineering, University of Colorado, Campus Box 424, Boulder, CO 80309-0424, USA; e-mail: randall.rice@colorado.edu

Xavier Ronot, E.P.H.E., Institut Albert Bonniot, 38706 La Tronche Cedex, France; e-mail: Xavier.Ronot@ujf-grenoble.fr

Roland Rzehak, Institut für Festkörperphysik, Forschungszentrum Jülich, 52425 Jülich, Germany; e-mail: R.Rzehak@fz-juelich.de

Ralf Sambeth, Forschungszentrum Jülich, 52425 Jülich, Germany; e-mail: r.sambeth@fz-juelich.de

Mark Alexander Schweitzer, Abteilung Wissenschaftliches Rechnen und Numerische Simulation, , Institut für Angewandte Mathematik, Wegelerstr. 6, 53115 Bonn, Germany; e-mail: schweitz@iam.uni-bonn.de

Udo Seifert, Max-Planck-Institut für Kolloid- und Grenzflächenforschung, MPI-KG, 14424 Potsdam, Germany: e-mail: useifert@mpikg-golm.mpg.de

Angélique Stéphanou, Laboratoire TIMC-IMAG, UMR CNRS 5525, Faculté de Médecine de Grenoble, 38706 La Tronche Cedex, France; e-mail: Angelique.Stephanou@imag.fr

Angela Stevens, Max-Planck-Institute for Mathematics in the Sciences (MIS), Inselstr. 22-26, 04103 Leipzig, Germany; e-mail: stevens@mis.mpg.de

Philippe Tracqui, Laboratoire TIMC-IMAG, UMR CNRS 5525, Faculté de Médecine de Grenoble, 38706 La Tronche Cedex, France; e-mail: Philippe.Tracqui@imag.fr

Tamás Vicsek, Dept. of Biological Physics, Eötvös University, Pázmány P. stny. 1A, 1117 Budapest, Hungary; e-mail: vicsek@bangel.elte.hu

Stefan Wuertz, Departement of Civil and Environmental Engineering, University of California, Davis, USA; e-mail: swuertz@ucdavis.edu

Michael Wurzel, Abteilung Theoretische Biologie, Botanisches Institut der Universität Bonn, Kirschallee 1, 53115 Bonn, Germany; e-mail: michael-wurzel@web.de

Haoyu Yu, Physiome Sciences, 307 College Road East, Princeton, New Jersey 08540, USA

Walter Zimmermann, Theoretische Physik, Universität des Saarlandes, 66041 Saarbrücken, Germany; e-mail: wz@lusi.uni-sb.de

Index

actin 46, 70, 76, 98, 114
actin polymerization 46, 98
actin-myosin network 80, 99
activated platelet 146
adaptive grid refinement 160
adaptivity 164
adiabatic mapping 29, 34
advection diffusion equation 147
affine motion model 94, 104
aggregation 134, 136, 143, 221, 226, 230
algebraic multigrid 165
amoeba, swarm dynamics 221
amphiphilic surfactant 8, 10
animal tissue cell, locomotion 75
assembly of virus capsids 9

bcl complex 34
bead-spring model 46
biased random walk 253, 255, 257, 261
biased random walk equation 144
biased random walk model 251
bifurcation, parameter 101
biofilm ix, 142, 155, 161
biofilm growth 161
biological wastewater treatment 155
biomass growth 157
birds, swarm dynamics 221
bird swarm 234f

blastula formation 185
bleb 82f
Brownian dynamics simulation 55, 62
Brownian dynamics simulation, artificial inertia for 55
Brownian dynamics simulation, neighbor list 58
brush border effect 267
bulk viscosity 47

calmodulin 36
CAMLab 23
cAMP gradient 133
capsid assembly 7, 9, 14
capsid simulation 10
CAsim 161
cell alignment 205f
cell collision 134
cell concentration 243
cell cortex contraction 107
cell deformation and motility 47
cell mechanical response 135
cell migration 132
cell morphological changes 104
cell object 127f, 130
cell population 185
cell protrusion 92, 96
cell response 130
cell sorting 173, 179, 184, 190
cell surface model 47
cell tensegrity 126, 134
cell uni-axial compression 130
cell-based discretization 164
cell-cell interaction 134, 163
cell-centered grid 160
cellular automata 161, 172, 174ff, 195, 212
cellular interaction 143
CHARMM 22, 23
chemoattractant 132, 223
chemotaxis 143, 223, 225ff, 253, 262f, 272
Chorin's projection method 159

citrate synthase 31
coagulation dynamics 141
coagulation factor 148ff
collagen gel 143
collective behavior 8
collective cell motion 143
collective movement 230
compression waves 236
confocal laser scanning microscopy (CLSM) 156
conformation, of proteins 29
conformational variability 29
contact response 222
contact surface 231, 240
continuum limit 236
contractile fluid dynamics 46
contraction 82
contraction, cell 243
convection-diffusion equation 159
cooperative behavior 143
critical exponents 217
crypt 176, 186, 188
crypt fission 185
cytomechanical model 98
cytomechanics 130
cytoskeleton 81

data hierarchization 165
database, macromolecular motions 29
Delaunay triangulation 238
detachment 163
Dictyostelium 47, 223, 239
Dictyostelium discoideum, aggregation 132, 261, 271, 273
differential perception 232
differential response 239
diffusion 151, 253
diffusion limit 255
diffusion-reaction equation 159
dilatational viscosity 47, 224
dilemma of different scales 156
displacement, maximum 31, 36
distorted geometry 39
dynamic response 223, 226

eden model 172
elastic bending 46
elastic modulus 133
elasticity 224, 233
elasticity of cell 132
electrostatic potential 21, 23ff
elliptic or parabolic system xii
energetic barrier 29
energy minimization 4, 22f
erythrocyte 131
Eulerian modeling 141
Experimentally Validated Model (EVAM) 19
extracellular polymeric substances 163

F-actin 98
ferromagnetic model 206
fibrin 145, 148, 150
fibroblast 92, 243
Fick's Law 159
finite difference method 23
finite difference scheme 159
Finite Element Method (FEM) 245
finite volume discretization 159
Finitely Extensible Nonlinear Elastic (FENE) spring 52
flock 206
flow 151, 157f
fluctuation-dissipation theorem 51
fluid-structure interaction 163
fluorescently labeled lectin 163
foam 193
force field 22
format, URML 37
formats, video 37
fractal dimension 171f

fragment 29, 31
friction force 223
full approximation scheme 165
full multigrid 165

Galerkin operators 160
Galilean invariance 207
gastrulation 181, 185
generative grammar 194
glycerol kinase 34
gOpenMol 23
gradient perception 228

Hamiltonian function 206
haptotaxis 262f, 270
hemophilia 151f
hexokinase 31
hierarchy of cells 165
hinge 32
hinge axis 36
hinge location 36
homogenization 162
homology modeling 21
homology technique 4
human interleukin 5, 34
human keratinocyte 82
hybrid continuum-discrete evolution model 142
hybrid discrete-continuum (HDC) model of angiogenesis 262
hybrid discrete-continuum (HDC) model of *Dictyostelium* aggregation 273
hybrid discrete-continuum (HDC) model of invasion 269
hybrid discrete-continuum (HDC) technique 251, 253, 255ff, 261, 264, 267, 275
hybrid method xii
hybrid model 251
hydrodynamic evolution equation x
hydrodynamic interaction, method of 54, 56
hydrodynamic steering effect 63
hydrostatic pressure 80
hyperbolic transport equation xii

immersed boundary (IB) method 141, 146
individual-based model 142, 169
individual-based process 255f, 265, 270
interaction energy 178, 182
interaction field 222ff, 231
interaction force 222f
interaction tensor 231f
interactive movement 221
interpolation 29, 36
interpolation engine 37
interpolation, Cartesian 34
invasion and metastasis 268f

keratinocyte 82, 224

laboratory reactor 157
Lagrangian modeling 141
Langevin dynamics 3
Langevin equation xii
Langevin-type equation 77
lattice model 142, 170
lattice growth model 170
Law of Mass Action 151
Lennard-Jones potential 53
liftase 20
Lindenmeyer system 194
linker 32
loss modulus $G''(\omega)$ 79
low-resolution model 9

MEAD 22f

mean-field approximation 216
mean-field description 215
mechanical force 130
mechanical interaction 126
mechanical stress 163
metropolis algorithm 178, 183
micelle formation 12
micelle simulation 10
micropipet aspiration 113
migrating birds 236
molecular dynamics 4, 7, 30, 62
molecular machine 4
molecular pharmaceutics ix
momental ellipse 83
monod kinetic 159
monolayer, endothelial or epithelial 239
Monte Carlo simulation 178, 182, 196
Monte Carlo technique 3
morph movie 29
morph server 31
morph, optimal 37
motility assay 70
motion, allosteric 33
motion, "gold-standard" 33
motion, complex 33
motion, macromolecular 29
motion, non-allosteric 33
motion, protein 30
motion, rigid-body 30
motion, shear 32
motion, unclassifiable 33
motor protein 70
multicellular system 169
multiculture biofilm 163
multigrid 160, 165
multiparticle model x, 46, 76, 80, 169
multiparticle model, actin-myosin system 76
multiparticle model of cellular motility 80
multi-phase flow dynamics with free boundaries 47
multiple sequence alignment 21
myosin 70, 77
myxobacteria, swarm dynamics 221

Nast++ 159
Navier-Stokes equation 141, 146, 158
Navier-Stokes equation, compressible 237f
nearest-neighbor interaction 230
nearest-neighbor response 233
neutrophil 113
NK-lysin (NKL) 19f
nonequilibrium process 8
nonlinear Boltzmann equation 215
nonlinear boundary condition 163
nutrient transport 158

Occam's razor 15
octree 164
off-lattice model 180
optical flow 94, 103
optical tweezer 131
order parameter 209
Ornstein-Uhlenbeck, differential equation 222
oscillating magnetic bead microrheometer 79
oscillatory morphology 102
Oseen tensor 53

packing 29
packing classification 32
parallelization 159
partial differential equation (PDE) 142
particle tracking 157
pattern formation 4

perception scope 225, 230
phase transition 209, 212, 215f
physical-object-oriented (POO) approach 126
plasma membrane motion 46
platelet activation 146, 151
platelet aggregation 145
platelet coagulation 145
platelet count 152f
point-based discretization 164
Poisson equation 159
Poisson-Boltzmann equation 23
polarity 82
polarity map 96
polarization 46
polymer fluid 46
polymer network 46
polymers, bead-spring model for 50
polymers, blob models for 61
polymers, drag coefficient of 59
polymers, dumbbell model for 63
polymers, elongation of 60
polymers, excluded volume effect in 52
polymers, finite extensibility of 52
polymers, hydrodynamic interaction 53
polymers, partial draining in 59
polymers, relaxation time of 63
postgreSQL 33
Potts model 177, 196
Protein Databank 33
protein flexibility 31
protonation 21ff
pulsatile behavior 104

quadtree 164
quasi-equivalence 9
quasi-stationary approach 162

random motility 263
reactive process 157
refolding, partial 33
rigid body dynamics 10
rotation 36
Rouse model 52

saposin 19
saposin fold 20
screw axis 36
script language 161
self-assembly 4, 7f, 11
self-sustained oscillation 100
semi-flexible polymer 46
shear viscosity 224
sieve-fit 30, 36
simulation model, scaling xi
size classification 31
slime mold aggregation 132
sphingolipid 20
sphingomyelin 24ff
stability analysis, linear 215
stability analysis, mean-field 216
staggered grid 159
standardized coordinate system 29
statistics, standardized 37
stochastic equation 222
stochastic hyperbolic-parabolic free boundary problem 237
storage modulus $G'(\omega)$ 79
structural mechanic equation 163
subunit 29, 31
superposition, RMS 36
swarm dynamics 143, 221, 230
swarming 143

tensegrity 47
tension field 224
tensiotaxis 143, 224ff, 230f
thrombin 148, 150, 152
ToPLign 22
transition probability 253ff

transmission of signals 223
transport process 157
triangulation 189
tumor 170ff, 184
tumor-induced angiogenesis 262, 265
two-phase flow model 47

unclassifiable motion 33
uniform flow 58

velocity, correlation function 211
vesicle 63
virus capsid, assembly 9
viscoelastic chain 236
viscoelastic evolution equation x
viscoelastic fluid dynamics 46
viscoelastic medium 143
viscoelastic porous medium 163
viscoelastic substratum 143, 224, 230
viscoelasticity 181, 188
viscosity 233, 239
visualization 11, 29
Voronoi neighborhood 232
Voronoi polyhedron 232
Voronoi surface 231, 240
Voronoi tessellation 176f, 184, 231, 238f
wound healing 184

X-PLOR 36

Color plates

Color plates

Color plate 1

a)

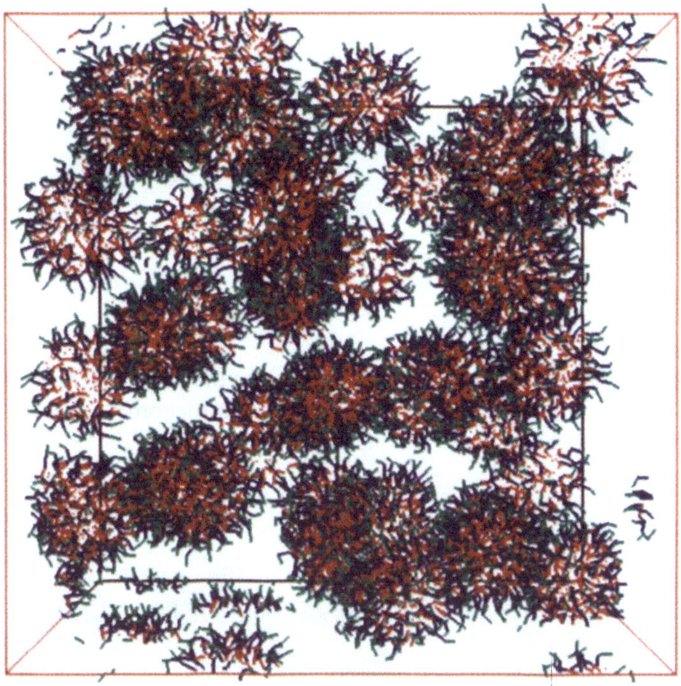

b)

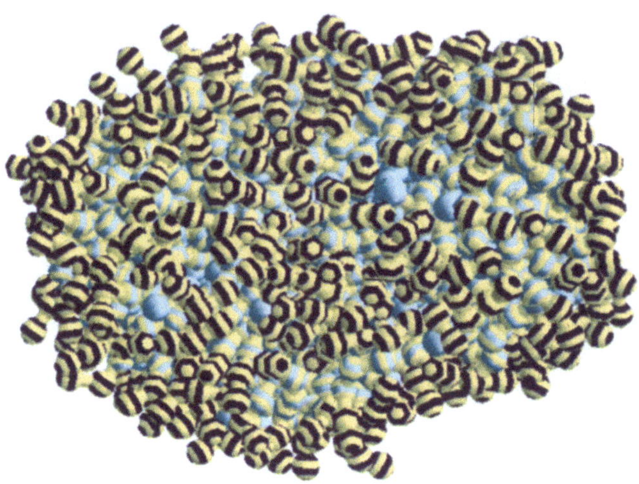

Fig. 2 from page 12.
(a) Surfactant system ($c = 0.05$) after 5×10^5 time steps showing chains and solute (because of periodic boundaries, some micelles wrap around); (b) the largest micelle extracted from Fig. 2(a); the color scheme distinguishes between the surfactant heads (dark blue with yellow stripes) and tails (light blue and yellow), and the solute (light blue).

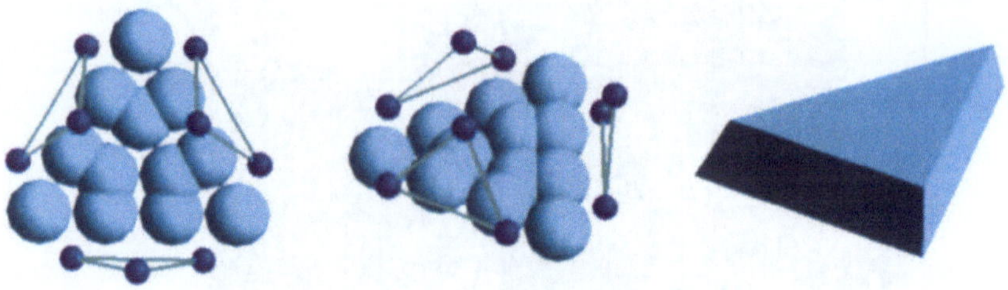

Fig. 1 from page 11.
Views of the structure of the capsid subunit showing the large spheres that produce excluded volume and the small spheres denoting attraction sites; an alternative view shows the faces of the equivalent triangular block.

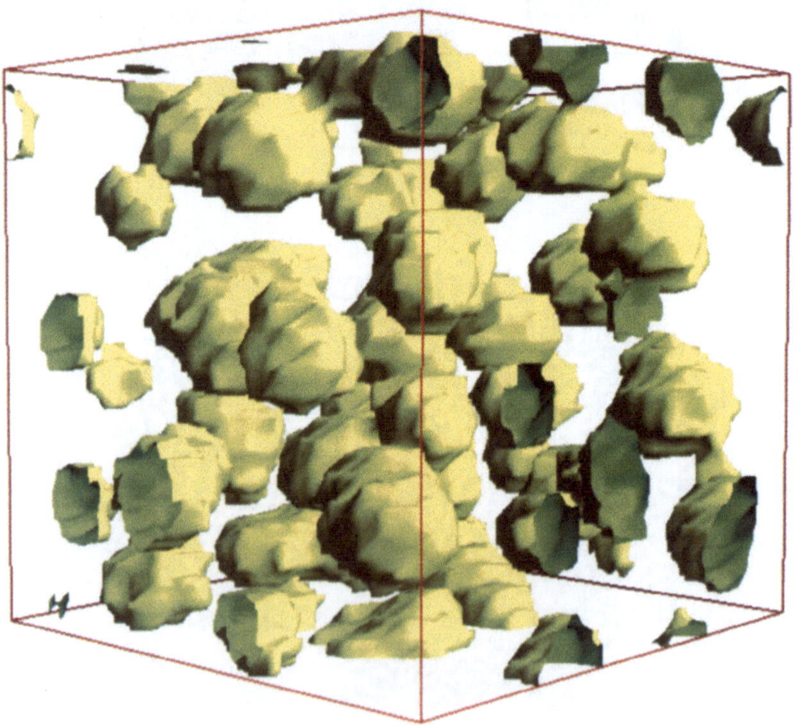

Fig. 3 from page 13.
Surfactant system (c = 0.025) after 5 x 10^5 time steps showing the micelle boundaries (the fact that some surfaces appear open is an artifact due to periodic boundaries).

Color plate 3

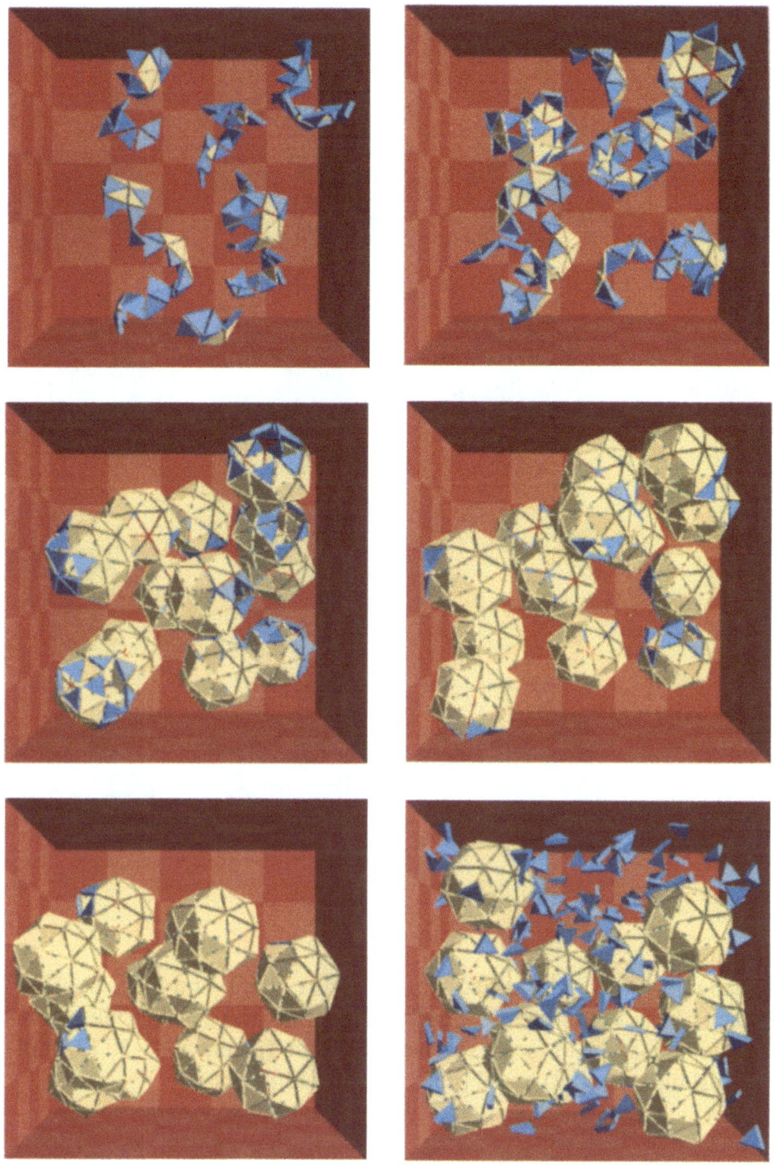

Fig. 5 from page 15.
Stages in the capsid self-assembly process; unattached monomers are omitted for clarity, except in the final image (fully bonded subunits are shown in yellow, all others are in blue).

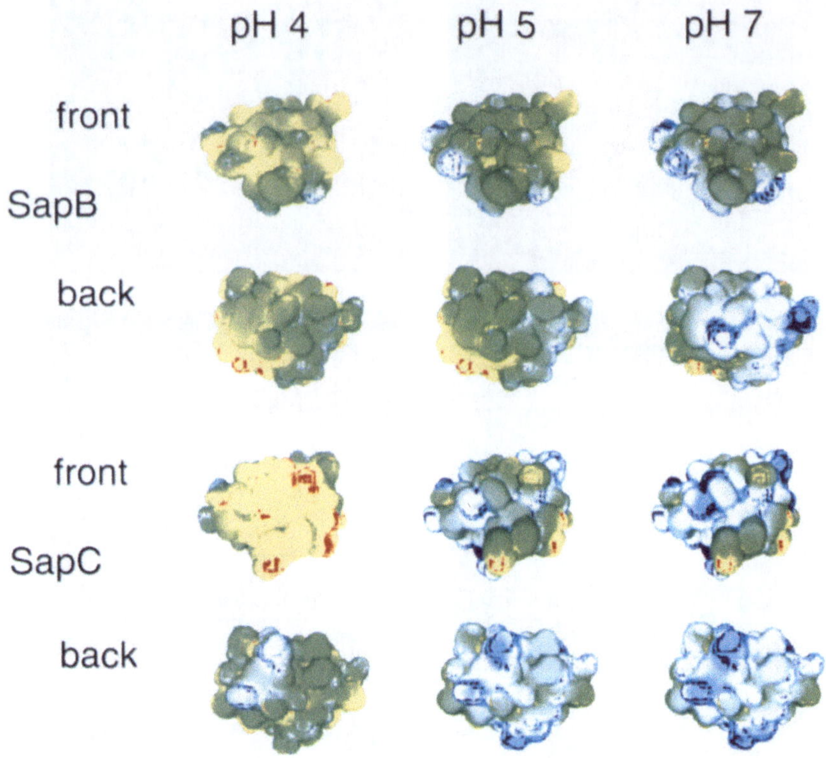

Fig. 4 from page 25.
Electrostatic potential at the surfaces of EVAMs of SapB and SapC at pH 4, 5 and 7. For each of the two proteins at each pH, „front“ and „back“ sides (arbitrarily defined) are shown. Color scale reaches from blue (potential ≤ -20 kJ/(mol e)) to green (vanishing potential) to red (potential ≥ 20 kJ/(mol e).

Color plate 5

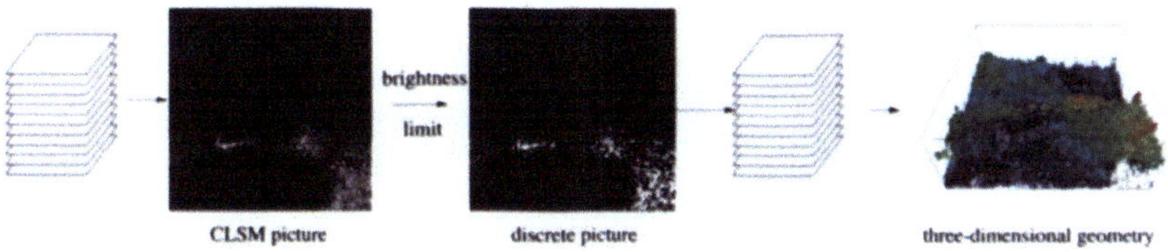

Fig. 1 from page 158.
Construction of a 3D biofilm geometry from 2D grayscale CLSM images.

a)

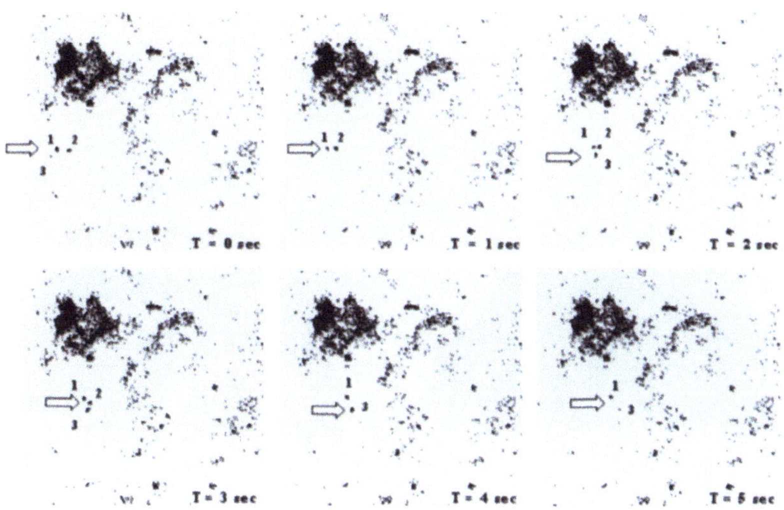

b)

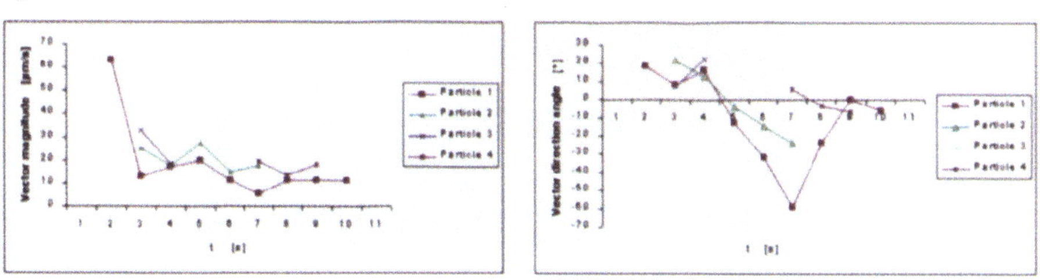

Fig. 2 from page 158.
Time-resolved measurements: sequence of snapshots tracking particles in a horizontal slice through a monoculture biofilm (a) and experimentally captured local velocities of four tracked particles (b).

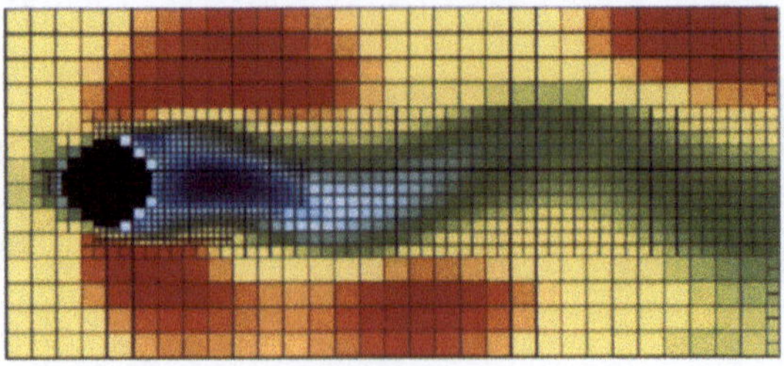

Fig. 4 (c) from page 161.
Efficient simulation on an adaptively refined grid (flow round a circular obstacle). Color indicates flow velocities, from low (in blue) to high (in red).

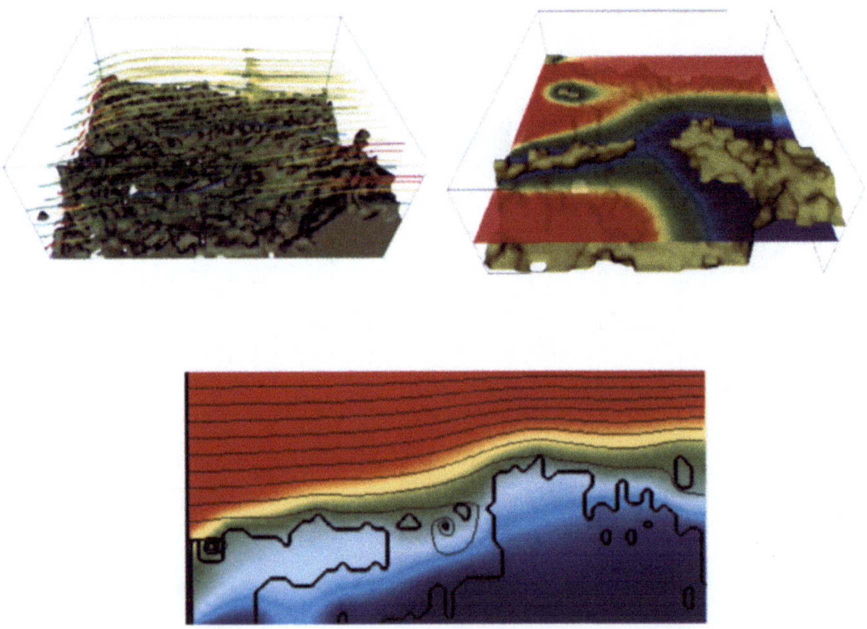

Fig. 5 from page 161.
(a) Streamlines, (b) nutrient concentration in the fluid phase, and (c) nutrient concentration in a vertical cut through an experimentally defined section (512 x 512 x 50 μm) of a four-day old E. coli DH5α monoculture biofilm. Colors in (b) and (c) indicate nutrient concentration from low (in blue) to high (in red).

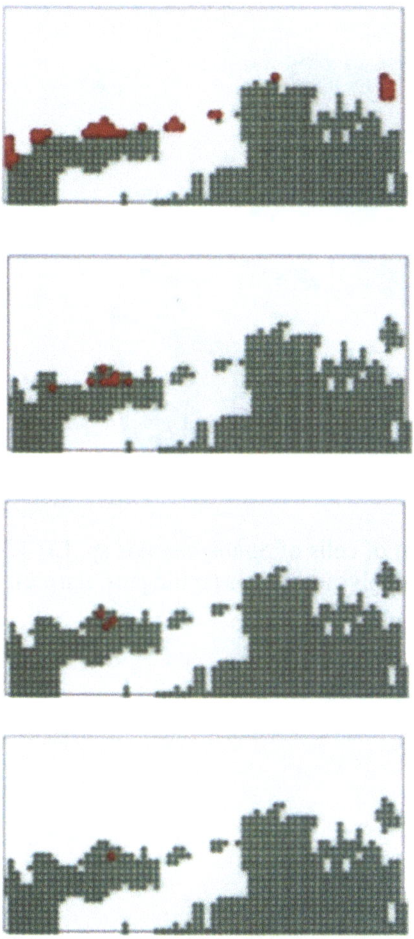

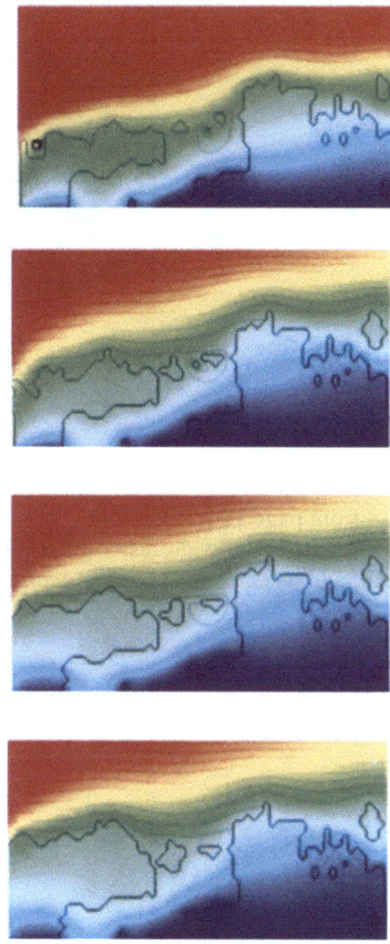

Fig. 6 from page 162.
CAsim: cellular automaton iterations for one growth step in the 2D biofilm model (red: cells ready for dividing).

Fig 7 (b) from page 162.
Simulated growth of a vertical cut through a single-species biofilm starting from an experimentally defined geometry (color indicates nutrient concentration).

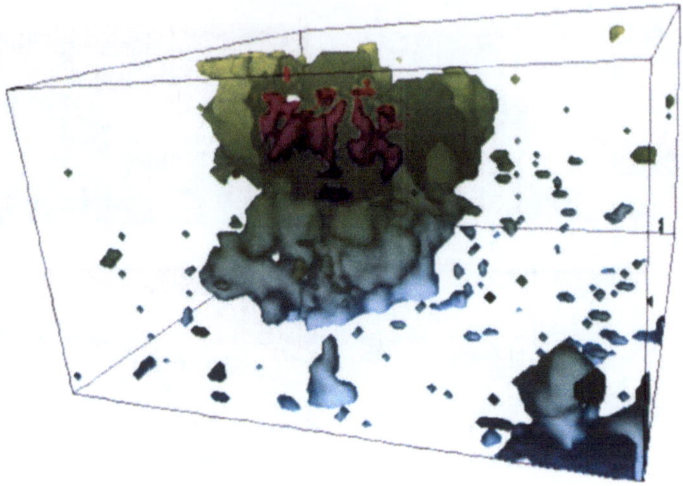

Fig. 8 from page 164.
Experimentally defined cluster (30 x 30 µm) of cells of *Sphingomonas* sp. LB 126 (green) and ist embedded *Sphingomonas* exopolysaccharides (sphingans, purple).

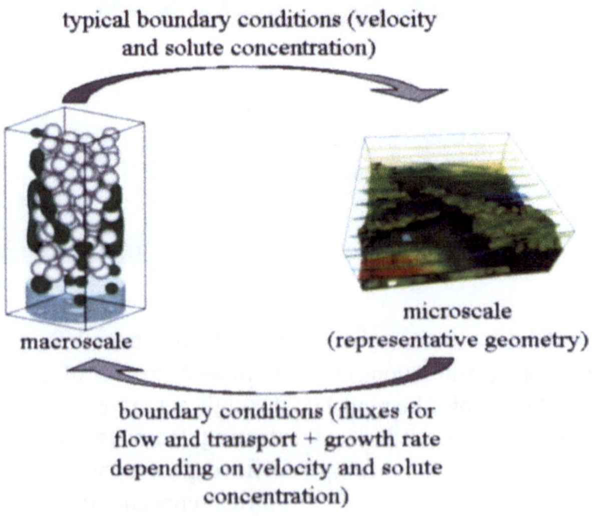

Fig. 9 from page 164.
Schematic illustration of the coupling between micro- and macroscale computations for the convection-diffusion problem.

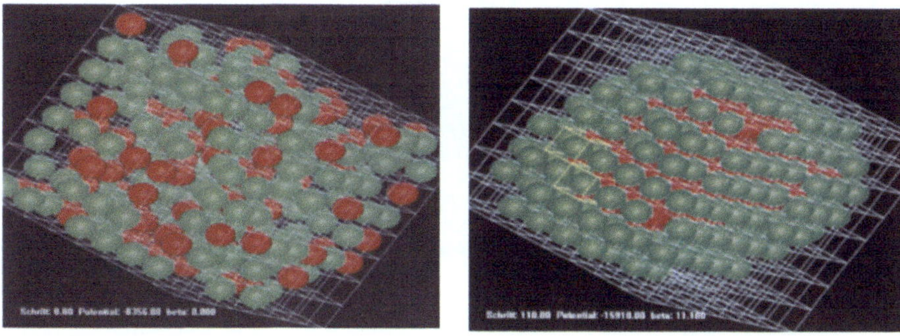

Fig. 3 from page 174.
Cell sorting on a square lattice with two cell types A (red), B (green) in d = 3 dimensions in a computer simulation assuming periodic boundary conditions in x,y- and z-direction. (For more details see caption in text).

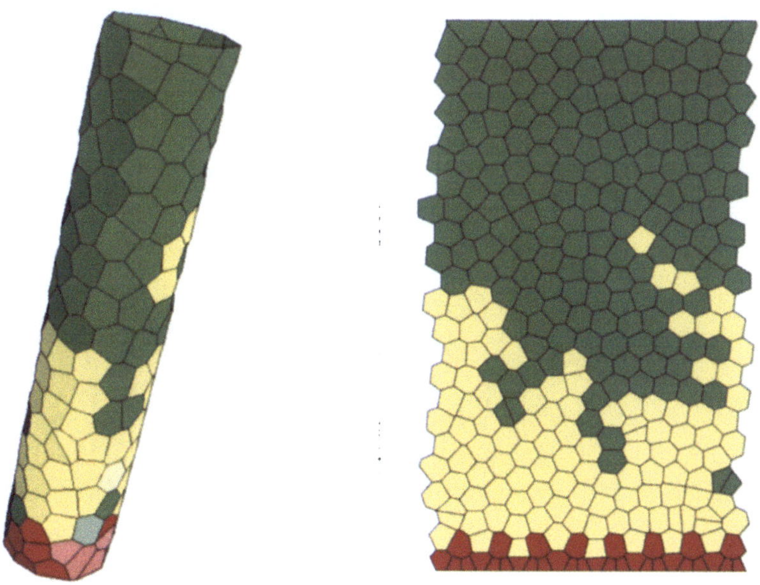

Fig. 4 from page 176.
Model design for studying the cell dynamics in an intestinal crypt (represented by a cylinder) using a Voronoi diagram. The paneth cells are red, S and T_r-cells are yellow and mature cells are green.

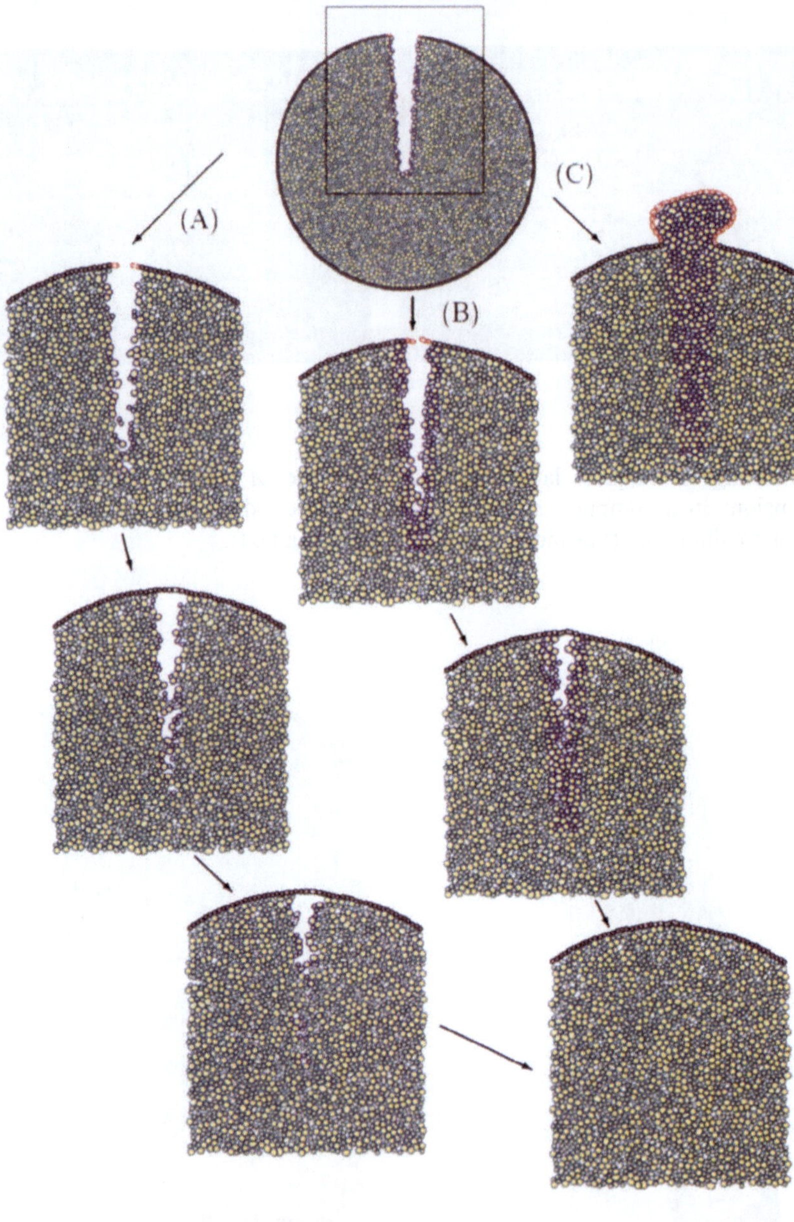

Fig. 10 from page 186.
Regeneration after a cut-like lesion of a 2D model tissue, composed of epithelial cells surrounding fibroblasts. Cells may switch from proliferating (p) to differentiated, non-proliferating (n) state. Epithelia are marked pink, with black (n) or red (p) boundaries, fibroblasts in yellow, with gray (n) or blue (p) boundaries.

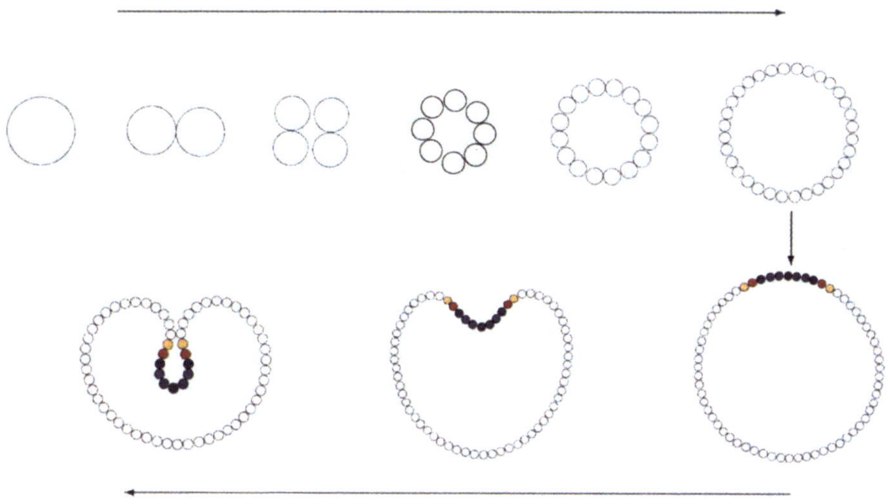

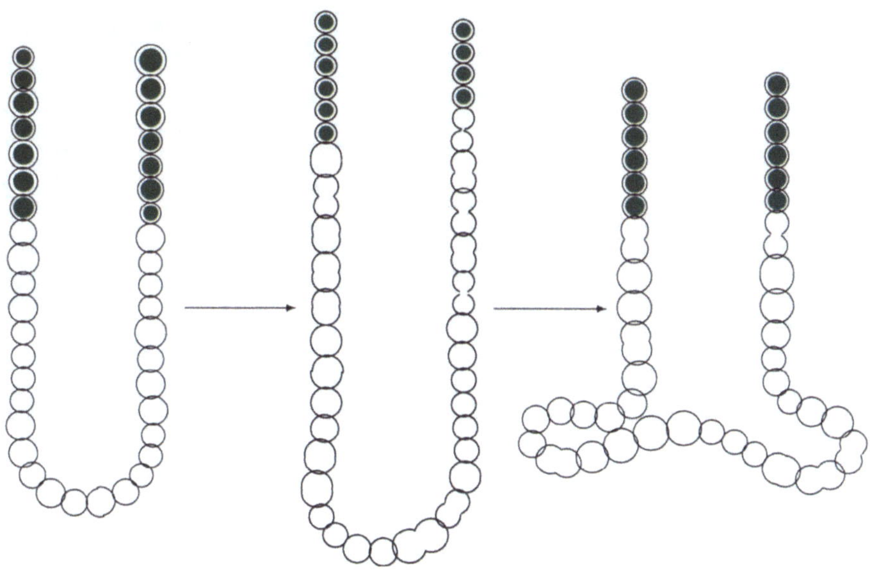

Fig. 10 from page 186.
Upper panel: Blastulation and gastrulation in a 2D model of *synapta digita* or sea urchin. Blue cells have a large negative spontaneous curvature, yellow cells have a small negative spontaneous curvature, hollow circles have a zero spontaneous curvature. *Lower panel*: Typical fission scenario in a crypt. Green cells denote mature cells, hollow circles denote cells capable to divide. (For more details see caption in text).

Color plate 12

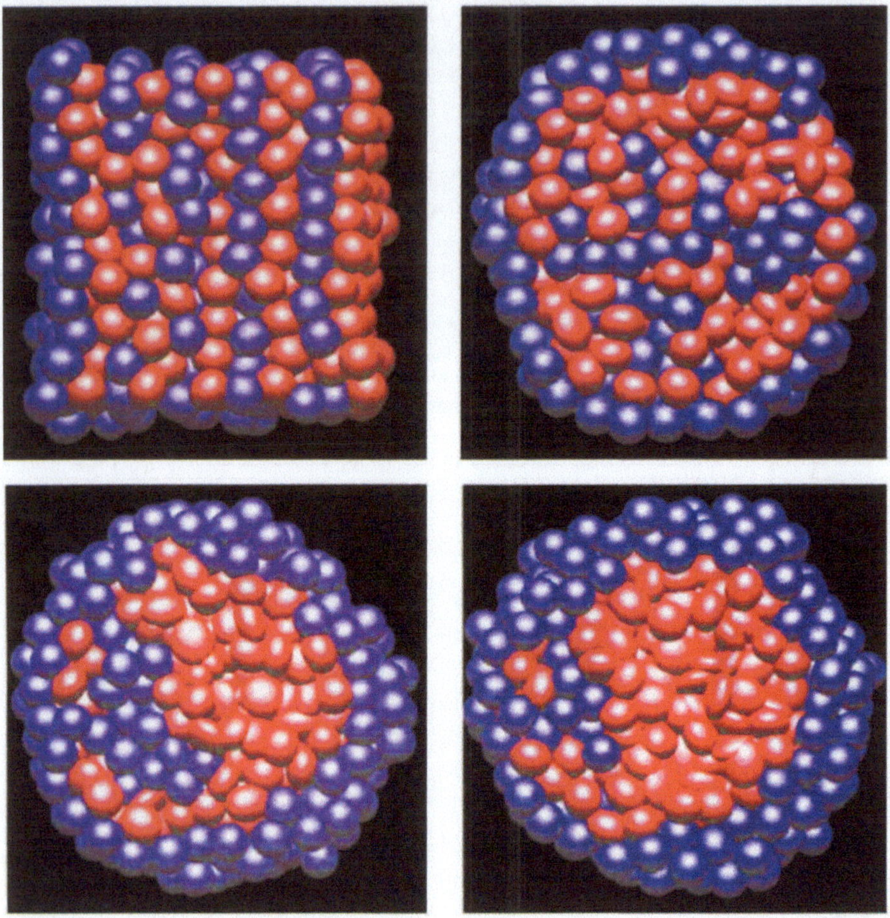

Fig 13 from page 190.
Typical time series of a cell-sorting process with the model of Palsson and Othmer [76] (pictures with courtesy by E. Palsson). *Upper left:* inital state: alternating layers of *A* (red) and *B* (blue) cells (see text for describing the interaction energies) *Upper right, lower left*: intermediate states. *Lower right*: final state, where *A* cells are surrounded by *B* cells.

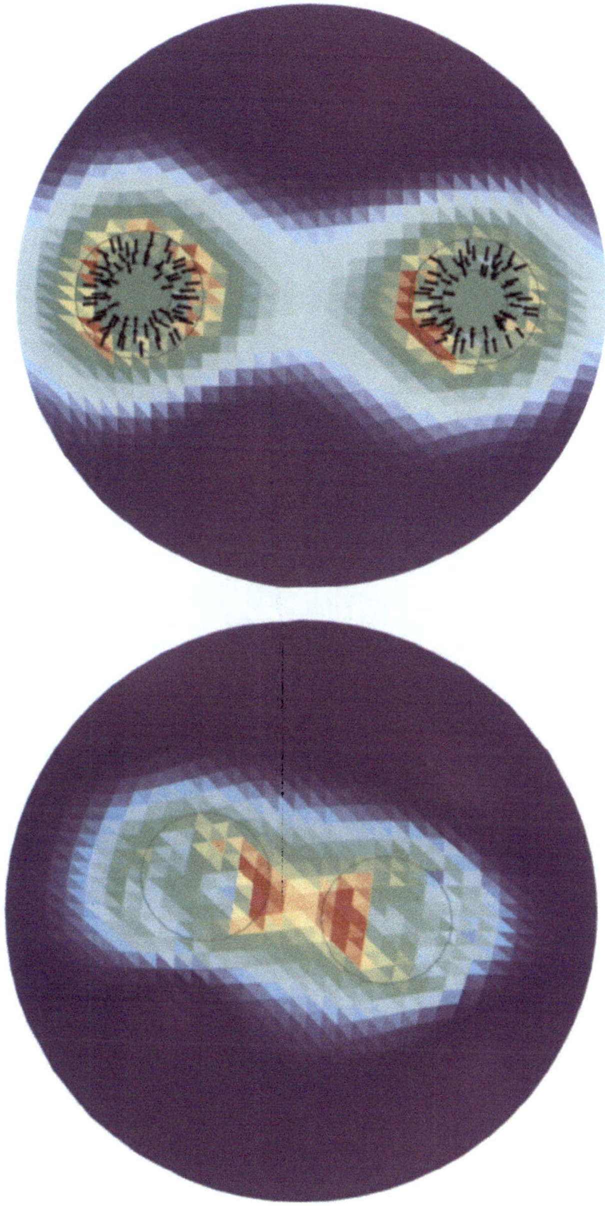

Fig. 4 from page 232.
Simulation snapshots of two cells migrating and interacting on a viscoelastic polymer substrate in the 2D tensiotaxis model (Example II). Arrows indicate forces acting on the substrate at adhesion sites. They are confined to the outer rim (lamella region) of the cell. Colors code tension in the polymer matrix (see eq. (9)).

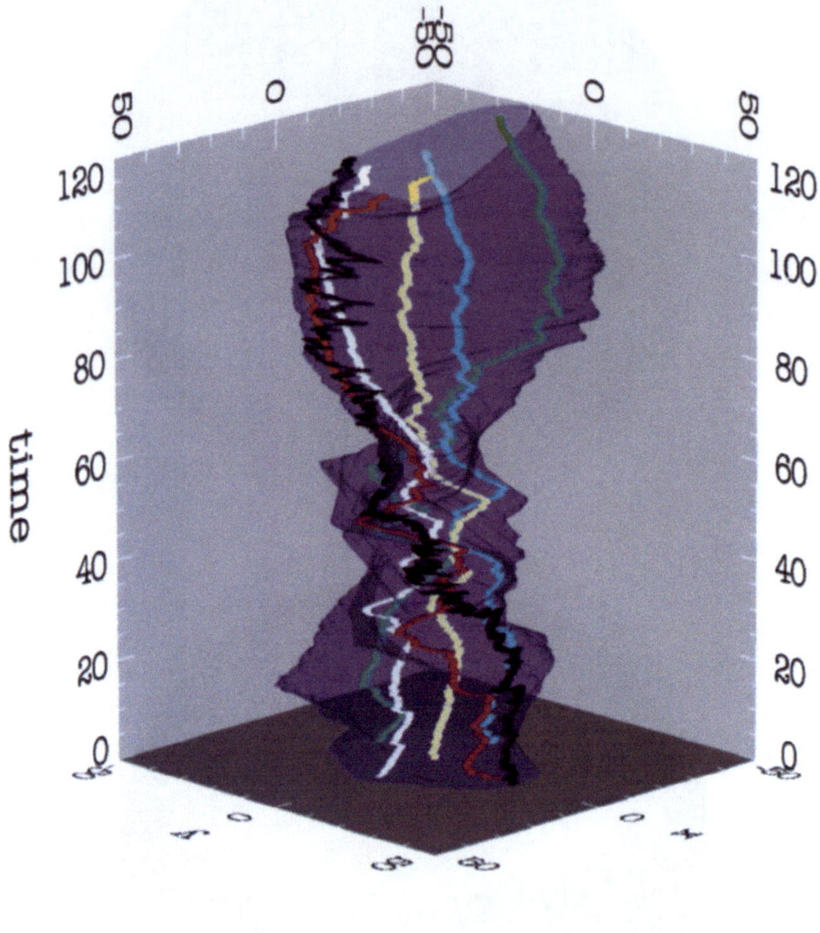

Fig. 7 from page 235.
Time-space plot of a complete bird swarm simulation (with fixed swarm center, time $t = 0 - 120$ s, from bottom to top). The light blue surface depicts the swarm boundary in the x-y plane changing in time; the black curve on the boundary indicates the mean moving direction of the swarm (viewed from the center); curves in red, blue, green, white and yellow represent the relative positions of five selected birds. For further discussion see text.

Fig. 3 from page 268.
Spatiotemporal evolution of a 3D capillary network from numerical simulations with the HDC technique. The figure shows the evolution of the capillary network (red capillaries) towards a spherical tumor (yellow spheroid) for the times $t = 1, 3, 7$, and 10 days.

Color plate 16

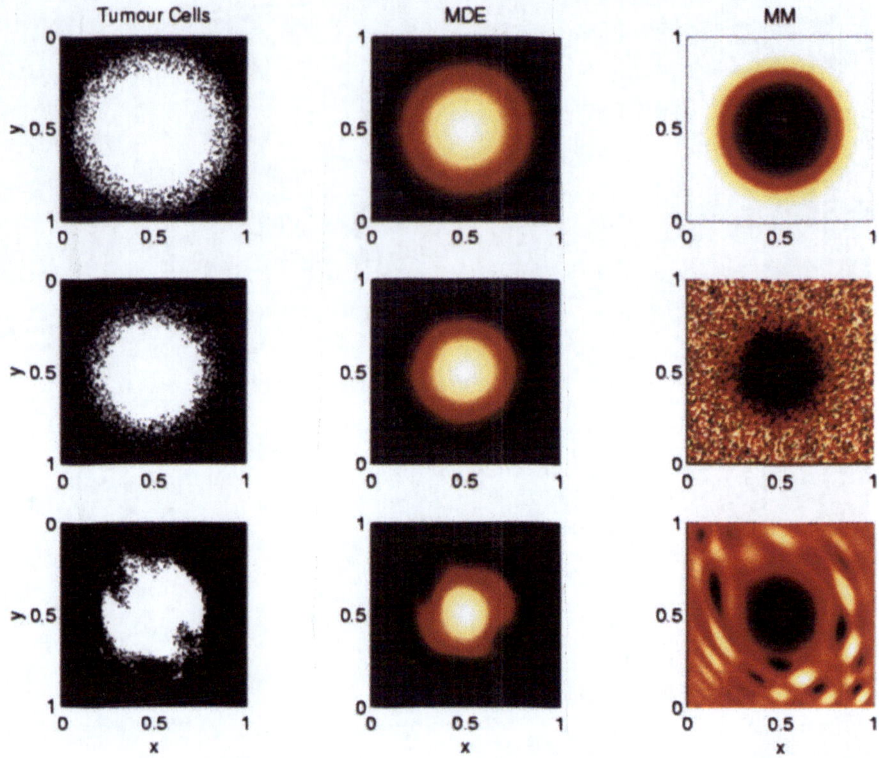

Fig. 4 from page 272.
Spatial distribution at t = 4000 time steps, of a tumor (column one) producing MDE (column two) and subsequently invading three different MM distributions (column three): (i) homogeneous, first row; (ii) random, second row; (iii) heterogeneous, third row. Colors in MDE and MM code for concentrations, from low (black) to high (yellow).

MIX
Papier aus verantwortungsvollen Quellen
Paper from responsible sources
FSC® C105338

If you have any concerns about our products,
you can contact us on
ProductSafety@springernature.com

In case Publisher is established outside the EU,
the EU authorized representative is:
Springer Nature Customer Service Center GmbH
Europaplatz 3, 69115 Heidelberg, Germany

Printed by Libri Plureos GmbH
in Hamburg, Germany